Universitext

T0155877

Springer
Berlin
Heidelberg
New York
Hong Kong
London
Milan
Paris
Tokyo

M. Gross • D. Huybrechts • D. Joyce

Calabi-Yau Manifolds and Related Geometries

Lectures at a Summer School in Nordfjordeid, Norway, June 2001

Springer

Mark Gross
University of Warwick
Mathematics Institute
Coventry CV4 7AL
United Kingdom
e-mail: mgross@maths.warwick.ac.uk

Dominic Joyce
Lincoln College
Oxford OX1 3DR
United Kingdom
e-mail: dominic.joyce@lincoln.ox.ac.uk

Daniel Huybrechts
Université Paris 7 Denis Diderot
Institut de Mathématiques
Topologie et Géometrie Algébriques
2, place Jussieu
75251 Paris Cedex 05
e-mail: huybrech@math.jussieu.fr

Editors:

Geir Ellingsrud
Universtity of Oslo
Department of Mathematics
P.O. Box 1053 Blindern
0316 Oslo, Norway
e-mail: ellingsr@math.uio.no

Kristian Ranestad
Universtity of Oslo
Department of Mathematics
P.O. Box 1053 Blindern
0316 Oslo, Norway
e-mail: ranestad@math.uio.no

Loren Olson
Universtity of Tromsø
Institute of Mathematics and Statistics
9037 Tromsø, Norway
e-mail: loren@math.uit.no

Stein A. Strømme
Universtity of Bergen
Department of Mathematics
5008 Bergen, Norway
e-mail: stromme@mi.uib.no

Library of Congress Cataloging-in-Publication Data applied for

A catalog record for this book is available from the Library of Congress.

Bibliographic information published by Die Deutsche Bibliothek
Die Deutsche Bibliothek lists this publication in the Deutsche Nationalbibliografie;
detailed bibliographic data is available in the Internet at http://dnb.ddb.de

ISBN 3-540-44059-3 Springer-Verlag Berlin Heidelberg New York

Mathematics Subject Classification (2000): 14J32, 32Q25, 53C26, 53C29, 53C38

Springer-Verlag Berlin Heidelberg New York
a member of BertelsmannSpringer Science+Business Media GmbH

http://www.springer.de

© Springer-Verlag Berlin Heidelberg 2003
Printed in Germany

Cover design: design & production, Heidelberg
Typesetting by the authors using a TeX macro package
Printed on acid-free paper   SPIN 10873463   46/3142ck-5 4 3 2 1 0

Preface

Each summer since 1996, algebraic geometers and algebraists in Norway have organised a summer school in Nordfjordeid, a small place in the western part of Norway. In addition to the beauty of the place, located between the mountains, close to the fjord and not far from the Norway's largest glacier, a reason for going there is that Sophus Lie was born and spent his few first years in Nordfjordeid, so it has a flavour of both the exotic and pilgrimage. It is also convenient: the municipality of Eid has created a conference centre named after Sophus Lie, aimed at attracting activities to fill the summer term of the local boarding school.

The summer schools are a joint effort of the four universities in Norway — the universities of Tromsø, Trondheim, Bergen and Oslo. They are primarily meant as training for Norwegian graduate students, but have over the years attracted increasing numbers of students from other parts of the world, adding the value of being international to the schools.

The themes of the schools have been varied, but build around some central topics in contemporary mathematics. The format of the school has by now become tradition — three international experts giving independent, but certainly connected, series of talks with exercise sessions in the evening, over five or six days.

In 2001 the organising committee consisted of Stein Arild Strømme from the University of Bergen, Loren Olson from the University of Tromsø, and Kristian Ranestad and Geir Ellingsrud from the University of Oslo. We wanted to make a summer school giving the students insight in some of the new interactions between differential and algebraic geometry. The three topics we finally chose, Riemannian holonomy and calibrated geometry, Calabi-Yau manifolds and mirror symmetry, and Compact hyperkähler manifolds, are parts of the fascinating current development of mathematics, and we think they illustrate well the modern interplay between differential and algebraic geometry.

We were fortunate enough to get positive answers when we asked Dominic Joyce, Mark Gross and Daniel Huybrechts to give the courses, and we are thankful for the great job the three lecturers did, both on stage in Nordfjordeid and by writing up the nice notes which have now developed into this book.

Geir Ellingsrud

Contents

Preface V

Part I. Riemannian Holonomy Groups and Calibrated Geometry
Dominic Joyce

1 Introduction... 3
2 Introduction to Holonomy Groups............................. 4
3 Berger's Classification of Holonomy Groups 10
4 Kähler Geometry and Holonomy 14
5 The Calabi Conjecture....................................... 20
6 The Exceptional Holonomy Groups............................. 26
7 Introduction to Calibrated Geometry 31
8 Calibrated Submanifolds in $\mathbb{R}^n$ 36
9 Constructions of SL m-folds in $\mathbb{C}^m$ 41
10 Compact Calibrated Submanifolds 50
11 Singularities of Special Lagrangian m-folds 57
12 The SYZ Conjecture, and SL Fibrations 63

Part II. Calabi–Yau Manifolds and Mirror Symmetry
Mark Gross

13 Introduction.. 71
14 The Classical Geometry of Calabi–Yau Manifolds 72
15 Kähler Moduli and Gromov–Witten Invariants................. 93
16 Variation and Degeneration of Hodge Structures 102
17 A Mirror Conjecture 121
18 Mirror Symmetry in Practice 122
19 The Strominger–Yau–Zaslow Approach to Mirror Symmetry 140

Part III. Compact Hyperkähler Manifolds
Daniel Huybrechts

20 Introduction... 163
21 Holomorphic Symplectic Manifolds 164
22 Deformations of Complex Structures 172
23 The Beauville–Bogomolov Form 177
24 Cohomology of Compact Hyperkähler Manifolds 190
25 Twistor Space and Moduli Space............................. 196

26 Projectivity of Hyperkähler Manifolds 203
27 Birational Hyperkähler Manifolds 213
28 The (Birational) Kähler Cone 221

References 227

Index 237

Part I

Riemannian Holonomy Groups
and Calibrated Geometry

Dominic Joyce

1 Introduction

The *holonomy group* Hol(g) of a Riemannian n-manifold (M, g) is a global invariant which measures the constant tensors on the manifold. It is a Lie subgroup of SO(n), and for generic metrics Hol(g) = SO(n). If Hol(g) is a proper subgroup of SO(n) then we say g has *special holonomy*. Metrics with special holonomy are interesting for a number of different reasons. They include *Kähler metrics* with holonomy U(m), which are the most natural class of metrics on complex manifolds, *Calabi–Yau manifolds* with holonomy SU(m), and *hyperkähler manifolds* with holonomy Sp(m).

Calibrated submanifolds are a class of k-dimensional submanifolds N of a Riemannian manifold (M, g) defined using a closed k-form φ on M called a *calibration*. Calibrated submanifolds are automatically *minimal submanifolds*. Manifolds with special holonomy (M, g) generally come equipped with one or more natural calibrations φ, which then define interesting classes of submanifolds in M. One such class is *special Lagrangian submanifolds* (*SL m-folds*) of Calabi–Yau manifolds.

This part of the book is an expanded version of a course of 8 lectures given at Njordfjordeid in June 2001. The first half of the course discussed Riemannian holonomy groups, focussing in particular on Kähler and Calabi–Yau manifolds. The second half discussed calibrated geometry and calibrated submanifolds of manifolds with special holonomy, focussing in particular on SL m-folds of Calabi–Yau m-folds. The final lecture surveyed research on the *SYZ Conjecture*, which explains Mirror Symmetry between Calabi–Yau 3-folds using special Lagrangian fibrations, and so made contact with Mark Gross' lectures.

I have retained this basic format, so that the first half §1–§6 is on Riemannian holonomy, and the second half §7–§12 on calibrated geometry, finishing with the SYZ Conjecture. The principal aim of the first half is to provide a firm grounding in Kähler and Calabi–Yau geometry from the differential geometric point of view, to serve as background for the more advanced, algebro-geometric material discussed in Parts II and III below. Therefore I have treated other subjects such as the exceptional holonomy groups fairly briefly.

In the second half I shall concentrate mainly on SL m-folds in $\mathbb{C}^m$ and Calabi–Yau m-folds. This is partly because of the focus of the book on Calabi–Yau manifolds and the link with Mirror Symmetry, partly because of my own research interests, and partly because more work has been done on special Lagrangian geometry than on other interesting classes of calibrated submanifolds, so there is simply more to say.

This is not intended as an even-handed survey of a field, but is biassed in favour of my own interests, and areas of research I want to promote in future. Therefore my own publications appear more often than they deserve, whilst more significant work is omitted, through my oversight or ignorance. I apologize to other authors who feel left out.

This is particularly true in the second half. Much of sections 9, 11 and 12 is an account of my own research programme into the singularities of SL m-folds. It therefore has a provisional, unfinished quality, with some conjectural material. My excuse for including it in a book is that I believe that a proper understanding of SL m-folds and their singularities will lead to exciting new discoveries — new invariants of Calabi–Yau 3-folds, and new chapters in the Mirror Symmetry story which are obscure at present, but are hinted at in the Kontsevich Mirror Symmetry proposal and the SYZ Conjecture. So I would like to get more people interested in this area.

We begin in §2 with some background from Differential Geometry, and define holonomy groups of connections and of Riemannian metrics. Section 3 explains Berger's classification of holonomy groups of Riemannian manifolds. Section 4 discusses Kähler geometry and Ricci curvature of Kähler manifolds and defines Calabi–Yau manifolds, and §5 sketches the proof of the Calabi Conjecture, and how it is used to construct examples of Calabi–Yau and hyperkähler manifolds via Algebraic Geometry. Section 6 surveys the exceptional holonomy groups G_2 and Spin(7).

The second half begins in §7 with an introduction to calibrated geometry. Section 8 covers general properties of calibrated submanifolds in $\mathbb{R}^n$, and §9 construction of examples of SL m-folds in $\mathbb{C}^m$. Section 10 discusses compact calibrated submanifolds in special holonomy manifolds, and §11 the singularities of SL m-folds. Finally, §12 briefly introduces String Theory and Mirror Symmetry, explains the SYZ Conjecture, and summarizes some research on the singularities of special Lagrangian fibrations.

2 Introduction to Holonomy Groups

We begin by giving some background from differential and Riemannian geometry, principally to establish notation, and move on to discuss connections on vector bundles, parallel transport, and the definition of holonomy groups. Some suitable reading for this section is my book [113, §2–§3].

2.1 Tensors and Forms

Let M be a smooth n-dimensional manifold, with tangent bundle TM and cotangent bundle T^*M. Then TM and T^*M are *vector bundles* over M. If E is a vector bundle over M, we use the notation $C^\infty(E)$ for the vector space of smooth sections of E. Elements of $C^\infty(TM)$ are called *vector fields*, and elements of $C^\infty(T^*M)$ are called 1-*forms*. By taking tensor products of the vector bundles TM and T^*M we obtain the bundles of *tensors* on M. A *tensor* T on M is a smooth section of a bundle $\bigotimes^k TM \otimes \bigotimes^l T^*M$ for some $k, l \in \mathbb{N}$.

It is convenient to write tensors using the *index notation*. Let U be an open set in M, and $(x^1, \ldots, x^n)$ coordinates on U. Then at each point $x \in U$,

$\frac{\partial}{\partial x^1}, \ldots, \frac{\partial}{\partial x^n}$ are a basis for $T_x U$. Hence, any vector field v on U may be uniquely written $v = \sum_{a=1}^{n} v^a \frac{\partial}{\partial x^a}$ for some smooth functions $v^1, \ldots, v^n$: $U \to \mathbb{R}$. We denote v by v^a, which is understood to mean the collection of n functions $v^1, \ldots, v^n$, so that a runs from 1 to n.

Similarly, at each $x \in U$, $dx^1, \ldots, dx^n$ are a basis for $T_x^* U$. Hence, any 1-form α on U may be uniquely written $\alpha = \sum_{b=1}^{n} \alpha_b dx^b$ for some smooth functions $\alpha_1, \ldots, \alpha_n : U \to \mathbb{R}$. We denote α by α_b, where b runs from 1 to n. In the same way, a general tensor T in $C^\infty(\bigotimes^k TM \otimes \bigotimes^l T^* M)$ is written $T^{a_1 \ldots a_k}_{b_1 \ldots b_l}$, where

$$T = \sum_{\substack{1 \le a_i \le n,\ 1 \le i \le k \\ 1 \le b_j \le n,\ 1 \le j \le l}} T^{a_1 \ldots a_k}_{b_1 \ldots b_l} \frac{\partial}{\partial x^{a_1}} \otimes \cdots \frac{\partial}{\partial x^{a_k}} \otimes dx^{b_1} \otimes \cdots \otimes dx^{b_l}.$$

The kth exterior power of the cotangent bundle $T^* M$ is written $\Lambda^k T^* M$. Smooth sections of $\Lambda^k T^* M$ are called k-forms, and the vector space of k-forms is written $C^\infty(\Lambda^k T^* M)$. They are examples of tensors. In the index notation they are written $T_{b_1 \ldots b_k}$, and are antisymmetric in the indices $b_1, \ldots, b_k$. The exterior product $\wedge$ and the exterior derivative d are important natural operations on forms. If α is a k-form and β an l-form then $\alpha \wedge \beta$ is a $(k+l)$-form and $d\alpha$ a $(k+1)$-form, which are given in index notation by

$$(\alpha \wedge \beta)_{a_1 \ldots a_{k+l}} = \alpha_{[a_1 \ldots a_k} \beta_{a_{k+1} \ldots a_{k+l}]} \quad \text{and} \quad (d\alpha)_{a_1 \ldots a_{k+1}} = \frac{\partial}{\partial x^{[a_1}} \alpha_{a_2 \ldots a_{k+1}]},$$

where $[\cdots]$ denotes antisymmetrization over the enclosed group of indices.

Let v, w be vector fields on M. The Lie bracket $[v, w]$ of v and w is another vector field on M, given in index notation by

$$[v, w]^a = v^b \frac{\partial w^a}{\partial x^b} - w^b \frac{\partial v^a}{\partial x^b}. \tag{1}$$

Here we have used the Einstein summation convention, that is, the repeated index b on the right hand side is summed from 1 to n. The important thing about this definition is that it is independent of choice of coordinates $(x^1, \ldots, x^n)$.

2.2 Connections on Vector Bundles and Curvature

Let M be a manifold, and $E \to M$ a vector bundle. A connection ∇^E on E is a linear map $\nabla^E : C^\infty(E) \to C^\infty(E \otimes T^* M)$ satisfying the condition

$$\nabla^E(\alpha\, e) = \alpha \nabla^E e + e \otimes d\alpha,$$

whenever $e \in C^\infty(E)$ is a smooth section of E and α is a smooth function on M.

If ∇^E is such a connection, $e \in C^\infty(E)$, and $v \in C^\infty(TM)$ is a vector field, then we write $\nabla^E_v e = v \cdot \nabla^E e \in C^\infty(E)$, where '$\cdot$' contracts together the

TM and T^*M factors in v and $\nabla^E e$. Then if $v \in C^\infty(TM)$ and $e \in C^\infty(E)$ and α, β are smooth functions on M, we have

$$\nabla^E_{\alpha v}(\beta e) = \alpha\beta\nabla^E_v e + \alpha(v \cdot \beta)e.$$

Here $v \cdot \beta$ is the Lie derivative of β by v. It is a smooth function on M, and could also be written $v \cdot \mathrm{d}\beta$.

There exists a unique, smooth section $R(\nabla^E) \in C^\infty(\mathrm{End}(E) \otimes \Lambda^2 T^*M)$ called the *curvature* of ∇^E, that satisfies the equation

$$R(\nabla^E) \cdot (e \otimes v \wedge w) = \nabla^E_v \nabla^E_w e - \nabla^E_w \nabla^E_v e - \nabla^E_{[v,w]}e \qquad (2)$$

for all $v, w \in C^\infty(TM)$ and $e \in C^\infty(E)$, where $[v, w]$ is the Lie bracket of v, w.

Here is one way to understand the curvature of ∇^E. Define $v_i = \partial/\partial x^i$ for $i = 1, \ldots, n$. Then v_i is a vector field on U, and $[v_i, v_j] = 0$. Let e be a smooth section of E. Then we may interpret $\nabla^E_{v_i} e$ as a kind of *partial derivative* $\partial e/\partial x^i$ of e. Equation (2) then implies that

$$R(\nabla^E) \cdot (e \otimes v_i \wedge v_j) = \frac{\partial^2 e}{\partial x^i \partial x^j} - \frac{\partial^2 e}{\partial x^j \partial x^i}. \qquad (3)$$

Thus, *the curvature $R(\nabla^E)$ measures how much partial derivatives in E fail to commute*.

Now let ∇ be a connection on the tangent bundle TM of M, rather than a general vector bundle E. Then there is a unique tensor $T = T^a_{bc}$ in $C^\infty(TM \otimes \Lambda^2 T^*M)$ called the *torsion* of ∇, satisfying

$$T \cdot (v \wedge w) = \nabla_v w - \nabla_w v - [v, w] \quad \text{for all } v, w \in C^\infty(TM).$$

A connection ∇ with zero torsion is called *torsion-free*. Torsion-free connections have various useful properties, so we usually restrict attention to torsion-free connections on TM.

A connection ∇ on TM extends naturally to connections on all the bundles of tensors $\bigotimes^k TM \otimes \bigotimes^l T^*M$ for $k, l \in \mathbb{N}$, which we will also write ∇. That is, we can use ∇ to differentiate not just vector fields, but any tensor on M.

2.3 Parallel Transport and Holonomy Groups

Let M be a manifold, $E \to M$ a vector bundle over M, and ∇^E a connection on E. Let $\gamma : [0, 1] \to M$ be a smooth curve in M. Then the pull-back $\gamma^*(E)$ of E to $[0, 1]$ is a vector bundle over $[0, 1]$ with fibre $E_{\gamma(t)}$ over $t \in [0, 1]$, where E_x is the fibre of E over $x \in M$. The connection ∇^E pulls back under γ to give a connection on $\gamma^*(E)$ over $[0, 1]$.

Definition 2.1 Let M be a manifold, E a vector bundle over M, and ∇^E a connection on E. Suppose $\gamma : [0, 1] \to M$ is (piecewise) smooth, with $\gamma(0) = x$

and $\gamma(1) = y$, where $x, y \in M$. Then for each $e \in E_x$, there exists a unique smooth section s of $\gamma^*(E)$ satisfying $\nabla^E_{\dot\gamma(t)} s(t) = 0$ for $t \in [0,1]$, with $s(0) = e$. Define $P_\gamma(e) = s(1)$. Then $P_\gamma : E_x \to E_y$ is a well-defined linear map, called the *parallel transport map*.

We use parallel transport to define the *holonomy group* of ∇^E.

Definition 2.2 Let M be a manifold, E a vector bundle over M, and ∇^E a connection on E. Fix a point $x \in M$. We say that γ is a *loop based at x* if $\gamma : [0,1] \to M$ is a piecewise-smooth path with $\gamma(0) = \gamma(1) = x$. The parallel transport map $P_\gamma : E_x \to E_x$ is an invertible linear map, so that P_γ lies in $\mathrm{GL}(E_x)$, the group of invertible linear transformations of E_x. Define the *holonomy group* $\mathrm{Hol}_x(\nabla^E)$ of ∇^E based at x to be

$$\mathrm{Hol}_x(\nabla^E) = \big\{ P_\gamma : \gamma \text{ is a loop based at } x \big\} \subset \mathrm{GL}(E_x). \tag{4}$$

The holonomy group has the following important properties.

- It is a *Lie subgroup* of $\mathrm{GL}(E_x)$. To show that $\mathrm{Hol}_x(\nabla^E)$ is a subgroup of $\mathrm{GL}(E_x)$, let γ, δ be loops based at x, and define loops $\gamma\delta$ and γ^{-1} by

$$\gamma\delta(t) = \begin{cases} \delta(2t) & t \in [0, \tfrac{1}{2}] \\ \gamma(2t-1) & t \in [\tfrac{1}{2}, 1] \end{cases} \quad \text{and} \quad \gamma^{-1}(t) = \gamma(1-t) \quad \text{for } t \in [0,1].$$

  Then $P_{\gamma\delta} = P_\gamma \circ P_\delta$ and $P_{\gamma^{-1}} = P_\gamma^{-1}$, so $\mathrm{Hol}_x(\nabla^E)$ is closed under products and inverses.
- It is *independent of basepoint $x \in M$*, in the following sense. Let $x, y \in M$, and let $\gamma : [0,1] \to M$ be a smooth path from x to y. Then $P_\gamma : E_x \to E_y$, and $\mathrm{Hol}_x(\nabla^E)$ and $\mathrm{Hol}_y(\nabla^E)$ satisfy $\mathrm{Hol}_y(\nabla^E) = P_\gamma \mathrm{Hol}_x(\nabla^E) P_\gamma^{-1}$.
  Suppose E has fibre $\mathbb{R}^k$, so that $\mathrm{GL}(E_x) \cong \mathrm{GL}(k, \mathbb{R})$. Then we may regard $\mathrm{Hol}_x(\nabla^E)$ as a subgroup of $\mathrm{GL}(k, \mathbb{R})$ defined up to conjugation, and it is then independent of basepoint x.
- If M is simply-connected, then $\mathrm{Hol}_x(\nabla^E)$ is connected. To see this, note that any loop γ based at x can be continuously shrunk to the constant loop at x. The corresponding family of parallel transports is a continuous path in $\mathrm{Hol}_x(\nabla^E)$ joining P_γ to the identity.

The holonomy group of a connection is closely related to its curvature. Here is one such relationship. As $\mathrm{Hol}_x(\nabla^E)$ is a Lie subgroup of $\mathrm{GL}(E_x)$, it has a *Lie algebra* $\mathfrak{hol}_x(\nabla^E)$, which is a Lie subalgebra of $\mathrm{End}(E_x)$. It can be shown that the curvature $R(\nabla^E)_x$ at x lies in the linear subspace $\mathfrak{hol}_x(\nabla^E) \otimes \Lambda^2 T_x^* M$ of $\mathrm{End}(E_x) \otimes \Lambda^2 T_x^* M$. Thus, *the holonomy group of a connection places a linear restriction upon its curvature*.

Now let ∇ be a connection on TM. Then from §2.2, ∇ extends to connections on all the tensor bundles $\bigotimes^k TM \otimes \bigotimes^l T^*M$. We call a tensor S on M *constant* if $\nabla S = 0$. The constant tensors on M are determined by the holonomy group $\mathrm{Hol}(\nabla)$.

Theorem 2.3 *Let M be a manifold, and ∇ a connection on TM. Fix $x \in M$, and let $H = \mathrm{Hol}_x(\nabla)$. Then H acts naturally on the tensor powers $\bigotimes^k T_x M \otimes \bigotimes^l T_x^* M$. Suppose $S \in C^\infty(\bigotimes^k TM \otimes \bigotimes^l T^*M)$ is a constant tensor. Then $S|_x$ is fixed by the action of H on $\bigotimes^k T_x M \otimes \bigotimes^l T_x^* M$. Conversely, if $S|_x \in \bigotimes^k T_x M \otimes \bigotimes^l T_x^* M$ is fixed by H, it extends to a unique constant tensor $S \in C^\infty(\bigotimes^k TM \otimes \bigotimes^l T^*M)$.*

The main idea in the proof is that if S is a constant tensor and $\gamma : [0,1] \to M$ is a path from x to y, then $P_\gamma(S|_x) = S|_y$. Thus, constant tensors are invariant under parallel transport.

2.4 Riemannian Metrics and the Levi-Civita Connection

Let g be a *Riemannian metric* on M. We refer to the pair (M, g) as a *Riemannian manifold*. Here g is a tensor in $C^\infty(S^2 T^*M)$, so that $g = g_{ab}$ in index notation with $g_{ab} = g_{ba}$. There exists a unique, torsion-free connection ∇ on TM with $\nabla g = 0$, called the *Levi-Civita connection*, which satisfies

$$2g(\nabla_u v, w) = u \cdot g(v, w) + v \cdot g(u, w) - w \cdot g(u, v)$$
$$+ g([u, v], w) - g([v, w], u) - g([u, w], v)$$

for all $u, v, w \in C^\infty(TM)$. This result is known as the *fundamental theorem of Riemannian geometry*.

The curvature $R(\nabla)$ of the Levi-Civita connection is a tensor $R^a{}_{bcd}$ on M. Define $R_{abcd} = g_{ae} R^e{}_{bcd}$. We shall refer to both $R^a{}_{bcd}$ and R_{abcd} as the *Riemann curvature* of g. The following theorem gives a number of symmetries of R_{abcd}. Equations (6) and (7) are known as the *first* and *second Bianchi identities*, respectively.

Theorem 2.4 *Let (M, g) be a Riemannian manifold, ∇ the Levi-Civita connection of g, and R_{abcd} the Riemann curvature of g. Then*

$$R_{abcd} = -R_{abdc} = -R_{bacd} = R_{cdab}, \tag{5}$$
$$R_{abcd} + R_{adbc} + R_{acdb} = 0, \tag{6}$$
$$\text{and} \quad \nabla_e R_{abcd} + \nabla_c R_{abde} + \nabla_d R_{abec} = 0. \tag{7}$$

Let (M, g) be a Riemannian manifold, with Riemann curvature $R^a{}_{bcd}$. The *Ricci curvature* of g is $R_{ab} = R^c{}_{acb}$. It is a component of the full Riemann curvature, and satisfies $R_{ab} = R_{ba}$. We say that g is *Einstein* if $R_{ab} = \lambda g_{ab}$ for some constant $\lambda \in \mathbb{R}$, and *Ricci-flat* if $R_{ab} = 0$. Einstein and Ricci-flat metrics are of great importance in mathematics and physics.

2.5   Riemannian Holonomy Groups

Let (M, g) be a Riemannian manifold. We define the *holonomy group* $\mathrm{Hol}_x(g)$ of g to be the holonomy group $\mathrm{Hol}_x(\nabla)$ of the Levi-Civita connection ∇ of g, as in §2.3. Holonomy groups of Riemannian metrics, or *Riemannian holonomy groups*, have stronger properties than holonomy groups of connections on arbitrary vector bundles. We shall explore some of these.

Firstly, note that g is a *constant tensor* as $\nabla g = 0$, so g is invariant under $\mathrm{Hol}(g)$ by Theorem 2.3. That is, $\mathrm{Hol}_x(g)$ lies in the subgroup of $\mathrm{GL}(T_xM)$ which preserves $g|_x$. This subgroup is isomorphic to $O(n)$. Thus, $\mathrm{Hol}_x(g)$ may be regarded as a *subgroup of $O(n)$ defined up to conjugation*, and it is then independent of $x \in M$, so we will often write it as $\mathrm{Hol}(g)$, dropping the basepoint x.

Secondly, the holonomy group $\mathrm{Hol}(g)$ constrains the Riemann curvature of g, in the following way. The Lie algebra $\mathfrak{hol}_x(\nabla)$ of $\mathrm{Hol}_x(\nabla)$ is a vector subspace of $T_xM \otimes T_x^*M$. From §2.3, we have $R^a{}_{bcd}|_x \in \mathfrak{hol}_x(\nabla) \otimes \Lambda^2 T_x^*M$.

Use the metric g to identify $T_xM \otimes T_x^*M$ and $\otimes^2 T_x^*M$, by equating $T^a{}_b$ with $T_{ab} = g_{ac}T^c{}_b$. This identifies $\mathfrak{hol}_x(\nabla)$ with a vector subspace of $\otimes^2 T_x^*M$ that we will write as $\mathfrak{hol}_x(g)$. Then $\mathfrak{hol}_x(g)$ lies in $\Lambda^2 T_x^*M$, and $R_{abcd}|_x \in \mathfrak{hol}_x(g) \otimes \Lambda^2 T_x^*M$. Applying the symmetries (5) of R_{abcd}, we have:

Theorem 2.5 *Let (M, g) be a Riemannian manifold with Riemann curvature R_{abcd}. Then R_{abcd} lies in the vector subspace $S^2 \mathfrak{hol}_x(g)$ in $\Lambda^2 T_x^*M \otimes \Lambda^2 T_x^*M$ at each $x \in M$.*

Combining this theorem with the Bianchi identities, (6) and (7), gives strong restrictions on the curvature tensor R_{abcd} of a Riemannian metric g with a prescribed holonomy group $\mathrm{Hol}(g)$. These restrictions are the basis of the classification of Riemannian holonomy groups, which will be explained in §3.

2.6   Exercises

2.1 Let M be a manifold and u, v, w be vector fields on M. The *Jacobi identity* for the Lie bracket of vector fields is

$$[u, [v, w]] + [v, [w, u]] + [w, [u, v]] = 0.$$

Prove the Jacobi identity in coordinates $(x^1, \ldots, x^n)$ on a coordinate patch U. Use the coordinate expression (1) for the Lie bracket of vector fields.

2.2 In §2.3 we explained that if M is a manifold, $E \to M$ a vector bundle and ∇^E a connection, then $\mathrm{Hol}(\nabla^E)$ is connected when M is simply-connected. If M is not simply-connected, what is the relationship between the fundamental group $\pi_1(M)$ and $\mathrm{Hol}(\nabla^E)$?

2.3 Work out your own proof of Theorem 2.3.

3   Berger's Classification of Holonomy Groups

Next we describe Berger's classification of Riemannian holonomy groups, and briefly discuss the possibilities in the classification. A reference for the material of this section is [113, §3]. Berger's original paper is [18], but owing to language and notation most will now find it difficult to read.

3.1   Reducible Riemannian Manifolds

Let (P, g) and (Q, h) be Riemannian manifolds with positive dimension, and $P \times Q$ the product manifold. Then at each (p, q) in $P \times Q$ we have $T_{(p,q)}(P \times Q) \cong T_p P \oplus T_q Q$. Define the *product metric* $g \times h$ on $P \times Q$ by $g \times h|_{(p,q)} = g|_p + h|_q$ for all $p \in P$ and $q \in Q$. We call $(P \times Q, g \times h)$ a *Riemannian product*.

A Riemannian manifold (M, g') is said to be *(locally) reducible* if every point has an open neighbourhood isometric to a Riemannian product $(P \times Q, g \times h)$, and *irreducible* if it is not locally reducible. It is easy to show that the holonomy of a product metric $g \times h$ is the product of the holonomies of g and h.

Proposition 3.1 *If $(P \times Q, g \times h)$ is the product of Riemannian manifolds (P, g), (Q, h), then $\mathrm{Hol}(g \times h) = \mathrm{Hol}(g) \times \mathrm{Hol}(h)$.*

Here is a kind of converse to this.

Theorem 3.2 *Let M be an n-manifold, and g an irreducible Riemannian metric on M. Then the representation of $\mathrm{Hol}(g)$ on $\mathbb{R}^n$ is irreducible.*

To prove the theorem, suppose $\mathrm{Hol}(g)$ acts reducibly on $\mathbb{R}^n$, so that $\mathbb{R}^n$ is the direct sum of representations $\mathbb{R}^k$, $\mathbb{R}^l$ of $\mathrm{Hol}(g)$ with $k, l > 0$. Using parallel transport, one can define a splitting $TM = E \oplus F$, where E, F are vector subbundles with fibres $\mathbb{R}^k, \mathbb{R}^l$. These vector subbundles are *integrable*, so locally $M \cong P \times Q$ with $E = TP$ and $F = TQ$. One can then show that the metric on M is the product of metrics on P and Q, so that g is locally reducible.

3.2   Symmetric Spaces

Next we discuss Riemannian symmetric spaces.

Definition 3.3 A Riemannian manifold (M, g) is said to be a *symmetric space* if for every point $p \in M$ there exists an isometry $s_p : M \to M$ that is an involution (that is, s_p^2 is the identity), such that p is an isolated fixed point of s_p.

Examples include $\mathbb{R}^n$, spheres $\mathcal{S}^n$, projective spaces $\mathbb{CP}^m$ with the Fubini–Study metric, and so on. Symmetric spaces have a transitive group of isometries.

Proposition 3.4 *Let (M, g) be a connected, simply-connected symmetric space. Then g is complete. Let G be the group of isometries of (M, g) generated by elements of the form $s_q \circ s_r$ for $q, r \in M$. Then G is a connected Lie group acting transitively on M. Choose $p \in M$, and let H be the subgroup of G fixing p. Then H is a closed, connected Lie subgroup of G, and M is the homogeneous space G/H.*

Because of this, symmetric spaces can be classified completely using the theory of Lie groups. This was done in 1925 by Élie Cartan. From Cartan's classification one can quickly deduce the list of holonomy groups of symmetric spaces.

A Riemannian manifold (M, g) is called *locally symmetric* if every point has an open neighbourhood isometric to an open set in a symmetric space, and *nonsymmetric* if it is not locally symmetric. It is a surprising fact that Riemannian manifolds are locally symmetric if and only if they have *constant curvature*.

Theorem 3.5 *Let (M, g) be a Riemannian manifold, with Levi-Civita connection ∇ and Riemann curvature R. Then (M, g) is locally symmetric if and only if $\nabla R = 0$.*

3.3 Berger's Classification

In 1955, Berger proved the following result.

Theorem 3.6 (Berger) *Suppose M is a simply-connected manifold of dimension n, and that g is a Riemannian metric on M, that is irreducible and nonsymmetric. Then exactly one of the following seven cases holds.*

(i) $\mathrm{Hol}(g) = \mathrm{SO}(n)$,
(ii) $n = 2m$ with $m \geqslant 2$, and $\mathrm{Hol}(g) = \mathrm{U}(m)$ in $\mathrm{SO}(2m)$,
(iii) $n = 2m$ with $m \geqslant 2$, and $\mathrm{Hol}(g) = \mathrm{SU}(m)$ in $\mathrm{SO}(2m)$,
(iv) $n = 4m$ with $m \geqslant 2$, and $\mathrm{Hol}(g) = \mathrm{Sp}(m)$ in $\mathrm{SO}(4m)$,
(v) $n = 4m$ with $m \geqslant 2$, and $\mathrm{Hol}(g) = \mathrm{Sp}(m)\,\mathrm{Sp}(1)$ in $\mathrm{SO}(4m)$,
(vi) $n = 7$ and $\mathrm{Hol}(g) = G_2$ in $\mathrm{SO}(7)$, *or*
(vii) $n = 8$ and $\mathrm{Hol}(g) = \mathrm{Spin}(7)$ in $\mathrm{SO}(8)$.

Notice the three simplifying assumptions on M and g: that M is simply-connected, and g is irreducible and nonsymmetric. Each condition has consequences for the holonomy group $\mathrm{Hol}(g)$.

- As M is simply-connected, $\mathrm{Hol}(g)$ is connected, from §2.3.
- As g is irreducible, $\mathrm{Hol}(g)$ acts irreducibly on $\mathbb{R}^n$ by Theorem 3.2.
- As g is nonsymmetric, $\nabla R \not\equiv 0$ by Theorem 3.5.

The point of the third condition is that there are some holonomy groups H which can *only* occur for metrics g with $\nabla R = 0$, and these holonomy groups are excluded from the theorem.

One can remove the three assumptions, at the cost of making the list of holonomy groups much longer. To allow g to be symmetric, we must include the holonomy groups of Riemannian symmetric spaces, which are known from Cartan's classification. To allow g to be reducible, we must include all products of holonomy groups already on the list. To allow M not simply-connected, we must include non-connected Lie groups whose identity components are already on the list.

Berger proved that the groups on his list were the only possibilities, but he did not show whether the groups actually do occur as holonomy groups. It is now known (but this took another thirty years to find out) that all of the groups on Berger's list do occur as the holonomy groups of irreducible, nonsymmetric metrics.

3.4 A Sketch of the Proof of Berger's Theorem

Let (M, g) be a Riemannian n-manifold with M simply-connected and g irreducible and nonsymmetric, and let $H = \text{Hol}(g)$. Then it is known that H is a closed, connected Lie subgroup of $\text{SO}(n)$. The classification of such subgroups follows from the classification of Lie groups. Berger's method was to take the list of all closed, connected Lie subgroups H of $\text{SO}(n)$, and apply two tests to each possibility to find out if it could be a holonomy group. The only groups H which passed both tests are those in the Theorem 3.6.

Berger's tests are algebraic and involve the curvature tensor. Suppose R_{abcd} is the Riemann curvature of a metric g with $\text{Hol}(g) = H$, and let $\mathfrak{h}$ be the Lie algebra of H. Then Theorem 2.4 shows that $R_{abcd} \in S^2\mathfrak{h}$, and the first Bianchi identity (6) applies.

If $\mathfrak{h}$ has large codimension in $\mathfrak{so}(n)$, then the vector space $\mathfrak{R}^H$ of elements of $S^2\mathfrak{h}$ satisfying (6) will be small, or even zero. But the *Ambrose–Singer Holonomy Theorem* shows that $\mathfrak{R}^H$ must be big enough to generate $\mathfrak{h}$, in a certain sense. For many of the candidate groups H this does not hold, and so H cannot be a holonomy group. This is the first test.

Now $\nabla_e R_{abcd}$ lies in $(\mathbb{R}^n)^* \otimes \mathfrak{R}^H$, and also satisfies the second Bianchi identity (7). Frequently these requirements imply that $\nabla R = 0$, so that g is locally symmetric. Therefore we may exclude such H, and this is Berger's second test.

3.5 The Groups on Berger's List

Here are some brief remarks about each group on Berger's list.

(i) $\text{SO}(n)$ is the holonomy group of generic Riemannian metrics.

(ii) Riemannian metrics g with $\mathrm{Hol}(g) \subseteq \mathrm{U}(m)$ are called *Kähler metrics*. Kähler metrics are a natural class of metrics on complex manifolds, and generic Kähler metrics on a given complex manifold have holonomy $\mathrm{U}(m)$.

(iii) Metrics g with $\mathrm{Hol}(g) = \mathrm{SU}(m)$ are called *Calabi–Yau metrics*. Since $\mathrm{SU}(m)$ is a subgroup of $\mathrm{U}(m)$, all Calabi–Yau metrics are Kähler. If g is Kähler and M is simply-connected, then $\mathrm{Hol}(g) \subseteq \mathrm{SU}(m)$ if and only if g is Ricci-flat. Thus Calabi–Yau metrics are locally more or less the same as Ricci-flat Kähler metrics.

(iv) Metrics g with $\mathrm{Hol}(g) = \mathrm{Sp}(m)$ are called *hyperkähler*. As $\mathrm{Sp}(m) \subseteq \mathrm{SU}(2m) \subset \mathrm{U}(2m)$, hyperkähler metrics are Ricci-flat and Kähler.

(v) Metrics g with holonomy group $\mathrm{Sp}(m)\mathrm{Sp}(1)$ for $m \geqslant 2$ are called *quaternionic Kähler*. (Note that quaternionic Kähler metrics are not in fact Kähler.) They are Einstein, but not Ricci-flat.

(vi) and (vii) The holonomy groups G_2 and $\mathrm{Spin}(7)$ are called the *exceptional holonomy groups*. Metrics with these holonomy groups are Ricci-flat.

The groups can be understood in terms of the four *division algebras*: the *real numbers* $\mathbb{R}$, the *complex numbers* $\mathbb{C}$, the *quaternions* $\mathbb{H}$, and the *octonions* or *Cayley numbers* $\mathbb{O}$.

- $\mathrm{SO}(n)$ is a group of automorphisms of $\mathbb{R}^n$.
- $\mathrm{U}(m)$ and $\mathrm{SU}(m)$ are groups of automorphisms of $\mathbb{C}^m$
- $\mathrm{Sp}(m)$ and $\mathrm{Sp}(m)\mathrm{Sp}(1)$ are automorphism groups of $\mathbb{H}^m$.
- G_2 is the automorphism group of $\mathrm{Im}\,\mathbb{O} \cong \mathbb{R}^7$. $\mathrm{Spin}(7)$ is a group of automorphisms of $\mathbb{O} \cong \mathbb{R}^8$.

Here are three ways in which we can gather together the holonomy groups on Berger's list into subsets with common features.

- The *Kähler holonomy groups* are $\mathrm{U}(m)$, $\mathrm{SU}(m)$ and $\mathrm{Sp}(m)$. Any Riemannian manifold with one of these holonomy groups is a Kähler manifold, and thus a complex manifold.
- The *Ricci-flat holonomy groups* are $\mathrm{SU}(m)$, $\mathrm{Sp}(m)$, G_2 and $\mathrm{Spin}(7)$. Any metric with one of these holonomy groups is Ricci-flat. This follows from the effect of holonomy on curvature discussed in §2.5 and §3.4: if H is one of these holonomy groups and R_{abcd} any curvature tensor lying in $S^2\mathfrak{h}$ and satisfying (6), then R_{abcd} has zero Ricci component.
- The *exceptional holonomy groups* are G_2 and $\mathrm{Spin}(7)$. They are the exceptional cases in Berger's classification, and they are rather different from the other holonomy groups.

3.6 Exercises

3.1 Work out your own proofs of Proposition 3.1 and (harder) Theorem 3.2.

3.2 Suppose that (M, g) is a simply-connected Ricci-flat Kähler manifold of complex dimension 4. What are the possibilities for $\mathrm{Hol}(g)$?
[You may use the fact that the only simply-connected Ricci-flat symmetric spaces are $\mathbb{R}^n$, $n \in \mathbb{N}$.]

4   Kähler Geometry and Holonomy

We now focus our attention on Kähler geometry, and the Ricci curvature of Kähler manifolds. This leads to the definition of *Calabi–Yau manifolds*, compact Ricci-flat Kähler manifolds with holonomy $\mathrm{SU}(m)$. A reference for this section is my book [113, §4, §6].

4.1   Complex Manifolds

We begin by defining *complex manifolds* M. The usual definition of complex manifolds involves an atlas of complex coordinate patches covering M, whose transition functions are holomorphic. However, for our purposes we need a more differential geometric definition, involving a tensor J on M called a *complex structure*.

Let M be a real manifold of dimension $2m$. An *almost complex structure* J on M is a tensor J_a^b on M satisfying $J_a^b J_b^c = -\delta_a^c$. For each vector field v on M define Jv by $(Jv)^b = J_a^b v^a$. Then $J^2 = -1$, so J gives each tangent space $T_p M$ the structure of a *complex vector space*.

We can associate a tensor $N = N_{bc}^a$ to J, called the *Nijenhuis tensor*, which satisfies

$$N_{bc}^a v^b w^c = \big([v, w] + J\big([Jv, w] + [v, Jw]\big) - [Jv, Jw]\big)^a$$

for all vector fields v, w on M, where $[,]$ is the Lie bracket of vector fields. The almost complex structure J is called a *complex structure* if $N \equiv 0$. A *complex manifold* (M, J) is a manifold M with a complex structure J.

Here is why this is equivalent to the usual definition. A smooth function $f : M \to \mathbb{C}$ is called *holomorphic* if $J_a^b (\mathrm{d}f)_b \equiv i(\mathrm{d}f)_a$ on M. These are called the *Cauchy–Riemann equations*. It turns out that the Nijenhuis tensor N is the obstruction to the existence of holomorphic functions. If $N \equiv 0$ there are many holomorphic functions locally, enough to form a set of holomorphic coordinates around every point.

4.2   Kähler Manifolds

Let (M, J) be a complex manifold, and let g be a Riemannian metric on M. We call g a *Hermitian metric* if $g(v, w) = g(Jv, Jw)$ for all vector fields v, w on M, or $g_{ab} = J_a^c J_b^d g_{cd}$ in index notation. When g is Hermitian, define the *Hermitian form* ω of g by $\omega(v, w) = g(Jv, w)$ for all vector fields v, w on

M, or $\omega_{ac} = J_a^b g_{bc}$ in index notation. Then ω is a (1,1)-form, and we may reconstruct g from ω by $g(v,w) = \omega(v, Jw)$.

A Hermitian metric g on a complex manifold (M, J) is called *Kähler* if one of the following three equivalent conditions holds:

(i) $d\omega = 0$,
(ii) $\nabla J = 0$, or
(iii) $\nabla \omega = 0$,

where ∇ is the Levi-Civita connection of g. We then call (M, J, g) a *Kähler manifold*, and ω the *Kähler form*. Kähler metrics are a natural and important class of metrics on complex manifolds.

By parts (ii) and (iii), if g is Kähler then J and ω are *constant tensors* on M. Thus by Theorem 2.3, the holonomy group $\mathrm{Hol}(g)$ must preserve a complex structure J_0 and 2-form ω_0 on $\mathbb{R}^{2m}$. The subgroup of $\mathrm{O}(2m)$ preserving J_0 and ω_0 is $\mathrm{U}(m)$, so $\mathrm{Hol}(g) \subseteq \mathrm{U}(m)$. So we prove:

Proposition 4.1 *A metric g on a $2m$-manifold M is Kähler with respect to some complex structure J on M if and only if $\mathrm{Hol}(g) \subseteq \mathrm{U}(m) \subset \mathrm{O}(2m)$.*

Note here that if $\mathrm{Hol}(g) \subseteq \mathrm{U}(m)$ then by Theorem 2.3 there exists a constant tensor J on M which is pointwise equivalent to the complex structure J_0 on $\mathbb{C}^m$. Therefore $J_a^b J_b^c = -\delta_a^c$, so J is an *almost complex structure* in the sense of §4.1. Moreover, as ∇ is torsion-free the *Nijenhuis tensor* N of J is a component of $\nabla J = 0$, so $N = 0$ and J is integrable. That is, if $\mathrm{Hol}(g) \subseteq \mathrm{U}(m)$ then M automatically carries an *integrable* complex structure J.

4.3 Kähler Potentials

Let (M, J) be a complex manifold. We have seen that to each Kähler metric g on M there is associated a closed real (1,1)-form ω, called the Kähler form. Conversely, if ω is a closed real (1,1)-form on M, then ω is the Kähler form of a Kähler metric if and only if ω is *positive*, that is, $\omega(v, Jv) > 0$ for all nonzero vectors v.

Now there is an easy way to manufacture closed real (1,1)-forms, using the ∂ and $\bar{\partial}$ operators on M. If $\phi : M \to \mathbb{R}$ is smooth, then $i\partial\bar{\partial}\phi$ is a closed real (1,1)-form, and every closed real (1,1)-form may be locally written in this way. Therefore, every Kähler metric g on M may be described locally by a function $\phi : M \to \mathbb{R}$ called a *Kähler potential*, such that the Kähler form ω satisfies $\omega = i\partial\bar{\partial}\phi$.

However, in general one cannot write $\omega = i\partial\bar{\partial}\phi$ globally on M, because $i\partial\bar{\partial}\phi$ is *exact*, but ω is usually not exact (never, if M is compact). Thus we are led to consider the *de Rham cohomology class* $[\omega]$ of ω in $H^2(M, \mathbb{R})$. We call $[\omega]$ the *Kähler class* of g. If two Kähler metrics g, g' on M lie in the same Kähler class, then they differ by a Kähler potential.

Proposition 4.2 *Let (M, J) be a compact complex manifold, and let g, g' be Kähler metrics on M with Kähler forms ω, ω'. Suppose that $[\omega] = [\omega'] \in H^2(M, \mathbb{R})$. Then there exists a smooth, real function ϕ on M such that $\omega' = \omega + i\partial\bar{\partial}\phi$. This function ϕ is unique up to the addition of a constant.*

Note also that if ω is the Kähler form of a fixed Kähler metric g and ϕ is sufficiently small in C^2, then $\omega' = \omega + i\partial\bar{\partial}\phi$ is the Kähler form of another Kähler metric g' on M, in the same Kähler class as g. This implies that if there exists one Kähler metric on M, then there exists an infinite-dimensional family — Kähler metrics are very abundant.

4.4 Ricci Curvature and the Ricci Form

Let (M, J, g) be a Kähler manifold, with Ricci curvature R_{ab}. Define the *Ricci form* ρ by $\rho_{ac} = J_a^b R_{bc}$. Then it turns out that $\rho_{ac} = -\rho_{ca}$, so that ρ is a 2-form. Furthermore, it is a remarkable fact that ρ is a *closed, real* (1, 1)-*form*. Note also that the Ricci curvature can be recovered from ρ by the formula $R_{ab} = \rho_{ac} J_b^c$.

To explain this, we will give an explicit expression for the Ricci form. Let $(z_1, \ldots, z_m)$ be holomorphic coordinates on an open set U in M. Define a smooth function $f : U \to (0, \infty)$ by

$$\omega^m = f \cdot \frac{(-1)^{m(m-1)/2} i^m m!}{2^m} \cdot dz_1 \wedge \cdots \wedge dz_m \wedge d\bar{z}_1 \wedge \cdots \wedge d\bar{z}_m. \qquad (8)$$

Here the constant factor ensures that f is positive, and gives $f \equiv 1$ when ω is the standard Hermitian form on $\mathbb{C}^m$. Then it can be shown that

$$\rho = -i\partial\bar{\partial}(\log f) \quad \text{on } U, \qquad (9)$$

so that ρ is indeed a closed real (1,1)-form.

Using some algebraic geometry, we can interpret this. The *canonical bundle* $K_M = \Lambda^{(m,0)} T^* M$ is a holomorphic line bundle over M. The Kähler metric g on M induces a metric on K_M, and the combination of metric and holomorphic structure induces a connection ∇^K on K_M. The curvature of this connection is a closed 2-form with values in the Lie algebra $u(1)$, and identifying $u(1) \cong \mathbb{R}$ we get a closed 2-form, which is the Ricci form.

Thus the Ricci form ρ may be understood as the curvature 2-form of a connection ∇^K on the canonical bundle K_M. So by characteristic class theory we may identify the de Rham cohomology class $[\rho]$ of ρ in $H^2(M, \mathbb{R})$: it satisfies

$$[\rho] = 2\pi\, c_1(K_M) = 2\pi\, c_1(M), \qquad (10)$$

where $c_1(M)$ is the first Chern class of M in $H^2(M, \mathbb{Z})$. It is a topological invariant depending on the homotopy class of the (almost) complex structure J.

4.5 Calabi–Yau Manifolds

Here is our definition of Calabi–Yau manifold. Note that a different definition of Calabi–Yau manifolds, Definition 14.1, will be used in Part II.

Definition 4.3 Let $m \geqslant 2$. A *Calabi–Yau m-fold* is a quadruple (M, J, g, Ω) such that (M, J) is a compact m-dimensional complex manifold, g a Kähler metric on (M, J) with holonomy group $\mathrm{Hol}(g) = \mathrm{SU}(m)$, and Ω a nonzero constant $(m, 0)$-form on M called the *holomorphic volume form*, which satisfies

$$\omega^m/m! = (-1)^{m(m-1)/2}(i/2)^m \Omega \wedge \bar{\Omega}, \tag{11}$$

where ω is the Kähler form of g. The constant factor in (11) is chosen to make $\mathrm{Re}\,\Omega$ a *calibration*.

Readers are warned that there are several *different* definitions of Calabi–Yau manifolds in use in the literature. Ours is unusual in regarding Ω as part of the given structure. Some authors define a Calabi–Yau m-fold to be a compact Kähler manifold (M, J, g) with holonomy $\mathrm{SU}(m)$. We shall show that one can associate a holomorphic volume form Ω to such (M, J, g) to make it Calabi–Yau in our sense, and Ω is unique up to phase.

Lemma 4.4 *Let* (M, J, g) *be a compact Kähler manifold with* $\mathrm{Hol}(g) = \mathrm{SU}(m)$. *Then* M *admits a holomorphic volume form* Ω, *unique up to change of phase* $\Omega \mapsto e^{i\theta}\Omega$, *such that* (M, J, g, Ω) *is a Calabi–Yau manifold.*

Proof. Let (M, J, g) be compact and Kähler with $\mathrm{Hol}(g) = \mathrm{SU}(m)$. Now the holonomy group $\mathrm{SU}(m)$ preserves the standard metric g_0 and Kähler form ω_0 on $\mathbb{C}^m$, and an $(m, 0)$-form Ω_0 given by

$$g_0 = |dz_1|^2 + \cdots + |dz_m|^2, \quad \omega_0 = \frac{i}{2}(dz_1 \wedge d\bar{z}_1 + \cdots + dz_m \wedge d\bar{z}_m),$$
$$\text{and} \quad \Omega_0 = dz_1 \wedge \cdots \wedge dz_m.$$

Thus, by Theorem 2.3 there exist corresponding constant tensors g, ω (the Kähler form), and Ω on (M, J, g). Since ω_0 and Ω_0 satisfy

$$\omega_0^m/m! = (-1)^{m(m-1)/2}(i/2)^m \Omega_0 \wedge \bar{\Omega}_0$$

on $\mathbb{C}^m$, it follows that ω and Ω satisfy (11) at each point, so (M, J, g, Ω) is Calabi–Yau. It is easy to see that Ω is unique up to change of phase. □

Suppose (M, J, g, Ω) is a Calabi–Yau m-fold. Then Ω is a constant section of the canonical bundle K_M. As Ω is constant, it is holomorphic. Thus the canonical bundle K_M admits a nonvanishing holomorphic section, so (M, J) has *trivial canonical bundle*, with first Chern class $c_1(M) = 0$.

Further, the connection ∇^K on K_M must be *flat*. However, from §4.4 the curvature of ∇^K is the Ricci form ρ. Therefore $\rho \equiv 0$, and g is Ricci-flat.

That is, Calabi–Yau m-folds are automatically *Ricci-flat*. More generally, the following proposition explains the relationship between the Ricci curvature and holonomy group of a Kähler metric.

Proposition 4.5 *Let* (M, J, g) *be a Kähler m-fold with* $\mathrm{Hol}(g) \subseteq \mathrm{SU}(m)$. *Then g is Ricci-flat. Conversely, let (M, J, g) be a Ricci-flat Kähler m-fold. If M is simply-connected or K_M is trivial, then* $\mathrm{Hol}(g) \subseteq \mathrm{SU}(m)$.

In the last part, M simply-connected implies that K_M is trivial for Ricci-flat Kähler manifolds, but not vice versa.

4.6   Hyperkähler Manifolds

The *quaternions* are the associative algebra $\mathbb{H} = \langle 1, i_1, i_2, i_3 \rangle \cong \mathbb{R}^4$, with multiplication given by

$$i_1 i_2 = -i_2 i_1 = i_3, \;\; i_2 i_3 = -i_3 i_2 = i_1, \;\; i_3 i_1 = -i_1 i_3 = i_2, \;\; i_1^2 = i_2^2 = i_3^2 = -1.$$

The holonomy group $\mathrm{Sp}(m)$ is the group of $m \times m$ matrices A over $\mathbb{H}$ satisfying $A\bar{A}^T = I$, where $x \mapsto \bar{x}$ is the conjugation on $\mathbb{H}$ defined by $\bar{x} = x_0 - x_1 i_1 - x_2 i_2 - x_3 i_3$ when $x = x_0 + x_1 i_1 + x_2 i_2 + x_3 i_3$.

Now $\mathrm{Sp}(m)$ acts on $\mathbb{H}^m = \mathbb{R}^{4m}$, regarded as column matrices of quaternions, preserving the Euclidean metric and also three complex structures J_1, J_2, J_3, induced by right multiplication of $\mathbb{H}^m$ by i_1, i_2, i_3. If $a_1, a_2, a_3 \in \mathbb{R}$ with $a_1^2 + a_2^2 + a_3^2 = 1$ then $a_1 J_1 + a_2 J_2 + a_3 J_3$ is also a complex structure on $\mathbb{R}^{4m}$ preserved by $\mathrm{Sp}(m)$.

Thus, if (M, g) is a Riemannian $4m$-manifold and g has holonomy $\mathrm{Sp}(m)$, then there exists an S^2 family of constant complex structures $a_1 J_1 + a_2 J_2 + a_3 J_3$ on M for $a_1^2 + a_2^2 + a_3^2 = 1$, by Theorem 2.3, each compatible with g.

Definition 4.6 A Riemannian $4m$-manifold (M, g) is called *hyperkähler* if $\mathrm{Hol}(g) = \mathrm{Sp}(m)$. A hyperkähler manifold comes naturally equipped with complex structures J_1, J_2, J_3 with Kähler forms $\omega_1, \omega_2, \omega_3$, such that $\nabla J_j = \nabla \omega_j = 0$ for $j = 1, 2, 3$, where ∇ is the Levi-Civita connection of g. Furthermore, if $a_1, a_2, a_3 \in \mathbb{R}$ with $a_1^2 + a_2^2 + a_3^2 = 1$ then $a_1 J_1 + a_2 J_2 + a_3 J_3$ is a complex structure on M, and g is Kähler with respect to it, with Kähler form $a_1 \omega_1 + a_2 \omega_2 + a_3 \omega_3$.

Thus, if $\mathrm{Hol}(g) = \mathrm{Sp}(m)$ then g is Kähler in many different ways, which is why we call g hyperkähler. Note that in this book the term 'hyperkähler' is reserved for metrics g with $\mathrm{Hol}(g) = \mathrm{Sp}(m)$, but that some other authors use hyperkähler to mean $\mathrm{Hol}(g) \subseteq \mathrm{Sp}(m)$.

Sometimes it is helpful to treat all the S^2 family of complex structures on an equal footing, and sometimes it is helpful to choose one and work with that. Let us choose $J = J_1$ as our distinguished complex structure. Then the complex 2-form $\omega_{\mathbb{C}} = \omega_2 + i\omega_3$ is a *holomorphic symplectic form* w.r.t. J, which makes (M, J) into a *holomorphic symplectic manifold*.

Definition 4.7 Let (M, J) be a complex manifold of complex dimension $2m$. A *holomorphic symplectic structure* on M is a closed $(2,0)$-form $\omega_{\mathbb{C}}$ such that the mth wedge product $\omega_{\mathbb{C}}^m$ is nonzero at every point. We call $(M, J, \omega_{\mathbb{C}})$ a *holomorphic symplectic manifold*.

Note that as $\omega_{\mathbb{C}}$ is closed we have $\partial \omega_{\mathbb{C}} = \bar{\partial} \omega_{\mathbb{C}} = 0$, so that $\omega_{\mathbb{C}}$ is *holomorphic*. Also, $\Omega = \omega_{\mathbb{C}}^m$ is a nonvanishing holomorphic section of the *canonical bundle* $K_M = \Lambda^{2m,0} T^* M$, and so is a *holomorphic volume form*. Thus, holomorphic symplectic manifolds have *trivial canonical bundle*, and $c_1(M) = 0$.

Here is one reason this is important. A hyperkähler manifold (M, g) comes with complex structures J_1, J_2, J_3 and Kähler forms $\omega_1, \omega_2, \omega_3$. Of these about half the data, J_1, ω_2 and ω_3, give M the structure of a *holomorphic symplectic manifold*. We can use *complex algebraic geometry* to study such manifolds, and perhaps even to construct J_1, ω_2, ω_3 explicitly.

As $\text{Sp}(m) \subseteq \text{SU}(2m)$, hyperkähler metrics are automatically Ricci-flat, from §4.5. In complex dimension 2 we have $\text{Sp}(1) = \text{SU}(2)$, so (compact) hyperkähler 4-manifolds and Calabi–Yau 2-folds coincide. But in higher dimensions the inclusion $\text{Sp}(m) \subset \text{SU}(2m)$ is proper. Many examples of noncompact hyperkähler manifolds are known, but rather fewer compact examples. Compact hyperkähler manifolds are discussed at length in Part III.

4.7 Exercises

4.1 Let $U \subseteq \mathbb{C}^m$ be simply-connected with coordinates $(z_1, \ldots, z_m)$, and g a Ricci-flat Kähler metric on U with Kähler form ω. Use equations (8) and (9) to show that there exists a holomorphic $(m, 0)$-form Ω on U satisfying

$$\omega^m / m! = (-1)^{m(m-1)/2} (i/2)^m \Omega \wedge \bar{\Omega}.$$

Hint: Write $\Omega = F \, dz_1 \wedge \cdots \wedge dz_m$ for some holomorphic function F. Use the fact that if f is a real function on a simply-connected subset U of $\mathbb{C}^m$ and $\partial \bar{\partial} f \equiv 0$, then f is the real part of a holomorphic function on U.

4.2 Let $\mathbb{C}^2$ have complex coordinates (z_1, z_2), and define $u = |z_1|^2 + |z_2|^2$. Let $f : [0, \infty) \to \mathbb{R}$ be a smooth function, and define a closed real $(1,1)$-form ω on $\mathbb{C}^2$ by $\omega = i \partial \bar{\partial} f(u)$.

(a) Calculate the conditions on f for ω to be the Kähler form of a Kähler metric g on $\mathbb{C}^2$.
(You can define g by $g(v, w) = \omega(v, Jw)$, and need to ensure that g is positive definite).

(b) Supposing g is a metric, calculate the conditions on f for g to be Ricci-flat. You should get an o.d.e. on f. If you can, solve this o.d.e., and write down the corresponding Kähler metrics in coordinates.

5   The Calabi Conjecture

The *Calabi Conjecture* specifies which closed (1,1)-forms on a compact complex manifold can be the Ricci form of a Kähler metric. It was posed by Calabi in 1954, and proved by Yau in 1976. We shall explain the conjecture and sketch its proof, and discuss its applications to Calabi–Yau and hyperkähler geometry.

The Calabi Conjecture *Let (M, J) be a compact, complex manifold, and g a Kähler metric on M, with Kähler form ω. Suppose that ρ' is a real, closed (1,1)-form on M with $[\rho'] = 2\pi\, c_1(M)$. Then there exists a unique Kähler metric g' on M with Kähler form ω', such that $[\omega'] = [\omega] \in H^2(M, \mathbb{R})$, and the Ricci form of g' is ρ'.*

Note that $[\omega'] = [\omega]$ says that g and g' are in the *same Kähler class*. An important application of the Calabi Conjecture is the construction of compact Ricci-flat Kähler manifolds, including *Calabi–Yau manifolds* and *hyperkähler manifolds*. A good general reference for this section is my book [113, §5–§7], which includes a proof of the Calabi Conjecture. The original proof of the conjecture is Yau [206].

5.1   Sketch of the Proof of the Calabi Conjecture

The Calabi Conjecture is proved by rewriting it as a second-order nonlinear elliptic p.d.e. upon a real function ϕ on M, and then showing that this p.d.e. has a unique solution. We first explain how to rewrite the Calabi Conjecture as a p.d.e.

Let (M, J) be a compact, complex manifold, and let g, g' be two Kähler metrics on M with Kähler forms ω, ω' and Ricci forms ρ, ρ'. Suppose g, g' are in the same Kähler class, so that $[\omega'] = [\omega] \in H^2(M, \mathbb{R})$. Define a smooth function $f : M \to \mathbb{R}$ by $(\omega')^m = e^f \omega^m$. Then from equations (8) and (9) of §4.4, we find that $\rho' = \rho - i\partial\bar{\partial}f$. Furthermore, as $[\omega'] = [\omega]$ in $H^2(M, \mathbb{R})$, we have $[\omega']^m = [\omega]^m$ in $H^{2m}(M, \mathbb{R})$, and thus $\int_M e^f \omega^m = \int_M \omega^m$.

Now suppose that we are given the real, closed (1,1)-form ρ' with $[\rho'] = 2\pi\, c_1(M)$, and want to construct a metric g' with ρ' as its Ricci form. Since $[\rho] = [\rho'] = 2\pi\, c_1(M)$, $\rho - \rho'$ is an *exact* real (1,1)-form, and so by the $\partial\bar{\partial}$-Lemma there exists a smooth function $f : M \to \mathbb{R}$ with $\rho - \rho' = i\partial\bar{\partial}f$. This f is unique up to addition of a constant, but the constant is fixed by requiring that $\int_M e^f \omega^m = \int_M \omega^m$. Thus we have proved:

Proposition 5.1 *Let (M, J) be a compact complex manifold, g a Kähler metric on M with Kähler form ω and Ricci form ρ, and ρ' a real, closed (1,1)-form on M with $[\rho'] = 2\pi\, c_1(M)$. Then there is a unique smooth function $f : M \to \mathbb{R}$ such that*

$$\rho' = \rho - i\partial\bar{\partial}f \quad and \quad \int_M e^f \omega^m = \int_M \omega^m, \tag{12}$$

and a Kähler metric g on M with Kähler form ω' satisfying $[\omega'] = [\omega]$ in $H^2(M, \mathbb{R})$ has Ricci form ρ' if and only if $(\omega')^m = e^f \omega^m$.

Thus we have transformed the Calabi Conjecture from seeking a metric g' with *prescribed Ricci curvature* ρ' to seeking a metric g' with *prescribed volume form* $(\omega')^m$. This is an important simplification, because the Ricci curvature depends on the second derivatives of g', but the volume form depends only on g' and not on its derivatives.

Now by Proposition 4.2, as $[\omega'] = [\omega]$ we may write $\omega' = \omega + i\partial\bar{\partial}\phi$ for ϕ a smooth real function on M, unique up to addition of a constant. We can fix the constant by requiring that $\int_M \phi \, dV_g = 0$. So, from Proposition 5.1 we deduce that the Calabi Conjecture is equivalent to:

The Calabi Conjecture (second version) *Let (M, J) be a compact, complex manifold, and g a Kähler metric on M, with Kähler form ω. Let f be a smooth real function on M satisfying $\int_M e^f \omega^m = \int_M \omega^m$. Then there exists a unique smooth real function ϕ such that*

(i) $\omega + i\partial\bar{\partial}\phi$ *is a positive* $(1,1)$-*form, that is, it is the Kähler form of some Kähler metric g',*

(ii) $\int_M \phi \, dV_g = 0$, *and*

(iii) $(\omega + i\partial\bar{\partial}\phi)^m = e^f \omega^m$ *on M.*

This reduces the Calabi Conjecture to a problem in analysis, that of showing that the nonlinear p.d.e. $(\omega + i\partial\bar{\partial}\phi)^m = e^f \omega^m$ has a solution ϕ for every suitable function f. To prove this second version of the Calabi Conjecture, Yau used the *continuity method*.

For each $t \in [0, 1]$, define $f_t = tf + c_t$, where c_t is the unique real constant such that $e^{c_t} \int_M e^{tf} \omega^m = \int_M \omega^m$. Then f_t depends smoothly on t, with $f_0 \equiv 0$ and $f_1 \equiv f$. Define S to be the set of $t \in [0, 1]$ such that there exists a smooth real function ϕ on M satisfying parts (i) and (ii) above, and also

(iii)' $(\omega + i\partial\bar{\partial}\phi)^m = e^{f_t}\omega^m$ *on M.*

The idea of the continuity method is to show that S is both *open* and *closed* in $[0, 1]$. Thus, S is a connected subset of $[0, 1]$, so $S = \emptyset$ or $S = [0, 1]$. But $0 \in S$, since as $f_0 \equiv 0$ parts (i), (ii) and (iii)' are satisfied by $\phi \equiv 0$. Thus $S = [0, 1]$. In particular, (i), (ii) and (iii)' admit a solution ϕ when $t = 1$. As $f_1 \equiv f$, this ϕ satisfies (iii), and the Calabi Conjecture is proved.

Showing that S is open is fairly easy, and was done by Calabi. It depends on the fact that (iii) is an *elliptic* p.d.e. — basically, the operator $\phi \mapsto (\omega + i\partial\bar{\partial}\phi)^m$ is rather like a nonlinear Laplacian — and uses only standard facts about elliptic operators.

However, showing that S is closed is much more difficult. One must prove that S contains its limit points. That is, if $(t_n)_{n=1}^{\infty}$ is a sequence in S converging to $t \in [0, 1]$ then there exists a sequence $(\phi_n)_{n=1}^{\infty}$ satisfying (i), (ii) and $(\omega + i\partial\bar{\partial}\phi_n)^m = e^{f_{t_n}}\omega^m$ for $n = 1, 2, \ldots$, and we need to show that $\phi_n \to \phi$

as $n \to \infty$ for some smooth real function ϕ satisfying (i), (ii) and (iii)', so that $t \in S$.

The thing you have to worry about is that the sequence $(\phi_n)_{n=1}^{\infty}$ might converge to some horrible non-smooth function, or might not converge at all. To prove this doesn't happen you need *a priori estimates* on the ϕ_n and all their derivatives. In effect, you need upper bounds on $|\nabla^k \phi_n|$ for all n and k, bounds which are allowed to depend on M, J, g, k and f_{t_n}, but not on n or ϕ_n. These a priori estimates were difficult to find, because the nonlinearities in ϕ of $(\omega + i \partial \bar{\partial} \phi)^m = e^f \omega^m$ are of a particularly nasty kind, and this is why it took so long to prove the Calabi Conjecture.

5.2   Compact Ricci-flat Kähler Manifolds

Our main interest in the Calabi Conjecture is that when $c_1(M) = 0$ we can take $\rho' \equiv 0$, and then g' is Ricci-flat. Thus, from the proof of the Calabi Conjecture we deduce:

Theorem 5.2 *Let (M, J) be a compact complex manifold admitting Kähler metrics, with $c_1(M) = 0$ in $H^2(M, \mathbb{R})$. Then there is a unique Ricci-flat Kähler metric in each Kähler class on M. The Ricci-flat Kähler metrics on M form a smooth family of dimension $h^{1,1}(M)$, isomorphic to the Kähler cone $\mathcal{K}$ of M.*

As a consequence of the *Cheeger–Gromoll splitting Theorem* [20, §6.G] we get the following result, taken from Besse [20, Cor. 6.67].

Theorem 5.3 *Suppose (M, g) is a compact Ricci-flat Riemannian manifold. Then M admits a finite cover isometric to $T^n \times N$, where T^n carries a flat metric and N is a compact, simply-connected Riemannian manifold.*

Using this we prove the following crude classification result for compact Ricci-flat Kähler manifolds, which will also be treated from a more algebro-geometric point of view in Theorem 14.15.

Theorem 5.4 *Let (M, J, g) be a compact Ricci-flat Kähler manifold. Then (M, J, g) admits a finite cover isomorphic to the product Kähler manifold*

$$(T^{2l} \times M_1 \times \cdots \times M_k, J_0 \times \cdots \times J_k, g_0 \times \cdots \times g_k),$$

where (T^{2l}, J_0, g_0) is a flat Kähler torus, and (M_j, J_j, g_j) is a compact, simply-connected, irreducible, Ricci-flat Kähler manifold for $j = 1, \ldots, k$. Let $m_j = \dim_{\mathbb{C}} M_j$. Then either $m_j \geqslant 2$ and $\mathrm{Hol}(g_j) = \mathrm{SU}(m_j)$, or $m_j \geqslant 4$ is even and $\mathrm{Hol}(g_j) = \mathrm{Sp}(m_j/2)$.

Proof. By Theorem 5.3, M admits a finite cover isometric to $T^n \times N$, where T^n is flat and N is simply-connected. As N is compact, it is complete. Now by a result of de Rham, a complete, simply-connected Riemannian manifold is a Riemannian product of *irreducible* Riemannian manifolds, as in §3.1. Thus we may write N as $(M_1 \times \cdots \times M_k, g_1 \times \cdots \times g_k)$, where each (M_j, g_j) is compact, simply-connected and irreducible. As the only Ricci-flat symmetric spaces are $\mathbb{R}^n$, the (M_j, g_j) are also nonsymmetric.

As g is Kähler, it follows easily that T^n and (M_j, g_j) are Kähler. Thus $n = 2l$ is even. Each (M_j, J_j, g_j) is a simply-connected, Ricci-flat Kähler manifold of complex dimension m_j, so $\mathrm{Hol}(g_j) \subseteq \mathrm{SU}(m_j)$ by Proposition 4.5. Finally, as M_j is simply-connected and g_j irreducible and nonsymmetric, $\mathrm{Hol}(g_j)$ is one of cases (i)–(vii) of Theorem 3.6. Since $\mathrm{Hol}(g_j) \subseteq \mathrm{SU}(m_j)$, the only possibilities are $\mathrm{Hol}(g_j) = \mathrm{SU}(m_j)$ for $m_j \geqslant 2$, or $\mathrm{Hol}(g_j) = \mathrm{Sp}(m_j/2)$ for $m_j \geqslant 4$ even. $\square$

If (M, J, g) is a compact Ricci-flat Kähler manifold with Levi-Civita connection ∇, using a *Weitzenbock formula* one can prove that if ξ is a $(p, 0)$-form on M then $\bar{\partial}^* \bar{\partial} \xi = \frac{1}{2} \nabla^* \nabla \xi$. Taking the L^2-inner product with ξ and integrating by parts shows that $\|\bar{\partial}\xi\|_{L^2}^2 = \frac{1}{2}\|\nabla\xi\|_{L^2}^2$. Thus $\bar{\partial}\xi = 0$ if and only if $\nabla\xi = 0$. That is, a $(p, 0)$-form ξ on M is holomorphic if and only if it is constant.

Now by Theorem 2.3, the constant tensors on M are determined by the holonomy group $\mathrm{Hol}(g)$. Also, the vector space of holomorphic $(p, 0)$-forms on M is the Dolbeault cohomology group $H^{p,0}(M)$. Thus, if (M, J, g) is a compact Ricci-flat Kähler manifold then $\mathrm{Hol}(g)$ determines the Dolbeault groups $H^{p,0}(M)$. In this way we prove:

Proposition 5.5 *Let (M, J, g) be a compact Ricci-flat Kähler m-fold. Then for $0 \leqslant p \leqslant m$, the Dolbeault group $H^{p,0}(M)$ is isomorphic to the subspace of $\Lambda^p (\mathbb{C}^m)^*$ invariant under $\mathrm{Hol}(g)$.*

5.3 Calabi–Yau 2-folds and K3 Surfaces

Recall from §3.5 that the *Kähler holonomy groups* are $\mathrm{U}(m)$, $\mathrm{SU}(m)$ and $\mathrm{Sp}(k)$. Calabi–Yau manifolds of complex dimension m have holonomy $\mathrm{SU}(m)$ for $m \geqslant 2$, and hyperkähler manifolds of complex dimension $2k$ have holonomy $\mathrm{Sp}(k)$ for $k \geqslant 1$. In complex dimension 2 these coincide, as $\mathrm{SU}(2) = \mathrm{Sp}(1)$. Because of this, Calabi–Yau 2-folds have special features which are not present in Calabi–Yau m-folds for $m \geqslant 3$.

Calabi–Yau 2-folds are very well understood, through the classification of compact complex surfaces [4,16]. A *K3 surface* is defined to be a compact, complex surface (X, J) with $h^{1,0}(X) = 0$ and trivial canonical bundle. All Calabi–Yau 2-folds are K3 surfaces, and conversely, every K3 surface (X, J) admits a family of Kähler metrics g making it into a Calabi–Yau 2-fold. All K3 surfaces (X, J) are diffeomorphic, sharing the same smooth 4-manifold X, which is simply-connected, with Betti numbers $b^2 = 22$, $b_+^2 = 3$, and $b_-^2 = 19$.

The moduli space $\mathscr{M}_{\mathrm{K3}}$ of K3 surfaces is a connected 20-dimensional singular complex manifold, which can be described very precisely via the 'Torelli

Theorems'. Some K3 surfaces are *algebraic*, that is, they can be embedded as complex submanifolds in $\mathbb{CP}^N$ for some N, and some are not. The set of algebraic K3 surfaces is a countable, dense union of 19-dimensional subvarieties in $\mathcal{M}_{K3}$. Each K3 surface (X, J) admits a real 20-dimensional family of Calabi–Yau metrics g, so the family of Calabi–Yau 2-folds (X, J, g) is a nonsingular 60-dimensional real manifold. For more information about K3 surfaces, see §21.1.

5.4 General Properties of Calabi–Yau m-folds for $m \geqslant 3$

Let (M, J, g, Ω) be a Calabi–Yau m-fold, in the sense of Definition 4.3. Then $\mathrm{Hol}(g) = \mathrm{SU}(m)$. Theorem 5.4 shows that M has a finite cover which is simply-connected, and so $\pi_1(M)$ is finite. Also Proposition 5.5 shows that $H^{p,0}(M)$ is $\mathbb{C}$ if $p = 0, m$ and 0 if $0 < p < m$. Thus we prove:

Proposition 5.6 *Let (M, J, g, Ω) be a Calabi–Yau m-fold with Hodge numbers $h^{p,q}$. Then M has finite fundamental group, $h^{0,0} = h^{m,0} = 1$ and $h^{p,0} = 0$ for $p \neq 0, m$.*

When m is even we can improve this result using spin geometry, following [113, §3.6].

Proposition 5.7 *Suppose (M, J, g, Ω) is a Calabi–Yau $2n$-fold. Then M is simply-connected.*

Proof. As M carries an $\mathrm{SU}(2n)$-structure, and $\mathrm{SU}(2n)$ is simply-connected, M has a natural *spin structure*. The corresponding *spin bundle* $S \to M$ splits into positive and negative spinors, $S \cong S_+ \oplus S_-$. The *positive Dirac operator* D_+ maps $D_+ : C^\infty(S_+) \to C^\infty(S_-)$. It is an elliptic operator, with index

$$\mathrm{ind}(D_+) = \dim \mathrm{Ker}(D_+) - \dim \mathrm{Coker}(D_+) = \hat{A}(M),$$

the $\hat{A}$-*genus* of M.

By a similar argument to the proof of Proposition 5.5, it turns out that spinors in $\mathrm{Ker}(D_+)$ or $\mathrm{Coker}(D_+)$ are necessarily constant. Thus applying Theorem 2.3 to the Levi-Civita connection on spinors shows that $\mathrm{Hol}(g)$ determines $\mathrm{Ker}(D_+)$ and $\mathrm{Coker}(D_+)$, and hence $\mathrm{ind}(D_+) = \hat{A}(M)$. Calculating the $\mathrm{SU}(2n)$-invariant spaces of spinors [113, Th. 3.6.5] shows that $\hat{A}(M) = 2$.

Proposition 5.6 shows that $\pi_1(M)$ is finite. Let $\tilde{M}$ be the universal cover of M, and let $d = |\pi_1(M)|$, so that $\pi : \tilde{M} \to M$ is a d-fold cover. Now $\tilde{M}$ is also a Calabi–Yau $2n$-fold, and so $\hat{A}(\tilde{M}) = 2$. But the $\hat{A}$-genus is a characteristic class, and so $\hat{A}(\tilde{M}) = d \cdot \hat{A}(M)$ by the behaviour of characteristic classes under coverings. As $\hat{A}(\tilde{M}) = \hat{A}(M) = 2$ we have $d = 1$, and so M is simply-connected.   □

When m is odd this argument fails as $\text{ind}(D_+) = 0$, and in fact there do exist non-simply-connected Calabi–Yau 3-folds. Let (M, J, g, Ω) be a Calabi–Yau m-fold for $m \geqslant 3$. Then $h^{2,0}(M) = 0$ by Proposition 5.6, and this has important consequences for the complex manifold (M, J).

It can be shown that a complex line bundle L over a compact Kähler manifold (M, J, g) admits a holomorphic structure if and only if $c_1(L)$ lies in $H^{1,1}(M) \subseteq H^2(M, \mathbb{C})$. But $H^2(M, \mathbb{C}) = H^{2,0}(M) \oplus H^{1,1}(M) \oplus H^{0,2}(M)$, and $H^{2,0}(M) = H^{0,2}(M) = 0$ as $h^{2,0}(M) = 0$. Thus $H^{1,1}(M) = H^2(M, \mathbb{C})$, and so *every* complex line bundle L over M admits a holomorphic structure.

Thus, Calabi–Yau m-folds for $m \geqslant 3$ are richly endowed with holomorphic line bundles. Using the *Kodaira Embedding Theorem* one can show that some of these holomorphic line bundles admit many holomorphic sections. By taking a line bundle with enough holomorphic sections (a *very ample* line bundle) we can construct an embedding of M in $\mathbb{CP}^N$ as a complex submanifold. So we prove:

Theorem 5.8 *Let (M, J, g, Ω) be a Calabi–Yau m-fold for $m \geqslant 3$. Then M is projective. That is, (M, J) is isomorphic as a complex manifold to a complex submanifold of $\mathbb{CP}^N$, and is an algebraic variety.*

This shows that Calabi–Yau manifolds (or at least, the complex manifolds underlying them) can be studied using *complex algebraic geometry*, which is the point of view of Part II. For a survey of constructions of examples of Calabi–Yau manifolds, see §14.3.

5.5 Compact Hyperkähler Manifolds

Let (M, g) be a compact Riemannian $4m$-manifold with $\text{Hol}(g) = \text{Sp}(m)$, that is, a *compact hyperkähler manifold*. Then as in §4.6 there exist natural complex structures J_1, J_2, J_3 and Kähler forms $\omega_1, \omega_2, \omega_3$. Let us choose $J = J_1$ as our distinguished complex structure, and write $\omega_c = \omega_2 + i\omega_3$. Then (M, J, g) is a Kähler manifold, and ω_c is a *holomorphic symplectic structure* on (M, J), in the sense of Definition 4.6.

As (M, J, g) is a compact Ricci-flat Kähler $2m$-fold, Proposition 5.5 shows that $H^{p,0}(M)$ is isomorphic to the subspace of $\Lambda^p(\mathbb{C}^{2m})^*$ invariant under $\text{Sp}(m)$. If $p = 2k$ for $0 \leqslant k \leqslant m$ this subspace is $\mathbb{C}$, generated by ω_c^k, and otherwise it is zero. Thus $h^{2k,0} = 1$ for $0 \leqslant k \leqslant m$ and $h^{p,0} = 0$ otherwise. Also, following Proposition 5.7 we may show that M is simply-connected; the only difference is that $\hat{A}(M) = m+1$ rather than 2, by [113, Th. 3.6.5]. Thus we prove:

Theorem 5.9 *Let (M, g) be a compact hyperkähler $4m$-manifold and J an associated complex structure, and $h^{p,q}$ the Hodge numbers of (M, J). Then M is simply-connected, $h^{2k,0} = 1$ for $0 \leqslant k \leqslant m$, and $h^{p,0} = 0$ otherwise.*

Since $h^{2,0}(M) = 1$, for hyperkähler manifolds the underlying complex manifold (M, J) generically has few holomorphic line bundles, and is generically not projective, in contrast to Theorem 5.8 for Calabi–Yau m-folds.

Definition 5.10 We say that a compact holomorphic symplectic manifold (M, J, ω_c) is *irreducible* if M is simply-connected and $H^{2,0}(M) = \langle \omega_c \rangle$.

We can use this to give a criterion for when a holomorphic symplectic manifold admits hyperkähler metrics, which will be useful in Theorem 23.5 of Part III.

Theorem 5.11 *Let (M, J, ω_c) be a compact irreducible holomorphic symplectic $2m$-manifold admitting Kähler metrics. Then each Kähler class contains a unique hyperkähler metric g with $\mathrm{Hol}(g) = \mathrm{Sp}(m)$. Conversely, if (M, g) is a compact hyperkähler manifold with holomorphic symplectic structure (J, ω_c), then (M, J, ω_c) is irreducible.*

Proof. If (M, J, ω_c) is a compact holomorphic symplectic $2m$-manifold then $\Omega = \omega_c^m$ is a holomorphic volume form on M and $c_1(M) = 0$, as in §4.6. So by Theorem 5.2, each Kähler class contains a unique Ricci-flat Kähler metric g.

The proof of Proposition 5.5 shows that $\nabla \omega_c = 0$, where ∇ is the Levi-Civita connection of g. Thus ω_c is a *constant tensor*, and is fixed by $\mathrm{Hol}(g)$ by Theorem 2.3. As J, g are also constant tensors we have $\mathrm{Hol}(g) \subseteq \mathrm{Sp}(m)$, since $\mathrm{Sp}(m)$ is the subgroup of $\mathrm{GL}(4m, \mathbb{R})$ fixing J, g and ω_c.

Suppose for a contradiction that (M, g) is *reducible*, in the sense of §3.1. As M is simply-connected by Definition 5.10, by a result of de Rham (M, g) splits as a Riemannian product of irreducible factors $(M_1, g_1) \times \ldots \times (M_k, g_k)$ for $k > 1$. Theorem 3.6 gives $\mathrm{Hol}(g_j) = \mathrm{Sp}(m_j)$, where $\dim_{\mathbb{R}} M_j = 4m_j$.

But then $h^{1,0}(M_j) = 0$ and $h^{2,0}(M_j) = 1$ by Theorem 5.9, so $h^{2,0}(M) = k > 1$. This contradicts $h^{2,0}(M) = 1$ in Definition 5.10, as (M, J, ω_c) is irreducible. Thus g is irreducible, and Theorem 3.6 implies that $\mathrm{Hol}(g) = \mathrm{Sp}(m)$, as we want.

Conversely, suppose (M, g) is a compact hyperkähler manifold with holomorphic symplectic structure (J, ω_c). Then Theorem 5.9 implies that M is simply-connected and $h^{2,0}(M) = 1$, so that $H^{2,0}(M) = \langle \omega_c \rangle$. Thus (M, J, ω_c) is irreducible. $\qquad \square$

6   The Exceptional Holonomy Groups

In Berger's classification of Riemannian holonomy groups, Theorem 3.6, the exceptional cases are the holonomy group G_2 in 7 dimensions, and $\mathrm{Spin}(7)$ in 8 dimensions. We now discuss these very briefly. A much more extensive treatment can be found in the author's book [113], which is the source of all the material in this section.

6.1   The Holonomy Group G_2

Let $(x_1, \ldots, x_7)$ be coordinates on $\mathbb{R}^7$. Write $dx_{ij\ldots l}$ for the exterior form $dx_i \wedge dx_j \wedge \cdots \wedge dx_l$ on $\mathbb{R}^7$. Define a metric g_0, a 3-form φ_0 and a 4-form $*\varphi_0$ on $\mathbb{R}^7$ by $g_0 = dx_1^2 + \cdots + dx_7^2$,

$$\begin{aligned}
\varphi_0 &= dx_{123} + dx_{145} + dx_{167} + dx_{246} - dx_{257} - dx_{347} - dx_{356} \text{ and} \\
*\varphi_0 &= dx_{4567} + dx_{2367} + dx_{2345} + dx_{1357} - dx_{1346} - dx_{1256} - dx_{1247}.
\end{aligned} \tag{13}$$

The subgroup of $GL(7, \mathbb{R})$ preserving φ_0 is the *exceptional Lie group* G_2. It also preserves g_0, $*\varphi_0$ and the orientation on $\mathbb{R}^7$. It is a compact, semisimple, 14-dimensional Lie group, a subgroup of $SO(7)$.

A G_2-*structure* on a 7-manifold M is a principal subbundle of the frame bundle of M, with structure group G_2. Each G_2-structure gives rise to a 3-form φ and a metric g on M, such that every tangent space of M admits an isomorphism with $\mathbb{R}^7$ identifying φ and g with φ_0 and g_0 respectively. By an abuse of notation, we will refer to (φ, g) as a G_2-structure.

Here are some basic facts about G_2-structures, [113, Prop. 10.1.3].

Proposition 6.1 *Let M be a 7-manifold and (φ, g) a G_2-structure on M. Then the following are equivalent:*

(i) Hol$(g) \subseteq G_2$, *and φ is the induced 3-form,*
(ii) $\nabla\varphi = 0$ *on M, where ∇ is the Levi-Civita connection of g, and*
(iii) $d\varphi = d^*\varphi = 0$ *on M.*

If (i)–(iii) hold, then g is Ricci-flat.

We call $\nabla\varphi$ the *torsion* of the G_2-structure (φ, g), and when $\nabla\varphi = 0$ the G_2-structure is *torsion-free*. A triple (M, φ, g) is called a G_2-*manifold* if M is a 7-manifold and (φ, g) a torsion-free G_2-structure on M. By Proposition 6.1, if (M, φ, g) is a G_2-manifold then Hol$(g) \subseteq G_2$. Here is a topological criterion [113, Prop. 10.2.2] to determine when Hol$(g) = G_2$.

Theorem 6.2 *Let (M, φ, g) be a compact G_2-manifold. Then Hol$(g) = G_2$ if and only if $\pi_1(M)$ is finite.*

By [113, Th. 10.4.4], metrics with holonomy G_2 on compact 7-manifolds occur in smooth, finite-dimensional moduli spaces.

Theorem 6.3 *Let M be a compact 7-manifold, $\mathscr{X}$ the family of torsion-free G_2-structures (φ, g) on M, and $\mathscr{D}$ the group of diffeomorphisms of M isotopic to the identity. Then $\mathscr{M} = \mathscr{X}/\mathscr{D}$ is a smooth manifold of dimension $b^3(M)$, and the projection $\pi : \mathscr{M} \to H^3(M, \mathbb{R})$ given by $(\varphi, g)\mathscr{D} \mapsto [\varphi]$ is a local diffeomorphism.*

6.2   The Holonomy Group Spin(7)

Let $\mathbb{R}^8$ have coordinates $(x_1, \ldots, x_8)$. Define a 4-form Ω_0 on $\mathbb{R}^8$ by

$$
\begin{aligned}
\Omega_0 = {}& dx_{1234} + dx_{1256} + dx_{1278} + dx_{1357} - dx_{1368} - dx_{1458} - dx_{1467} \\
& - dx_{2358} - dx_{2367} - dx_{2457} + dx_{2468} + dx_{3456} + dx_{3478} + dx_{5678}.
\end{aligned}
\tag{14}
$$

The subgroup of $GL(8, \mathbb{R})$ preserving Ω_0 is the holonomy group Spin(7). It also preserves the orientation on $\mathbb{R}^8$ and the Euclidean metric $g_0 = dx_1^2 + \cdots + dx_8^2$. It is a compact, semisimple, 21-dimensional Lie group, a subgroup of SO(8).

A Spin(7)-structure on an 8-manifold M gives rise to a 4-form Ω and a metric g on M, such that each tangent space of M admits an isomorphism with $\mathbb{R}^8$ identifying Ω and g with Ω_0 and g_0 respectively. By an abuse of notation we will refer to the pair (Ω, g) as a Spin(7)-*structure*.

Here are some basic facts about Spin(7)-structures, [113, Prop. 10.5.3].

Proposition 6.4 *Let M be an 8-manifold and (Ω, g) a Spin(7)-structure on M. Then the following are equivalent:*

   (i) $\mathrm{Hol}(g) \subseteq \mathrm{Spin}(7)$, and Ω is the induced 4-form,
   (ii) $\nabla \Omega = 0$ on M, where ∇ is the Levi-Civita connection of g, and
   (iii) $d\Omega = 0$ on M.

If (i)–(iii) *hold, then g is Ricci-flat.*

We call $\nabla \Omega$ the *torsion* of the Spin(7)-structure (Ω, g), and (Ω, g) *torsion-free* if $\nabla \Omega = 0$. A triple (M, Ω, g) is called a Spin(7)-*manifold* if M is an 8-manifold and (Ω, g) a torsion-free Spin(7)-structure on M. By Proposition 6.4, if (M, Ω, g) is a Spin(7)-manifold then $\mathrm{Hol}(g) \subseteq \mathrm{Spin}(7)$. Here is a topological criterion [113, Prop. 10.6.1] to determine when $\mathrm{Hol}(g) = \mathrm{Spin}(7)$.

Theorem 6.5 *Let (M, Ω, g) be a compact Spin(7)-manifold. Then $\mathrm{Hol}(g) = \mathrm{Spin}(7)$ if and only if M is simply-connected, and $b^3(M) + b^4_+(M) = b^2(M) + 2b^4_-(M) + 25$.*

By [113, Th. 10.7.1], holonomy Spin(7) metrics on compact 8-manifolds occur in smooth, finite-dimensional moduli spaces.

Theorem 6.6 *Let M be a compact 8-manifold, $\mathscr{X}$ the family of torsion-free Spin(7)-structures (Ω, g) on M, and $\mathscr{D}$ the group of diffeomorphisms of M isotopic to the identity. Then $\mathscr{M} = \mathscr{X}/\mathscr{D}$ is a smooth manifold of dimension $\hat{A}(M) + b^1(M) + b^4_-(M)$, and the projection $\pi : \mathscr{M} \to H^4(M, \mathbb{R})$ given by $(\Omega, g)\mathscr{D} \mapsto [\Omega]$ is an immersion.*

Here $\hat{A}(M)$ is the *$\hat{A}$-genus* of M. If M admits Spin(7)-structures then

$$
24\hat{A}(M) = -1 + b^1(M) - b^2(M) + b^3(M) + b^4_+(M) - 2b^4_-(M).
$$

If $\mathrm{Hol}(g) = \mathrm{Spin}(7)$ then $\hat{A}(M) = 1$ and $b^1(M) = 0$, so $\dim \mathcal{M} = 1 + b^4_-(M)$.
There are inclusions between $\mathrm{SU}(m), G_2$ and $\mathrm{Spin}(7)$:

$$
\begin{array}{ccccc}
\mathrm{SU}(2) & \longrightarrow & \mathrm{SU}(3) & \longrightarrow & G_2 \\
\downarrow & & \downarrow & & \downarrow \\
\mathrm{SU}(2) \times \mathrm{SU}(2) & \longrightarrow & \mathrm{SU}(4) & \longrightarrow & \mathrm{Spin}(7).
\end{array}
$$

These are significant in constructing examples of 7- and 8-manifolds with holonomy G_2 and $\mathrm{Spin}(7)$. For instance, if X is a Calabi–Yau 3-fold then $\mathbb{R} \times X$ and $\mathcal{S}^1 \times X$ have torsion-free G_2-structures, as $\mathrm{SU}(3) \hookrightarrow G_2$. So we can use Calabi Conjecture methods to produce torsion-free G_2- and $\mathrm{Spin}(7)$-structures, which can then be patched together using analysis.

6.3 Constructing Compact Manifolds with Exceptional Holonomy

For some time after Berger's classification, the exceptional holonomy groups remained a mystery. In 1987, Bryant [30] used the theory of exterior differential systems to show that locally there exist many metrics with these holonomy groups, and gave some explicit, incomplete examples. Then in 1989, Bryant and Salamon [33] found explicit, *complete* metrics with holonomy G_2 and $\mathrm{Spin}(7)$ on noncompact manifolds.

In 1994-5 the author constructed the first examples of metrics with holonomy G_2 and $\mathrm{Spin}(7)$ on *compact* manifolds [109–111]. Later this construction was made more powerful by the author [113], to give many more examples. We now briefly describe the method used in [109,110] and [113, §11–§12] to construct examples of compact 7-manifolds with holonomy G_2. It is based on the *Kummer construction* for Calabi–Yau metrics on the K3 surface, and may be divided into four steps.

Step 1. Let T^7 be the 7-torus and (φ_0, g_0) a flat G_2-structure on T^7. Choose a finite group Γ of isometries of T^7 preserving (φ_0, g_0). Then the quotient T^7/Γ is a singular, compact 7-manifold, an *orbifold*.

Step 2. For certain special groups Γ the singularities of T^7/Γ are locally modelled on $\mathbb{R}^3 \times \mathbb{C}^2/G$ or $\mathbb{R} \times \mathbb{C}^3/G$ for finite G in $\mathrm{SU}(2)$ or $\mathrm{SU}(3)$. Then there is a method to resolve the singularities of T^7/Γ in a natural way using complex geometry. Locally we replace $\mathbb{R}^3 \times \mathbb{C}^2/G$ or $\mathbb{R} \times \mathbb{C}^3/G$ by $\mathbb{R}^3 \times X$ or $\mathbb{R} \times Y$, where X, Y are *crepant resolutions* of $\mathbb{C}^2/G$ and $\mathbb{C}^3/G$. This gives a nonsingular, compact 7-manifold M, together with a map $\pi : M \to T^7/\Gamma$, the resolving map. We choose Γ so that $\pi_1(M)$ is finite.

Step 3. We write down a 1-parameter family of G_2-structures (φ_t, g_t) on M depending on $t \in (0, \epsilon)$. They are not torsion-free, but have small torsion when t is small. They are defined using a family of Calabi–Yau metrics g_t on the resolutions X, Y of $\mathbb{C}^2/G$ or $\mathbb{C}^3/G$ used to define M. The g_t satisfy asymptotic conditions at infinity in X, Y, and are constructed using Calabi Conjecture methods. As $t \to 0$, the G_2-structure (φ_t, g_t) converges to the singular G_2-structure $\pi^*(\varphi_0, g_0)$.

Step 4. We prove using analysis that for sufficiently small t, the G_2-structure (φ_t, g_t) on M, with small torsion, can be deformed to a G_2-structure $(\tilde{\varphi}_t, \tilde{g}_t)$, with zero torsion. Finally, Theorem 6.2 shows that $\tilde{g}_t$ is a metric with holonomy G_2 on the compact 7-manifold M.

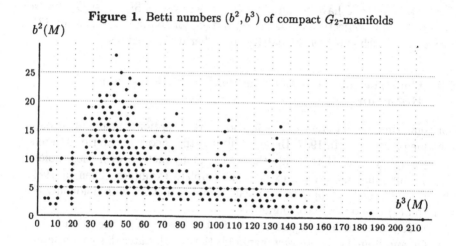

Figure 1. Betti numbers (b^2, b^3) of compact G_2-manifolds

By considering different groups Γ acting on T^7, and also by finding topo-logically distinct resolutions $M_1, \ldots, M_k$ of the same orbifold T^7/Γ, we can construct many compact Riemannian 7-manifolds with holonomy G_2. A good number of examples are given in [113, §12]. Figure 1 displays the Betti numbers of compact, simply-connected 7-manifolds with holonomy G_2 con-structed there. There are 252 different sets of Betti numbers.

Compact 8-manifolds with holonomy Spin(7) are constructed in [111] and [113, §13–§14] by a very similar method, involving resolving the singularities of orbifolds T^8/Γ. The examples in [113, §14] yield 52 different sets of Betti numbers, and rather more topologically distinct 8-manifolds.

There are also constructions of compact manifolds with exceptional holon-omy starting with a Calabi–Yau manifold or orbifold, rather than a flat torus T^7 or T^8. Kovalev [137] makes compact 7-manifolds with holonomy G_2 by finding two noncompact, asymptotically cylindrical Calabi–Yau 3-folds X_1, X_2, and gluing together $S^1 \times X_1$ and $S^1 \times X_2$ near infinity.

In the same spirit the author [112], [113, §15] constructs compact 8-manifolds with holonomy Spin(7) by starting with a Calabi–Yau 4-orbifold Y and resolving the singularities of $Y/\langle\sigma\rangle$, where $\sigma : Y \to Y$ is an antiholo-morphic involution satisfying certain conditions.

Table 1. Betti numbers (b^2, b^3, b^4) of compact Spin(7)-manifolds

$(4, 33, 200)$	$(3, 33, 202)$	$(2, 33, 204)$	$(1, 33, 206)$	$(0, 33, 208)$
$(1, 0, 908)$	$(0, 0, 910)$	$(1, 0, 1292)$	$(0, 0, 1294)$	$(1, 0, 2444)$
$(0, 0, 2446)$	$(0, 6, 3730)$	$(0, 0, 4750)$	$(0, 0, 11\,662)$	

The Betti numbers (b^2, b^3, b^4) of these examples are given in Table 1. Comparing them with those of the compact 8-manifolds constructed in [113, §14] by resolving torus orbifolds T^8/Γ, we see that in the examples constructed from Calabi–Yau 4-orbifolds the middle Betti number b^4 is much bigger, as much as $11\,662$ in one case.

6.4   Exercises

6.1 Show that the forms φ_0, $*\varphi_0$ in (13) are *generic* 3- and 4-forms on $\mathbb{R}^7$.

Hint: Consider the $GL(7, \mathbb{R})$-orbits of φ_0, $*\varphi_0$ in $\Lambda^k(\mathbb{R}^7)^*$.

6.2 Show that if (φ, g) is a G_2-structure then g is wholly determined by φ. Why should the equations $d\varphi = d^*\varphi = 0$ in part (iii) of Proposition 6.1 be regarded as a *nonlinear* equation on φ?

6.3 Generalize π in Theorem 6.3 to $\Pi : \mathscr{M} \to H^3(M, \mathbb{R}) \times H^4(M, \mathbb{R})$, given by $\Pi : (\varphi, g)\mathscr{D} \mapsto ([\varphi], [*\varphi])$. By Poincaré duality $H^4(M, \mathbb{R}) \cong H^3(M, \mathbb{R})^*$, and so $H^3(M, \mathbb{R}) \times H^4(M, \mathbb{R})$ has a natural symplectic structure. Show that the image of Π is a *Lagrangian submanifold* of $H^3(M, \mathbb{R}) \times H^4(M, \mathbb{R})$.

6.4 Show that the form Ω_0 of (14) is not a generic 4-form on $\mathbb{R}^8$.

7   Introduction to Calibrated Geometry

The theory of *calibrated geometry* was invented by Harvey and Lawson [93]. It concerns *calibrated submanifolds*, a special kind of *minimal submanifold* of a Riemannian manifold M, which are defined using a closed form on M called a *calibration*. It is closely connected with the theory of Riemannian holonomy groups because Riemannian manifolds with special holonomy usually come equipped with one or more natural calibrations.

Some references for this section are Harvey and Lawson [93, §I, §II], Harvey [92] and the author [113, §3.7]. Some background reading on minimal submanifolds and Geometric Measure Theory is Lawson [140] and Morgan [157].

7.1 Minimal Submanifolds

Let (M, g) be an n-dimensional Riemannian manifold, and N a compact k-dimensional submanifold of M. Regard N as an immersed submanifold (N, ι), with immersion $\iota : N \to M$. Using the metric g we can define the *volume* $\mathrm{Vol}(N)$ of N, by integration over N. We call N a *minimal submanifold* if its volume is stationary under small variations of the immersion $\iota : N \to M$. When $k = 1$, a curve in M is minimal if and only if it is a *geodesic*.

Let $\nu \to N$ be the normal bundle of N in M, so that $TM|_N = TN \oplus \nu$ is an orthogonal direct sum. The *second fundamental form* is a section B of $S^2 T^* N \otimes \nu$ such that whenever v, w are vector fields on M with $v|_N, w|_N$ sections of TN over N, then $B \cdot (v|_N \otimes w|_N) = \pi_\nu (\nabla_v w|_N)$, where '$\cdot$' contracts $S^2 T^* N$ with $TN \otimes TN$, ∇ is the Levi-Civita connection of g, and π_ν is the projection to ν in the splitting $TM|_N = TN \oplus \nu$.

The *mean curvature vector* κ of N is the trace of the second fundamental form B taken using the metric g on N. It is a section of the normal bundle ν. It can be shown by the Euler–Lagrange method that a submanifold N is minimal if and only if its mean curvature vector κ is zero. Note that this is a local condition. Therefore we can also define noncompact submanifolds N in M to be minimal if they have zero mean curvature. This makes sense even when N has infinite volume.

If $\iota : N \to M$ is a immersed submanifold, then the mean curvature κ of N depends on ι and its first and second derivatives, so the condition that N be minimal is a *second-order* equation on ι. Note that minimal submanifolds may not have minimal area, even amongst nearby homologous submanifolds. For instance, the equator in S^2 is minimal, but does not minimize length amongst lines of latitude.

The following argument is important in the study of minimal submanifolds. Let (M, g) be a compact Riemannian manifold, and α a nonzero homology class in $H_k(M, \mathbb{Z})$. We would like to find a compact, minimal immersed, k-dimensional submanifold N in M with homology class $[N] = \alpha$. To do this, we choose a minimizing sequence $(N_i)_{i=1}^\infty$ of compact submanifolds N_i with $[N_i] = \alpha$, such that $\mathrm{Vol}(N_i)$ approaches the infimum of volumes of submanifolds with homology class α as $i \to \infty$.

Pretend for the moment that the set of all closed k-dimensional submanifolds N with $\mathrm{Vol}(N) \leqslant C$ is a *compact* topological space. Then there exists a subsequence $(N_{i_j})_{j=1}^\infty$ which converges to some submanifold N, which is the minimal submanifold we want. In fact this does not work, because the set of submanifolds N does not have the compactness properties we need.

However, if we work instead with *rectifiable currents*, which are a measure-theoretic generalization of submanifolds, one can show that every integral homology class α in $H_k(M, \mathbb{Z})$ is represented by a minimal rectifiable current. One should think of rectifiable currents as a class of singular submanifolds, obtained by completing the set of nonsingular submanifolds with respect to some norm. They are studied in the subject of *Geometric Measure Theory*.

The question remains: how close are these minimal rectifiable currents to being submanifolds? For example, by a major result of Almgren [1] it is known that a k-dimensional minimal rectifiable current in a Riemannian n-manifold is an embedded submanifold except on a singular set of Hausdorff dimension at most $k - 2$. When $k = 2$ or $k = n - 1$ one can go further. In general, it is important to understand the possible singularities of such singular minimal submanifolds.

7.2 Calibrations and Calibrated Submanifolds

Let (M, g) be a Riemannian manifold. An *oriented tangent k-plane* V on M is a vector subspace V of some tangent space $T_x M$ to M with $\dim V = k$, equipped with an orientation. If V is an oriented tangent k-plane on M then $g|_V$ is a Euclidean metric on V, so combining $g|_V$ with the orientation on V gives a natural *volume form* vol_V on V, which is a k-form on V.

Now let φ be a closed k-form on M. We say that φ is a *calibration* on M if for every oriented k-plane V on M we have $\varphi|_V \leqslant \mathrm{vol}_V$. Here $\varphi|_V = \alpha \cdot \mathrm{vol}_V$ for some $\alpha \in \mathbb{R}$, and $\varphi|_V \leqslant \mathrm{vol}_V$ if $\alpha \leqslant 1$. Let N be an oriented submanifold of M with dimension k. Then each tangent space $T_x N$ for $x \in N$ is an oriented tangent k-plane. We say that N is a *calibrated submanifold* or *φ-submanifold* if $\varphi|_{T_x N} = \mathrm{vol}_{T_x N}$ for all $x \in N$.

All calibrated submanifolds are automatically *minimal submanifolds*. We prove this in the compact case, but it is true for noncompact submanifolds as well.

Proposition 7.1 *Let (M, g) be a Riemannian manifold, φ a calibration on M, and N a compact φ-submanifold in M. Then N is volume-minimizing in its homology class.*

Proof. Let $\dim N = k$, and let $[N] \in H_k(M, \mathbb{R})$ and $[\varphi] \in H^k(M, \mathbb{R})$ be the homology and cohomology classes of N and φ. Then

$$[\varphi] \cdot [N] = \int_{x \in N} \varphi|_{T_x N} = \int_{x \in N} \mathrm{vol}_{T_x N} = \mathrm{Vol}(N),$$

since $\varphi|_{T_x N} = \mathrm{vol}_{T_x N}$ for each $x \in N$, as N is a calibrated submanifold. If N' is any other compact k-submanifold of M with $[N'] = [N]$ in $H_k(M, \mathbb{R})$, then

$$[\varphi] \cdot [N] = [\varphi] \cdot [N'] = \int_{x \in N'} \varphi|_{T_x N'} \leqslant \int_{x \in N'} \mathrm{vol}_{T_x N'} = \mathrm{Vol}(N'),$$

since $\varphi|_{T_x N'} \leqslant \mathrm{vol}_{T_x N'}$ because φ is a calibration. The last two equations give $\mathrm{Vol}(N) \leqslant \mathrm{Vol}(N')$. Thus N is volume-minimizing in its homology class. $\square$

Now let (M, g) be a Riemannian manifold with a calibration φ, and let $\iota : N \to M$ be an immersed submanifold. Whether N is a φ-submanifold depends upon the tangent spaces of N. That is, it depends on ι and its first derivative. So, to be calibrated with respect to φ is a *first-order* equation on ι. But if N is calibrated then N is minimal, and we saw in §7.1 that to be minimal is a *second-order* equation on ι.

One moral is that the calibrated equations, being first-order, are often easier to solve than the minimal submanifold equations, which are second-order. So calibrated geometry is a fertile source of examples of minimal submanifolds.

7.3   Calibrated Submanifolds of $\mathbb{R}^n$

One simple class of calibrations is to take (M, g) to be $\mathbb{R}^n$ with the Euclidean metric, and φ to be a constant k-form on $\mathbb{R}^n$, such that $\varphi|_V \leqslant \mathrm{vol}_V$ for every oriented k-dimensional vector subspace $V \subseteq \mathbb{R}^n$. Each such φ defines a class of minimal k-submanifolds in $\mathbb{R}^n$. However, this class may be very small, or even empty. For instance, $\varphi = 0$ is a calibration on $\mathbb{R}^n$, but has no calibrated submanifolds.

For each constant calibration k-form φ on $\mathbb{R}^n$, define $\mathcal{F}_\varphi$ to be the set of oriented k-dimensional vector subspaces V of $\mathbb{R}^n$ such that $\varphi|_V = \mathrm{vol}_V$. Then an oriented submanifold N of $\mathbb{R}^n$ is a φ-submanifold if and only if each tangent space $T_x N$ lies in $\mathcal{F}_\varphi$. To be interesting, a calibration φ should define a fairly abundant class of calibrated submanifolds, and this will only happen if $\mathcal{F}_\varphi$ is reasonably large.

Define a *partial order* $\preceq$ on the set of constant calibration k-forms φ on $\mathbb{R}^n$ by $\varphi \preceq \varphi'$ if $\mathcal{F}_\varphi \subseteq \mathcal{F}_{\varphi'}$. A calibration φ is *maximal* if it is maximal with respect to this partial order. A maximal calibration φ is one in which $\mathcal{F}_\varphi$ is as large as possible.

It is an interesting problem to determine the maximal calibrations φ on $\mathbb{R}^n$. The symmetry group $G \subset \mathrm{O}(n)$ of a maximal calibration is usually quite large. This is because if $V \in \mathcal{F}_\varphi$ and $\gamma \in G$ then $\gamma \cdot V \in \mathcal{F}_\varphi$, that is, G acts on $\mathcal{F}_\varphi$. So if G is big we expect $\mathcal{F}_\varphi$ to be big too. Symmetry groups of maximal calibrations are often possible holonomy groups of Riemannian metrics, and the classification problem for maximal calibrations can be seen as in some ways parallel to the classification problem for Riemannian holonomy groups.

7.4   Calibrated Submanifolds and Special Holonomy

Next we explain the connection with Riemannian holonomy. Let $G \subset \mathrm{O}(n)$ be a possible holonomy group of a Riemannian metric. In particular, we can take G to be one of the holonomy groups $\mathrm{U}(m)$, $\mathrm{SU}(m)$, $\mathrm{Sp}(m)$, G_2 or $\mathrm{Spin}(7)$ from Berger's classification. Then G acts on the k-forms $\Lambda^k(\mathbb{R}^n)^*$ on $\mathbb{R}^n$, so we can look for G-invariant k-forms on $\mathbb{R}^n$.

Suppose φ_0 is a nonzero, G-invariant k-form on $\mathbb{R}^n$. By rescaling φ_0 we can arrange that for each oriented k-plane $U \subset \mathbb{R}^n$ we have $\varphi_0|_U \leqslant \mathrm{vol}_U$, and that $\varphi_0|_U = \mathrm{vol}_U$ for at least one such U. Thus $\mathcal{F}_{\varphi_0}$ is nonempty. Since

φ_0 is G-invariant, if $U \in \mathcal{F}_{\varphi_0}$ then $\gamma \cdot U \in \mathcal{F}_{\varphi_0}$ for all $\gamma \in G$. Generally this means that $\mathcal{F}_{\varphi_0}$ is 'reasonably large'.

Let M be a manifold of dimension n, and g a metric on M with Levi-Civita connection ∇ and holonomy group G. Then by Theorem 2.3 there is a k-form φ on M with $\nabla\varphi = 0$, corresponding to φ_0. Hence $d\varphi = 0$, and φ is closed. Also, the condition $\varphi_0|_U \leqslant \mathrm{vol}_U$ for all oriented k-planes U in $\mathbb{R}^n$ implies that $\varphi|_V \leqslant \mathrm{vol}_V$ for all oriented tangent k-planes V in M. Thus φ is a *calibration* on M.

At each point $x \in M$ the family of oriented tangent k-planes V with $\varphi|_V = \mathrm{vol}_V$ is isomorphic to $\mathcal{F}_{\varphi_0}$, which is 'reasonably large'. This suggests that locally there should exist many φ-submanifolds N in M, so the calibrated geometry of φ on (M, g) is nontrivial.

This gives us a general method for finding interesting calibrations on manifolds with reduced holonomy. Here are the most important examples of this.

- Let $G = \mathrm{U}(m) \subset \mathrm{O}(2m)$. Then G preserves a 2-form ω_0 on $\mathbb{R}^{2m}$. If g is a metric on M with holonomy $\mathrm{U}(m)$ then g is *Kähler* with complex structure J, and the 2-form ω on M associated to ω_0 is the *Kähler form* of g.

  One can show that ω is a calibration on (M, g), and the calibrated submanifolds are exactly the *holomorphic curves* in (M, J). More generally $\omega^k/k!$ is a calibration on M for $1 \leqslant k \leqslant m$, and the corresponding calibrated submanifolds are the complex k-dimensional submanifolds of (M, J).

- Let $G = \mathrm{SU}(m) \subset \mathrm{O}(2m)$. Compact manifolds (M, g) with holonomy $\mathrm{SU}(m)$ extend to *Calabi–Yau m-folds* (M, J, g, Ω), as in §4.5. The real part $\mathrm{Re}\,\Omega$ is a calibration on M, and the corresponding calibrated submanifolds are called *special Lagrangian submanifolds*.

- The group $G_2 \subset \mathrm{O}(7)$ preserves a 3-form φ_0 and a 4-form $*\varphi_0$ on $\mathbb{R}^7$. Thus a Riemannian 7-manifold (M, g) with holonomy G_2 comes with a 3-form φ and 4-form $*\varphi$, which are both calibrations. The corresponding calibrated submanifolds are called *associative 3-folds* and *coassociative 4-folds*.

- The group $\mathrm{Spin}(7) \subset \mathrm{O}(8)$ preserves a 4-form Ω_0 on $\mathbb{R}^8$. Thus a Riemannian 8-manifold (M, g) with holonomy $\mathrm{Spin}(7)$ has a 4-form Ω, which is a calibration. We call Ω-submanifolds *Cayley 4-folds*.

It is an important general principle that to each calibration φ on an n-manifold (M, g) with special holonomy we construct in this way, there corresponds a constant calibration φ_0 on $\mathbb{R}^n$. Locally, φ-submanifolds in M will look very like φ_0-submanifolds in $\mathbb{R}^n$, and have many of the same properties. Thus, to understand the calibrated submanifolds in a manifold with special holonomy, it is often a good idea to start by studying the corresponding calibrated submanifolds of $\mathbb{R}^n$.

In particular, singularities of φ-submanifolds in M will be locally modelled on singularities of φ_0-submanifolds in $\mathbb{R}^n$. (In the sense of Geometric Measure Theory, the *tangent cone* at a singular point of a φ-submanifold in M is a conical φ_0-submanifold in $\mathbb{R}^n$.) So by studying singular φ_0-submanifolds in $\mathbb{R}^n$, we may understand the singular behaviour of φ-submanifolds in M.

7.5   Exercises

7.1 The metric g and Kähler form ω on $\mathbb{C}^m$ are given by

$$g = |dz_1|^2 + \cdots + |dz_m|^2 \quad \text{and} \quad \omega = \frac{i}{2}(dz_1 \wedge d\bar{z}_1 + \cdots + dz_m \wedge d\bar{z}_m).$$

Show that a tangent 2-plane in $\mathbb{C}^m$ is calibrated w.r.t. ω if and only if it is a complex line in $\mathbb{C}^m$. (Harder) generalize to tangent $2k$-planes and $\frac{1}{k!}\omega^k$.

8   Calibrated Submanifolds in $\mathbb{R}^n$

In §7.4 we saw that Riemannian holonomy groups $G \subset O(n)$ yield interesting calibrations φ, both on $\mathbb{R}^n$ and on Riemannian n-manifolds (M, g) with $\mathrm{Hol}(g) = G$. The remainder of this part of the book will study *special Lagrangian m-folds* when $n = 2m$ and $G = \mathrm{SU}(m)$, and also *associative 3-folds* and *coassociative 4-folds* when $n = 7$ and $G = G_2$, and *Cayley 4-folds* when $n = 8$ and $G = \mathrm{Spin}(7)$.

We will concentrate mostly on special Lagrangian submanifolds, and then explain briefly how to generalize the material to the associative, coassociative and Cayley cases. This section discusses the basic theory of these submanifolds in $\mathbb{R}^n$.

8.1   Special Lagrangian Submanifolds in $\mathbb{C}^m$

We now discuss special Lagrangian submanifolds in $\mathbb{C}^m$. A reference for this section is Harvey and Lawson [93, §III.1–§III.2].

Definition 8.1 Let $\mathbb{C}^m \cong \mathbb{R}^{2m}$ have complex coordinates $(z_1, \ldots, z_m)$, and define a metric g, Kähler form ω and complex volume form Ω on $\mathbb{C}^m$ by

$$g = |dz_1|^2 + \cdots + |dz_m|^2, \quad \omega = \frac{i}{2}(dz_1 \wedge d\bar{z}_1 + \cdots + dz_m \wedge d\bar{z}_m),$$
$$\text{and} \quad \Omega = dz_1 \wedge \cdots \wedge dz_m. \tag{15}$$

Then $\mathrm{Re}\,\Omega$ and $\mathrm{Im}\,\Omega$ are real m-forms on $\mathbb{C}^m$. Let L be an oriented real submanifold of $\mathbb{C}^m$ of real dimension m. We call L a *special Lagrangian submanifold* of $\mathbb{C}^m$, or *SL m-fold* for short, if L is calibrated with respect to $\mathrm{Re}\,\Omega$, in the sense of §7.2.

In fact there is a more general definition involving a *phase* $e^{i\theta}$: if $\theta \in [0, 2\pi)$, we say that L is *special Lagrangian with phase* $e^{i\theta}$ if it is calibrated with respect to $\cos\theta\, \mathrm{Re}\, \Omega + \sin\theta\, \mathrm{Im}\, \Omega$. But we will not use this.

We shall identify the family $\mathcal{F}$ of tangent m-planes in $\mathbb{C}^m$ calibrated with respect to $\mathrm{Re}\, \Omega$. The subgroup of $\mathrm{GL}(2m, \mathbb{R})$ preserving g, ω and Ω is the Lie group $\mathrm{SU}(m)$ of complex unitary matrices with determinant 1. Define a real vector subspace U in $\mathbb{C}^m$ to be

$$U = \{(x_1, \ldots, x_m) : x_j \in \mathbb{R}\} \subset \mathbb{C}^m, \tag{16}$$

and let U have the usual orientation. Then U is calibrated w.r.t. $\mathrm{Re}\, \Omega$.

Furthermore, any oriented real vector subspace V in $\mathbb{C}^m$ calibrated w.r.t. $\mathrm{Re}\, \Omega$ is of the form $V = \gamma \cdot U$ for some $\gamma \in \mathrm{SU}(m)$. Therefore $\mathrm{SU}(m)$ acts transitively on $\mathcal{F}$. The stabilizer subgroup of U in $\mathrm{SU}(m)$ is the subset of matrices in $\mathrm{SU}(m)$ with real entries, which is $\mathrm{SO}(m)$. Thus $\mathcal{F} \cong \mathrm{SU}(m)/\mathrm{SO}(m)$, and we prove:

Proposition 8.2 *The family $\mathcal{F}$ of oriented real m-dimensional vector subspaces V in $\mathbb{C}^m$ with $\mathrm{Re}\, \Omega|_V = \mathrm{vol}_V$ is isomorphic to $\mathrm{SU}(m)/\mathrm{SO}(m)$, and has dimension $\frac{1}{2}(m^2 + m - 2)$.*

The dimension follows because $\dim \mathrm{SU}(m) = m^2 - 1$ and $\dim \mathrm{SO}(m) = \frac{1}{2}m(m-1)$. It is easy to see that $\omega|_U = \mathrm{Im}\, \Omega|_U = 0$. As $\mathrm{SU}(m)$ preserves ω and $\mathrm{Im}\, \Omega$ and acts transitively on $\mathcal{F}$, this gives $\omega|_V = \mathrm{Im}\, \Omega|_V = 0$ for any $V \in \mathcal{F}$. Conversely, if V is a real m-dimensional vector subspace of $\mathbb{C}^m$ and $\omega|_V = \mathrm{Im}\, \Omega|_V = 0$, then V lies in $\mathcal{F}$, with some orientation. This implies an alternative characterization of special Lagrangian submanifolds, [93, Cor. III.1.11]:

Proposition 8.3 *Let L be a real m-dimensional submanifold of $\mathbb{C}^m$. Then L admits an orientation making it into a special Lagrangian submanifold of $\mathbb{C}^m$ if and only if $\omega|_L \equiv 0$ and $\mathrm{Im}\, \Omega|_L \equiv 0$.*

Note that an m-dimensional submanifold L in $\mathbb{C}^m$ is called *Lagrangian* if $\omega|_L \equiv 0$. (This is a term from symplectic geometry, and ω is a symplectic structure.) Thus special Lagrangian submanifolds are Lagrangian submanifolds satisfying the extra condition that $\mathrm{Im}\, \Omega|_L \equiv 0$, which is how they get their name.

8.2 Special Lagrangian 2-folds in $\mathbb{C}^2$ and the Quaternions

The smallest interesting dimension, $m = 2$, is a special case. Let $\mathbb{C}^2$ have complex coordinates (z_1, z_2), complex structure I, and metric g, Kähler form ω and holomorphic 2-form Ω as in (15). Define real coordinates (x_0, x_1, x_2, x_3) on $\mathbb{C}^2 \cong \mathbb{R}^4$ by $z_0 = x_0 + ix_1$, $z_1 = x_2 + ix_3$. Then

$$g = dx_0^2 + \cdots + dx_3^2, \qquad \omega = dx_0 \wedge dx_1 + dx_2 \wedge dx_3,$$
$$\mathrm{Re}\, \Omega = dx_0 \wedge dx_2 - dx_1 \wedge dx_3 \quad \text{and} \quad \mathrm{Im}\, \Omega = dx_0 \wedge dx_3 + dx_1 \wedge dx_2.$$

Now define a *different* set of complex coordinates (w_1, w_2) on $\mathbb{C}^2 = \mathbb{R}^4$ by $w_1 = x_0 + ix_2$ and $w_2 = x_1 - ix_3$. Then $\omega - i\operatorname{Im}\Omega = dw_1 \wedge dw_2$.

But by Proposition 8.3, a real 2-submanifold $L \subset \mathbb{R}^4$ is special Lagrangian if and only if $\omega|_L \equiv \operatorname{Im}\Omega|_L \equiv 0$. Thus, L is special Lagrangian if and only if $(dw_1 \wedge dw_2)|_L \equiv 0$. But this holds if and only if L is a *holomorphic curve* with respect to the complex coordinates (w_1, w_2).

Here is another way to say this. There are *two different* complex structures I and J involved in this problem, associated to the two different complex coordinate systems (z_1, z_2) and (w_1, w_2) on $\mathbb{R}^4$. In the coordinates $(x_0, \ldots, x_3)$, I and J are given by

$$I\left(\tfrac{\partial}{\partial x_0}\right) = \tfrac{\partial}{\partial x_1}, \quad I\left(\tfrac{\partial}{\partial x_1}\right) = -\tfrac{\partial}{\partial x_0}, \quad I\left(\tfrac{\partial}{\partial x_2}\right) = \tfrac{\partial}{\partial x_3}, \quad I\left(\tfrac{\partial}{\partial x_3}\right) = -\tfrac{\partial}{\partial x_2},$$
$$J\left(\tfrac{\partial}{\partial x_0}\right) = \tfrac{\partial}{\partial x_2}, \quad J\left(\tfrac{\partial}{\partial x_1}\right) = -\tfrac{\partial}{\partial x_3}, \quad J\left(\tfrac{\partial}{\partial x_2}\right) = -\tfrac{\partial}{\partial x_0}, \quad J\left(\tfrac{\partial}{\partial x_3}\right) = \tfrac{\partial}{\partial x_1}.$$

The usual complex structure on $\mathbb{C}^2$ is I, but a 2-fold L in $\mathbb{C}^2$ is special Lagrangian if and only if it is holomorphic w.r.t. the alternative complex structure J. This means that special Lagrangian 2-folds are already very well understood, so we generally focus our attention on dimensions $m \geqslant 3$.

We can express all this in terms of the *quaternions* $\mathbb{H}$. The complex structures I, J anticommute, so that $IJ = -JI$, and $K = IJ$ is also a complex structure on $\mathbb{R}^4$, and $\langle 1, I, J, K \rangle$ is an algebra of automorphisms of $\mathbb{R}^4$ isomorphic to $\mathbb{H}$.

8.3 Special Lagrangian Submanifolds in $\mathbb{C}^m$ as Graphs

In symplectic geometry, there is a well-known way of manufacturing *Lagrangian* submanifolds of $\mathbb{R}^{2m} \cong \mathbb{C}^m$, which works as follows. Let $f : \mathbb{R}^m \to \mathbb{R}$ be a smooth function, and define

$$\Gamma_f = \left\{ \left(x_1 + i\tfrac{\partial f}{\partial x_1}(x_1, \ldots, x_m), \ldots, x_m + i\tfrac{\partial f}{\partial x_m}(x_1, \ldots, x_m) \right) : x_1, \ldots, x_m \in \mathbb{R} \right\}.$$

Then Γ_f is a smooth real m-dimensional submanifold of $\mathbb{C}^m$, with $\omega|_{\Gamma_f} \equiv 0$. Identifying $\mathbb{C}^m \cong \mathbb{R}^{2m} \cong \mathbb{R}^m \times (\mathbb{R}^m)^*$, we may regard Γ_f as the graph of the 1-form df on $\mathbb{R}^m$, so that Γ_f is the *graph of a closed 1-form*. Locally, but not globally, every Lagrangian submanifold arises from this construction.

Now by Proposition 8.3, a special Lagrangian m-fold in $\mathbb{C}^m$ is a Lagrangian m-fold L satisfying the additional condition that $\operatorname{Im}\Omega|_L \equiv 0$. We shall find the condition for Γ_f to be a special Lagrangian m-fold. Define the *Hessian* $\operatorname{Hess} f$ of f to be the $m \times m$ matrix $\left(\frac{\partial^2 f}{\partial x_i \partial x_j}\right)_{i,j=1}^m$ of real functions on $\mathbb{R}^m$. Then it is easy to show that $\operatorname{Im}\Omega|_{\Gamma_f} \equiv 0$ if and only if

$$\operatorname{Im}\det_{\mathbb{C}}\left(I_m + i\operatorname{Hess} f\right) \equiv 0 \quad \text{on } \mathbb{C}^m, \tag{17}$$

where I_m is the $m \times m$ identity matrix. This is a *nonlinear second-order elliptic partial differential equation* upon the function $f : \mathbb{R}^m \to \mathbb{R}$.

8.4   Local Discussion of Special Lagrangian Deformations

Suppose L_0 is a special Lagrangian submanifold in $\mathbb{C}^m$ (or, more generally, in some Calabi–Yau m-fold). What can we say about the family of *special Lagrangian deformations* of L_0, that is, the set of special Lagrangian m-folds L that are 'close to L_0' in a suitable sense? Essentially, deformation theory is one way of thinking about the question 'how many special Lagrangian submanifolds are there in $\mathbb{C}^m$?'.

Locally (that is, in small enough open sets), every special Lagrangian m-fold looks quite like $\mathbb{R}^m$ in $\mathbb{C}^m$. Therefore deformations of special Lagrangian m-folds should look like special Lagrangian deformations of $\mathbb{R}^m$ in $\mathbb{C}^m$. So, we would like to know what special Lagrangian m-folds L in $\mathbb{C}^m$ close to $\mathbb{R}^m$ look like.

As $\mathbb{R}^m$ is the graph Γ_f associated to the function $f \equiv 0$, a graph Γ_f will be close to $\mathbb{R}^m$ if the function f and its derivatives are small. But then Hess f is small, so we can approximate equation (17) by its *linearization*. Now

$$\operatorname{Im} \det_{\mathrm{c}}\big(I_m + i \operatorname{Hess} f\big) = \operatorname{Tr} \operatorname{Hess} f + \text{higher order terms}.$$

Thus, when the second derivatives of f are small, equation (17) reduces approximately to $\operatorname{Tr} \operatorname{Hess} f \equiv 0$. But $\operatorname{Tr} \operatorname{Hess} f = \frac{\partial^2 f}{(\partial x_1)^2} + \cdots + \frac{\partial^2 f}{(\partial x_m)^2} = \Delta f$, where Δ is the *Laplacian* on $\mathbb{R}^m$.

Hence, the small special Lagrangian deformations of $\mathbb{R}^m$ in $\mathbb{C}^m$ are approximately parametrized by small *harmonic functions* on $\mathbb{R}^m$. Actually, because adding a constant to f has no effect on Γ_f, this parametrization is degenerate. We can get round this by parametrizing instead by df, which is a closed and coclosed 1-form. This justifies the following:

Principle. *Small special Lagrangian deformations of a special Lagrangian m-fold L are approximately parametrized by closed and coclosed 1-forms α on L.*

This is the idea behind McLean's Theorem, Theorem 10.4 below.

We have seen using (17) that the deformation problem for special Lagrangian m-folds can be written as an *elliptic equation*. In particular, there are the same number of equations as functions, so the problem is neither overdetermined nor underdetermined. Therefore we do not expect special Lagrangian m-folds to be very few and very rigid (as would be the case if (17) were overdetermined), nor to be very abundant and very flabby (as would be the case if (17) were underdetermined).

If we think about Proposition 8.2 for a while, this may seem surprising. For the set $\mathcal{F}$ of special Lagrangian m-planes in $\mathbb{C}^m$ has dimension $\frac{1}{2}(m^2 + m - 2)$, but the set of all real m-planes in $\mathbb{C}^m$ has dimension m^2. So the special Lagrangian m-planes have codimension $\frac{1}{2}(m^2 - m + 2)$ in the set of all m-planes.

This means that the condition for a real m-submanifold L in $\mathbb{C}^m$ to be special Lagrangian is $\frac{1}{2}(m^2 - m + 2)$ real equations on each tangent space of L. However, the freedom to vary L is the sections of its normal bundle in $\mathbb{C}^m$, which is m real functions. When $m \geqslant 3$, there are more equations than functions, so we would expect the deformation problem to be *overdetermined*.

The explanation is that because ω is a *closed* 2-form, submanifolds L with $\omega|_L \equiv 0$ are much more abundant than would otherwise be the case. So the closure of ω is a kind of integrability condition necessary for the existence of many special Lagrangian submanifolds, just as the integrability of an almost complex structure is a necessary condition for the existence of many complex submanifolds of dimension greater than 1 in a complex manifold.

8.5   Associative and Coassociative Submanifolds of $\mathbb{R}^7$

As in §6.1, let g_0 be the Euclidean metric on $\mathbb{R}^7$, and define a 3-form φ_0 and 4-form $*\varphi_0$ on $\mathbb{R}^7$ by (13), so that g_0, φ_0 and $*\varphi_0$ are preserved by G_2 acting on $\mathbb{R}^7$. Then φ_0 and $*\varphi_0$ are both *calibrations* on $\mathbb{R}^7$. We define an *associative 3-fold* in $\mathbb{R}^7$ to be a 3-dimensional submanifold of $\mathbb{R}^7$ calibrated with respect to φ_0, and a *coassociative 4-fold* in $\mathbb{R}^7$ to be a 4-dimensional submanifold of $\mathbb{R}^7$ calibrated with respect to $*\varphi_0$.

Define an *associative 3-plane* to be an oriented 3-dimensional vector subspace V of $\mathbb{R}^7$ with $\varphi_0|_V = \mathrm{vol}_V$, and a *coassociative 4-plane* to be an oriented 4-dimensional vector subspace V of $\mathbb{R}^7$ with $*\varphi_0|_V = \mathrm{vol}_V$. By analogy with Proposition 8.2, we can prove:

Proposition 8.4 *The family $\mathcal{F}^3$ of associative 3-planes in $\mathbb{R}^7$ and the family $\mathcal{F}^4$ of coassociative 4-planes in $\mathbb{R}^7$ are both isomorphic to $G_2/SO(4)$, with dimension 8.*

There is also an analogue of Proposition 8.3 for coassociative 4-folds:

Proposition 8.5 *Let L be a real 4-dimensional submanifold of $\mathbb{R}^7$. Then L admits an orientation making it into a coassociative 4-fold of $\mathbb{R}^7$ if and only if $\varphi_0|_L \equiv 0$.*

The set of all 3-planes in $\mathbb{R}^7$ has dimension 12, and the set of associative 3-planes in $\mathbb{R}^7$ has dimension 8 by Proposition 8.4. Thus the associative 3-planes are of codimension 4 in the set of all 3-planes. This means that the condition for a 3-fold L in $\mathbb{R}^7$ to be associative is 4 equations on each tangent space. The freedom to vary L is the sections of its normal bundle in $\mathbb{R}^7$, which is 4 real functions.

Thus, the deformation problem for associative 3-folds involves 4 equations on 4 functions, so it is a *determined* problem. In fact, the relevant equation is *elliptic*, essentially the Dirac equation on L. This implies that the deformation theory of associative 3-folds is quite well-behaved.

For coassociative 4-folds, the deformation problem has 4 equations on 3 real functions, which is apparently overdetermined. But because φ_0 is closed, we can rewrite the problem as an elliptic equation, as we did for SL m-folds. So, the closure of φ_0 can be seen as an integrability condition for the existence of many coassociative 4-folds.

8.6   Cayley 4-folds in $\mathbb{R}^8$

As in §6.2, let g_0 be the Euclidean metric on $\mathbb{R}^8$, and define a 4-form Ω_0 on $\mathbb{R}^8$ by (14), so that g_0 and Ω_0 are preserved by Spin(7) acting on $\mathbb{R}^8$. Then Ω_0 is a *calibration* on $\mathbb{R}^8$. Submanifolds of $\mathbb{R}^8$ calibrated with respect to Ω_0 are called *Cayley 4-folds*.

Define a *Cayley 4-plane* to be an oriented 4-dimensional vector subspace V of $\mathbb{R}^8$ with $\Omega_0|_V = \mathrm{vol}_V$. Then one can prove the following analogue of Propositions 8.2 and 8.4.

Proposition 8.6 *The family $\mathcal{F}$ of Cayley 4-planes in $\mathbb{R}^8$ is isomorphic to* Spin(7)/K, *where* $K \cong (\mathrm{SU}(2) \times \mathrm{SU}(2) \times \mathrm{SU}(2))/\mathbb{Z}_2$ *is a Lie subgroup of* Spin(7). *The dimension of $\mathcal{F}$ is* 12.

The set of all 4-planes in $\mathbb{R}^8$ has dimension 16, so that the Cayley 4-planes have codimension 4. Thus, the deformation problem for a Cayley 4-fold L may be written as 4 real equations on 4 real functions, a determined problem. In fact this is an elliptic equation, essentially the positive Dirac equation upon L. Therefore the deformation theory of Cayley 4-folds is quite well-behaved.

8.7   Exercises

8.1 Find your own proofs of Propositions 8.2 and 8.3.

9   Constructions of SL m-folds in $\mathbb{C}^m$

We now describe five methods of constructing special Lagrangian m-folds in $\mathbb{C}^m$, drawn from papers by the author [124,115,116,122,123,117–119], Bryant [31], Castro and Urbano [40], Goldstein [69,68], Harvey [92, p. 139–143], Harvey and Lawson [93, §III], Haskins [94], Lawlor [139], Ma and Ma [146], McIntosh [154] and Sharipov [181]. These yield many examples of singular SL m-folds, and so hopefully will help in understanding what general singularities of SL m-folds in Calabi–Yau m-folds are like.

9.1   SL m-folds with Large Symmetry Groups

Here is a method used by Harvey and Lawson [93, §III.3], Haskins [94], Goldstein [69,68] and the author [124] to construct examples of SL m-folds in $\mathbb{C}^m$. The group $\mathrm{SU}(m) \ltimes \mathbb{C}^m$ acts on $\mathbb{C}^m$ preserving all the structure g, ω, Ω, so that it takes SL m-folds to SL m-folds in $\mathbb{C}^m$. Let G be a Lie subgroup of $\mathrm{SU}(m) \ltimes \mathbb{C}^m$ with Lie algebra $\mathfrak{g}$, and N a connected G-invariant SL m-fold in $\mathbb{C}^m$.

Since G preserves the symplectic form ω on $\mathbb{C}^m$, one can show that it has a *moment map* $\mu : \mathbb{C}^m \to \mathfrak{g}^*$. As N is Lagrangian, one can show that μ is constant on N, that is, $\mu \equiv c$ on N for some $c \in Z(\mathfrak{g}^*)$, the *centre* of $\mathfrak{g}^*$.

If the orbits of G in N are of codimension 1 (that is, dimension $m-1$), then N is a 1-parameter family of G-orbits $\mathcal{O}_t$ for $t \in \mathbb{R}$. After reparametrizing the variable t, it can be shown that the special Lagrangian condition is equivalent to an o.d.e. in t upon the orbits $\mathcal{O}_t$.

Thus, we can construct examples of cohomogeneity one SL m-folds in $\mathbb{C}^m$ by solving an o.d.e. in the family of $(m-1)$-dimensional G-orbits $\mathcal{O}$ in $\mathbb{C}^m$ with $\mu|_{\mathcal{O}} \equiv c$, for fixed $c \in Z(\mathfrak{g}^*)$. This o.d.e. usually turns out to be *integrable*.

Now suppose N is a *special Lagrangian cone* in $\mathbb{C}^m$, invariant under a subgroup $G \subset SU(m)$ which has orbits of dimension $m-2$ in N. In effect the symmetry group of N is $G \times \mathbb{R}_+$, where $\mathbb{R}_+$ acts by *dilations*, as N is a cone. Thus, in this situation too the symmetry group of N acts with cohomogeneity one, and we again expect the problem to reduce to an o.d.e.

One can show that $N \cap \mathcal{S}^{2m-1}$ is a 1-parameter family of G-orbits $\mathcal{O}_t$ in $\mathcal{S}^{2m-1} \cap \mu^{-1}(0)$ satisfying an o.d.e. By solving this o.d.e. we construct SL cones in $\mathbb{C}^m$. When $G = U(1)^{m-2}$, the o.d.e. has many *periodic solutions* which give large families of distinct SL cones on T^{m-1}. In particular, we can find many examples of SL T^2-cones in $\mathbb{C}^3$.

9.2 Evolution Equations for SL m-folds

The following method was used in [115,116] to construct examples of SL m-folds in $\mathbb{C}^m$. A related but less general method was used by Lawlor [139], and completed by Harvey [92, p. 139–143].

Let P be a real analytic $(m-1)$-dimensional manifold, and χ a nonvanishing real analytic section of $\Lambda^{m-1}TP$. Let $\{\phi_t : t \in \mathbb{R}\}$ be a 1-parameter family of real analytic maps $\phi_t : P \to \mathbb{C}^m$. Consider the o.d.e.

$$\left(\frac{\mathrm{d}\phi_t}{\mathrm{d}t}\right)^b = (\phi_t)_*(\chi)^{a_1 \ldots a_{m-1}} (\mathrm{Re}\,\Omega)_{a_1 \ldots a_{m-1} a_m} g^{a_m b}, \qquad (18)$$

using the index notation for (real) tensors on $\mathbb{C}^m$, where g^{ab} is the inverse of the Euclidean metric g_{ab} on $\mathbb{C}^m$.

It is shown in [115, §3] that if the ϕ_t satisfy (18) and $\phi_0^*(\omega) \equiv 0$, then $\phi_t^*(\omega) \equiv 0$ for all t, and $N = \{\phi_t(p) : p \in P,\ t \in \mathbb{R}\}$ is an SL m-fold in $\mathbb{C}^m$ wherever it is nonsingular. We think of (18) as an *evolution equation*, and N as the result of evolving a 1-parameter family of $(m-1)$-submanifolds $\phi_t(P)$ in $\mathbb{C}^m$.

Here is one way to understand this result. Suppose we are given $\phi_t : P \to \mathbb{C}^m$ for some t, and we want to find an SL m-fold N in $\mathbb{C}^m$ containing the $(m-1)$-submanifold $\phi_t(P)$. As N is Lagrangian, a necessary condition for this is that $\omega|_{\phi_t(P)} \equiv 0$, and hence $\phi_t^*(\omega) \equiv 0$ on P.

The effect of equation (18) is to flow $\phi_t(P)$ in the direction in which $\mathrm{Re}\,\Omega$ is 'largest'. The result is that $\mathrm{Re}\,\Omega$ is 'maximized' on N, given the initial conditions. But $\mathrm{Re}\,\Omega$ is maximal on N exactly when N is calibrated w.r.t.

Re Ω, that is, when N is special Lagrangian. The same technique also works for other calibrations, such as the associative and coassociative calibrations on $\mathbb{R}^7$, and the Cayley calibration on $\mathbb{R}^8$.

Now (18) evolves amongst the infinite-dimensional family of real analytic maps $\phi : P \to \mathbb{C}^m$ with $\phi^*(\omega) \equiv 0$, so it is an *infinite-dimensional* problem, and thus difficult to solve explicitly. However, there are *finite-dimensional* families C of maps $\phi : P \to \mathbb{C}^m$ such that evolution stays in C. This gives a *finite-dimensional* o.d.e., which can hopefully be solved fairly explicitly. For example, if we take G to be a Lie subgroup of $\mathrm{SU}(m) \ltimes \mathbb{C}^m$, P to be an $(m-1)$-dimensional homogeneous space G/H, and $\phi : P \to \mathbb{C}^m$ to be G-equivariant, we recover the construction of §9.1.

But there are also other possibilities for C which do not involve a symmetry assumption. Suppose P is a submanifold of $\mathbb{R}^n$, and χ the restriction to P of a linear or affine map $\mathbb{R}^n \to \Lambda^{m-1} \mathbb{R}^n$. (This is a strong condition on P and χ.) Then we can take C to be the set of restrictions to P of linear or affine maps $\mathbb{R}^n \to \mathbb{C}^m$.

For instance, set $m = n$ and let P be a quadric in $\mathbb{R}^m$. Then one can construct SL m-folds in $\mathbb{C}^m$ with few symmetries by evolving quadrics in Lagrangian planes $\mathbb{R}^m$ in $\mathbb{C}^m$. When P is a quadric cone in $\mathbb{R}^m$ this gives many SL cones on products of spheres $S^a \times S^b \times S^1$.

9.3 Ruled Special Lagrangian 3-folds

A 3-submanifold N in $\mathbb{C}^3$ is called *ruled* if it is fibred by a 2-dimensional family $\mathcal{F}$ of real lines in $\mathbb{C}^3$. A *cone* N_0 in $\mathbb{C}^3$ is called *two-sided* if $N_0 = -N_0$. Two-sided cones are automatically ruled. If N is a ruled 3-fold in $\mathbb{C}^3$, we define the *asymptotic cone* N_0 of N to be the two-sided cone fibred by the lines passing through 0 and parallel to those in $\mathcal{F}$.

Ruled SL 3-folds are studied by Harvey and Lawson [93, §III.3.C, §III.4.B], Bryant [31, §3] and the author [122]. Each (oriented) real line in $\mathbb{C}^3$ is determined by its *direction* in S^5 together with an orthogonal *translation* from the origin. Thus a ruled 3-fold N is determined by a 2-dimensional family of directions and translations.

The condition for N to be special Lagrangian turns out [122, §5] to reduce to two equations, the first involving only the direction components, and the second *linear* in the translation components. Hence, if a ruled 3-fold N in $\mathbb{C}^3$ is special Lagrangian, then so is its asymptotic cone N_0. Conversely, the ruled SL 3-folds N asymptotic to a given two-sided SL cone N_0 come from solutions of a linear equation, and so form a *vector space*.

Let N_0 be a two-sided SL cone, and let $\Sigma = N_0 \cap S^5$. Then Σ is a *Riemann surface*. Holomorphic vector fields on Σ give solutions to the linear equation (though not all solutions) [122, §6], and so yield new ruled SL 3-folds. In particular, each SL T^2-cone gives 2-dimensional family of ruled SL 3-folds, which are generically diffeomorphic to $T^2 \times \mathbb{R}$ as immersed 3-submanifolds.

9.4   Integrable Systems

Let N_0 be a special Lagrangian cone in $\mathbb{C}^3$, and set $\Sigma = N_0 \cap S^5$. As N_0 is calibrated, it is minimal in $\mathbb{C}^3$, and so Σ is minimal in S^5. That is, Σ is a *minimal Legendrian surface* in S^5. Let $\pi : S^5 \to \mathbb{CP}^2$ be the Hopf projection. One can also show that $\pi(\Sigma)$ is a *minimal Lagrangian surface* in $\mathbb{CP}^2$.

Regard Σ as a *Riemann surface*. Then the inclusions $\iota : \Sigma \to S^5$ and $\pi \circ \iota : \Sigma \to \mathbb{CP}^2$ are *conformal harmonic maps*. Now harmonic maps from Riemann surfaces into S^n and $\mathbb{CP}^m$ are an *integrable system*. There is a complicated theory for classifying them in terms of algebro-geometric 'spectral data', and finding 'explicit' solutions. In principle, this gives all harmonic maps from T^2 into S^n and $\mathbb{CP}^m$. So, the field of integrable systems offers the hope of a *classification* of all SL T^2-cones in $\mathbb{C}^3$.

For a good general introduction to this field, see Fordy and Wood [56]. Sharipov [181] and Ma and Ma [146] apply this integrable systems machinery to describe minimal Legendrian tori in S^5, and minimal Lagrangian tori in $\mathbb{CP}^2$, respectively, giving explicit formulae in terms of Prym theta functions. McIntosh [154] provides a more recent, readable, and complete discussion of special Lagrangian cones in $\mathbb{C}^3$ from the integrable systems perspective.

The families of SL T^2-cones constructed by U(1)-invariance in §9.1, and by evolving quadrics in §9.2, are both part of a more general, explicit, 'integrable systems' family of conformal harmonic maps $\mathbb{R}^2 \to S^5$ with Legendrian image, involving two commuting, integrable o.d.e.s, described in [123].

9.5   Analytic Construction of U(1)-invariant SL 3-folds in $\mathbb{C}^3$

Next we summarize the author's three papers [117–119], which study SL 3-folds N in $\mathbb{C}^3$ invariant under the U(1)-action

$$e^{i\theta} : (z_1, z_2, z_3) \mapsto (e^{i\theta} z_1, e^{-i\theta} z_2, z_3) \quad \text{for } e^{i\theta} \in U(1). \tag{19}$$

These three papers are briefly surveyed in [120]. Locally we can write N in the form

$$N = \big\{ (z_1, z_2, z_3) \in \mathbb{C}^3 : z_1 z_2 = v(x, y) + iy, \quad z_3 = x + iu(x, y),$$
$$|z_1|^2 - |z_2|^2 = 2a, \quad (x, y) \in S \big\}, \tag{20}$$

where S is a domain in $\mathbb{R}^2$, $a \in \mathbb{R}$ and $u, v : S \to \mathbb{R}$ are continuous.

Here we may take $|z_1|^2 - |z_2|^2 = 2a$ to be one of the equations defining N as $|z_1|^2 - |z_2|^2$ is the *moment map* of the U(1)-action (19), and so $|z_1|^2 - |z_2|^2$ is constant on any U(1)-invariant Lagrangian 3-fold in $\mathbb{C}^3$. Effectively (20) just means that we are choosing $x = \mathrm{Re}(z_3)$ and $y = \mathrm{Im}(z_1 z_2)$ as local coordinates on the 2-manifold $N/U(1)$. Then we find [117, Prop. 4.1]:

Proposition 9.1 *Let S, a, u, v and N be as above. Then*

(a) If $a = 0$, then N is a (possibly singular) SL 3-fold in $\mathbb{C}^3$ if u, v are differentiable and satisfy

$$\frac{\partial u}{\partial x} = \frac{\partial v}{\partial y} \quad \text{and} \quad \frac{\partial v}{\partial x} = -2(v^2 + y^2)^{1/2}\frac{\partial u}{\partial y}, \tag{21}$$

except at points $(x, 0)$ in S with $v(x, 0) = 0$, where u, v need not be differentiable. The singular points of N are those of the form $(0, 0, z_3)$, where $z_3 = x + iu(x, 0)$ for $(x, 0) \in S$ with $v(x, 0) = 0$.

(b) If $a \neq 0$, then N is a nonsingular SL 3-fold in $\mathbb{C}^3$ if and only if u, v are differentiable in S and satisfy

$$\frac{\partial u}{\partial x} = \frac{\partial v}{\partial y} \quad \text{and} \quad \frac{\partial v}{\partial x} = -2(v^2 + y^2 + a^2)^{1/2}\frac{\partial u}{\partial y}. \tag{22}$$

Now (21) and (22) are *nonlinear Cauchy–Riemann equations*. Thus, we may treat $u + iv$ as like a holomorphic function of $x + iy$. Many of the results in [117–119] are analogues of well-known results in elementary complex analysis.

In [117, Prop. 7.1] we show that solutions $u, v \in C^1(S)$ of (22) come from a potential $f \in C^2(S)$ satisfying a second-order quasilinear elliptic equation.

Proposition 9.2 *Let S be a domain in $\mathbb{R}^2$ and $u, v \in C^1(S)$ satisfy (22) for $a \neq 0$. Then there exists $f \in C^2(S)$ with $\frac{\partial f}{\partial y} = u$, $\frac{\partial f}{\partial x} = v$ and*

$$P(f) = \left(\left(\frac{\partial f}{\partial x}\right)^2 + y^2 + a^2\right)^{-1/2}\frac{\partial^2 f}{\partial x^2} + 2\frac{\partial^2 f}{\partial y^2} = 0. \tag{23}$$

This f is unique up to addition of a constant, $f \mapsto f + c$. Conversely, all solutions of (23) yield solutions of (22).

In the following result, a condensation of [117, Th. 7.6] and [118, Th.s 9.20 & 9.21], we prove existence and uniqueness for the *Dirichlet problem* for (23).

Theorem 9.3 *Suppose S is a strictly convex domain in $\mathbb{R}^2$ invariant under $(x, y) \mapsto (x, -y)$, and $\alpha \in (0, 1)$. Let $a \in \mathbb{R}$ and $\phi \in C^{3,\alpha}(\partial S)$. Then if $a \neq 0$ there exists a unique solution f of (23) in $C^{3,\alpha}(S)$ with $f|_{\partial S} = \phi$. If $a = 0$ there exists a unique $f \in C^1(S)$ with $f|_{\partial S} = \phi$, which is twice weakly differentiable and satisfies (23) with weak derivatives. Furthermore, the map $C^{3,\alpha}(\partial S) \times \mathbb{R} \to C^1(S)$ taking $(\phi, a) \mapsto f$ is continuous.*

Here a domain S in $\mathbb{R}^2$ is *strictly convex* if it is convex and the curvature of ∂S is nonzero at each point. Also domains are by definition compact, with smooth boundary, and $C^{3,\alpha}(\partial S)$ and $C^{3,\alpha}(S)$ are *Hölder spaces* of functions on ∂S and S. For more details see [117, 118].

Combining Propositions 9.1 and 9.2 and Theorem 9.3 gives existence and uniqueness for a large class of U(1)-invariant SL 3-folds in $\mathbb{C}^3$, with boundary conditions, and including *singular* SL 3-folds. It is interesting that this existence and uniqueness is *entirely unaffected* by singularities appearing in S°.

Here are some other areas covered in [117–119]. Examples of solutions u, v of (21) and (22) are given in [117, §5]. In [118] we give more precise statements on the regularity of singular solutions of (21) and (23). In [117, §6] and [119, §7] we consider the zeroes of $(u_1, v_1) - (u_2, v_2)$, where (u_j, v_j) are (possibly singular) solutions of (21) and (22).

We show that if $(u_1, v_1) \not\equiv (u_2, v_2)$ then the zeroes of $(u_1, v_1) - (u_2, v_2)$ in S° are *isolated*, with a positive integer *multiplicity*, and that the zeroes of $(u_1, v_1) - (u_2, v_2)$ in S° can be counted with multiplicity in terms of boundary data on ∂S. In particular, under some boundary conditions we can show $(u_1, v_1) - (u_2, v_2)$ has no zeroes in S°, so that the corresponding SL 3-folds do not intersect. This will be important in constructing U(1)-invariant SL fibrations in §12.5.

In [119, §9–§10] we study singularities of solutions u, v of (21). We show that either $u(x, -y) \equiv u(x, y)$ and $v(x, -y) \equiv -v(x, y)$, so that u, v are singular all along the x-axis, or else the singular points of u, v in S° are all *isolated*, with a positive integer *multiplicity*, and one of two *types*. We also show that singularities exist with every multiplicity and type, and multiplicity n singularities occur in codimension n in the family of all U(1)-invariant SL 3-folds.

9.6   Examples of Singular Special Lagrangian 3-folds in $\mathbb{C}^3$

We finish by describing four families of SL 3-folds in $\mathbb{C}^3$, as examples of the material of §9.1–§9.4. They have been chosen to illustrate different kinds of singular behaviour of SL 3-folds, and also to show how nonsingular SL 3-folds can converge to a singular SL 3-fold, to serve as a preparation for our discussion of singularities of SL m-folds in §11.

Our first example derives from Harvey and Lawson [93, §III.3.A], and is discussed in detail in [121, §3] and [114, §4].

Example 9.4 Define a subset L_0 in $\mathbb{C}^3$ by

$$L_0 = \left\{ (re^{i\theta_1}, re^{i\theta_2}, re^{i\theta_3}) : r \geqslant 0, \quad \theta_1, \theta_2, \theta_3 \in \mathbb{R}, \quad \theta_1 + \theta_2 + \theta_3 = 0 \right\}.$$

Then L_0 is a *special Lagrangian cone* on T^2. An alternative definition is

$$L_0 = \left\{ (z_1, z_2, z_3) \in \mathbb{C}^3 : |z_1| = |z_2| = |z_3|, \, \mathrm{Im}(z_1 z_2 z_3) = 0, \, \mathrm{Re}(z_1 z_2 z_3) \geqslant 0 \right\}.$$

Let $a > 0$, write $\mathcal{S}^1 = \{e^{i\theta} : \theta \in \mathbb{R}\}$, and define $\phi_a : \mathcal{S}^1 \times \mathbb{C} \to \mathbb{C}^3$ by

$$\phi_a : (e^{i\theta}, z) \mapsto \left((|z|^2 + 2a)^{1/2} e^{i\theta}, z, e^{-i\theta} \bar{z} \right).$$

Then ϕ_a is an *embedding*. Define $L_a = \text{Image}\,\phi_a$. Then L_a is a nonsingular special Lagrangian 3-fold in $\mathbb{C}^3$ diffeomorphic to $\mathcal{S}^1 \times \mathbb{R}^2$. An equivalent definition is

$$L_a = \big\{(z_1, z_2, z_3) \in \mathbb{C}^3 : |z_1|^2 - 2a = |z_2|^2 = |z_3|^2,$$
$$\text{Im}(z_1 z_2 z_3) = 0, \quad \text{Re}(z_1 z_2 z_3) \geqslant 0\big\}.$$

As $a \to 0_+$, the nonsingular SL 3-fold L_a converges to the singular SL cone L_0. Note that L_a is *asymptotic* to L_0 at infinity, and that $L_a = a^{1/2} L_1$ for $a > 0$, so that the L_a for $a > 0$ are all homothetic to each other. Also, each L_a for $a \geqslant 0$ is invariant under the T^2 subgroup of SU(3) acting by

$$(z_1, z_2, z_3) \mapsto (e^{i\theta_1} z_1, e^{i\theta_2} z_2, e^{i\theta_3} z_3) \text{ for } \theta_1, \theta_2, \theta_3 \in \mathbb{R} \text{ with } \theta_1 + \theta_2 + \theta_3 = 0,$$

and so fits into the framework of §9.1. By [117, Th. 5.1] the L_a may also be written in the form (20) for continuous $u, v : \mathbb{R}^2 \to \mathbb{R}$, as in §9.5.

Our second example is adapted from Harvey and Lawson [93, §III.3.B].

Example 9.5 For each $t > 0$, define

$$L_t = \big\{(e^{i\theta} x_1, e^{i\theta} x_2, e^{i\theta} x_3) : x_j \in \mathbb{R}, \quad \theta \in (0, \pi/3),$$
$$x_1^2 + x_2^2 + x_3^2 = t^2 (\sin 3\theta)^{-2/3}\big\}.$$

Then L_t is a nonsingular embedded SL 3-fold in $\mathbb{C}^3$ diffeomorphic to $\mathcal{S}^2 \times \mathbb{R}$. As $t \to 0_+$ it converges to the singular union L_0 of the two SL 3-planes

$$\Pi_1 = \big\{(x_1, x_2, x_3) : x_j \in \mathbb{R}\big\} \text{ and } \Pi_2 = \big\{(e^{i\pi/3} x_1, e^{i\pi/3} x_2, e^{i\pi/3} x_3) : x_j \in \mathbb{R}\big\},$$

which intersect at 0. Note that L_t is invariant under the action of the Lie subgroup SO(3) of SU(3), acting on $\mathbb{C}^3$ in the obvious way, so again this comes from the method of §9.1. Also L_t is asymptotic to L_0 at infinity.

Our third example is taken from [124, Ex. 9.4 & Ex. 9.5].

Example 9.6 Let a_1, a_2 be positive, coprime integers, and set $a_3 = -a_1 - a_2$. Let $c \in \mathbb{R}$, and define

$$L_c^{a_1, a_2} = \big\{(e^{ia_1\theta} x_1, e^{ia_2\theta} x_2, ie^{ia_3\theta} x_3) : \theta \in \mathbb{R}, \ x_j \in \mathbb{R}, \ a_1 x_1^2 + a_2 x_2^2 + a_3 x_3^2 = c\big\}.$$

Then $L_c^{a_1, a_2}$ is a special Lagrangian 3-fold, which comes from the 'evolving quadrics' construction of §9.2. It is also symmetric under the U(1)-action

$$(z_1, z_2, z_3) \mapsto (e^{ia_1\theta} z_1, e^{ia_2\theta} z_2, ie^{ia_3\theta} z_3) \quad \text{for } \theta \in \mathbb{R},$$

but this is not a necessary feature of the construction; these are just the easiest examples to write down.

When $c = 0$ and a_3 is odd, $L_0^{a_1,a_2}$ is an embedded special Lagrangian cone on T^2, with one singular point at 0. When $c = 0$ and a_3 is even, $L_0^{a_1,a_2}$ is two opposite embedded SL T^2-cones with one singular point at 0.

When $c > 0$ and a_3 is odd, $L_c^{a_1,a_2}$ is an embedded 3-fold diffeomorphic to a nontrivial real line bundle over the Klein bottle. When $c > 0$ and a_3 is even, $L_c^{a_1,a_2}$ is an embedded 3-fold diffeomorphic to $T^2 \times \mathbb{R}$. In both cases, $L_c^{a_1,a_2}$ is a *ruled* SL 3-fold, as in §9.3, since it is fibred by hyperboloids of one sheet in $\mathbb{R}^3$, which are ruled in two different ways.

When $c < 0$ and a_3 is odd, $L_c^{a_1,a_2}$ an immersed copy of $\mathcal{S}^1 \times \mathbb{R}^2$. When $c < 0$ and a_3 is even, $L_c^{a_1,a_2}$ two immersed copies of $\mathcal{S}^1 \times \mathbb{R}^2$.

All the singular SL 3-folds we have seen so far have been *cones* in $\mathbb{C}^3$. Our final example, taken from [116], has more complicated singularities which are not cones. They are difficult to describe in a simple way, so we will not say much about them. For more details, see [116].

Example 9.7 In [116, §5] the author constructed a family of maps $\Phi : \mathbb{R}^3 \rightarrow \mathbb{C}^3$ with special Lagrangian image $N = \text{Image}\,\Phi$. It is shown in [116, §6] that generic Φ in this family are immersions, so that N is nonsingular as an immersed SL 3-fold, but in codimension 1 in the family they develop isolated singularities.

Here is a rough description of these singularities, taken from [116, §6]. Taking the singular point to be at $\Phi(0,0,0) = 0$, one can write Φ as

$$\Phi(x,y,t) = \left(x + \tfrac{1}{4}g(\mathbf{u},\mathbf{v})t^2\right)\mathbf{u} + \left(y^2 - \tfrac{1}{4}|\mathbf{u}|^2 t^2\right)\mathbf{v} \\ + 2yt\,\mathbf{u} \times \mathbf{v} + O\left(x^2 + |xy| + |xt| + |y|^3 + |t|^3\right), \qquad (24)$$

where $\mathbf{u}, \mathbf{v}$ are linearly independent vectors in $\mathbb{C}^3$ with $\omega(\mathbf{u},\mathbf{v}) = 0$, and $\times : \mathbb{C}^3 \times \mathbb{C}^3 \rightarrow \mathbb{C}^3$ is defined by

$$(r_1, r_2, r_3) \times (s_1, s_2, s_3) = \tfrac{1}{2}(\bar{r}_2 \bar{s}_3 - \bar{r}_3 \bar{s}_2, \bar{r}_3 \bar{s}_1 - \bar{r}_1 \bar{s}_3, \bar{r}_1 \bar{s}_2 - \bar{r}_2 \bar{s}_1).$$

The next few terms in the expansion (24) can also be given very explicitly, but we will not write them down as they are rather complex, and involve further choices of vectors $\mathbf{w}, \mathbf{x}, \ldots$.

What is going on here is that the lowest order terms in Φ are a *double cover* of the special Lagrangian plane $\langle \mathbf{u}, \mathbf{v}, \mathbf{u} \times \mathbf{v} \rangle_{\mathbb{R}}$ in $\mathbb{C}^3$, *branched* along the real line $\langle \mathbf{u} \rangle_{\mathbb{R}}$. The branching occurs when $y = t = 0$. Higher order terms deviate from the 3-plane $\langle \mathbf{u}, \mathbf{v}, \mathbf{u} \times \mathbf{v} \rangle_{\mathbb{R}}$, and make the singularity isolated.

9.7 Exercises

The group of automorphisms of $\mathbb{C}^m$ preserving g, ω and Ω is $\mathrm{SU}(m) \ltimes \mathbb{C}^m$, where $\mathbb{C}^m$ acts by translations. Let G be a Lie subgroup of $\mathrm{SU}(m) \ltimes \mathbb{C}^m$, let $\mathfrak{g}$

be its Lie algebra, and let $\phi : \mathfrak{g} \to \text{Vect}(\mathbb{C}^m)$ be the natural map associating an element of $\mathfrak{g}$ to the corresponding vector field on $\mathbb{C}^m$.

A *moment map* for the action of G on $\mathbb{C}^m$ is a smooth map $\mu : \mathbb{C}^m \to \mathfrak{g}^*$, such that $\phi(x) \cdot \omega = x \cdot d\mu$ for all $x \in \mathfrak{g}$, and $\mu : \mathbb{C}^m \to \mathfrak{g}^*$ is equivariant with respect to the G-action on $\mathbb{C}^m$ and the coadjoint G-action on $\mathfrak{g}^*$. Moment maps always exist if G is compact or semisimple, and are unique up to the addition of a constant in the centre $Z(\mathfrak{g}^*)$ of $\mathfrak{g}^*$, that is, the G-invariant subspace of $\mathfrak{g}^*$.

9.1 Suppose L is a Lagrangian m-fold in $\mathbb{C}^m$ invariant under a Lie subgroup G in $\text{SU}(m) \ltimes \mathbb{C}^m$, with moment map μ. Show that $\mu \equiv c$ on L for some $c \in Z(\mathfrak{g}^*)$.

9.2 Let G be a Lie subgroup of $\text{SU}(m) \ltimes \mathbb{C}^m$ with moment map μ, and L a special Lagrangian m-fold in $\mathbb{C}^m$, not necessarily G-invariant. Show that $\mu|_L$ is a *harmonic* function $L \to \mathfrak{g}^*$.

9.3 Define a smooth map $f : \mathbb{C}^3 \to \mathbb{R}^3$ by

$$f(z_1, z_2, z_3) = \left(|z_1|^2 - |z_3|^2, |z_2|^2 - |z_3|^2, \text{Im}(z_1 z_2 z_3) \right).$$

For each $a, b, c \in \mathbb{R}^3$, define $N_{a,b,c} = f^{-1}(a, b, c)$. Then $N_{a,b,c}$ is a real 3-dimensional submanifold of $\mathbb{C}^3$, which may be singular.

(i) At $z = (z_1, z_2, z_3) \in \mathbb{C}^3$, determine $df|_z : \mathbb{C}^3 \to \mathbb{R}^3$. Find the conditions on z for $df|_z$ to be surjective.

  Now $N_{a,b,c}$ is nonsingular at $z \in N_{a,b,c}$ if and only if $df|_z$ is surjective. Hence determine which of the $N_{a,b,c}$ are singular, and find their singular points.

(ii) If z is a nonsingular point of $N_{a,b,c}$, then $T_z N_{a,b,c} = \text{Ker}\, df|_z$. Determine $\text{Ker}\, df|_z$ in this case, and show that it is a special Lagrangian 3-plane in $\mathbb{C}^3$.

  Hence prove that $N_{a,b,c}$ is a special Lagrangian 3-fold wherever it is nonsingular, and that $f : \mathbb{C}^3 \to \mathbb{R}^3$ is a *special Lagrangian fibration*.

(iii) Observe that $N_{a,b,c}$ is invariant under the Lie group $G = \text{U}(1)^2$, acting by

$$(e^{i\theta_1}, e^{i\theta_2}) : (z_1, z_2, z_3) \mapsto (e^{i\theta_1} z_1, e^{i\theta_2} z_2, e^{-i\theta_1 - i\theta_2} z_3).$$

  How is the form of f related to the ideas of Exercise 9.1? How might G-invariance have been used to construct the fibration f?

(iv) Describe the topology of $N_{a,b,c}$, distinguishing different cases according to the singularities.

10   Compact Calibrated Submanifolds

In this section we shall discuss *compact* calibrated submanifolds in special
holonomy manifolds, focussing mainly on special Lagrangian submanifolds in
Calabi–Yau manifolds. Here are three important questions which motivate
work in this area.

1. Let N be a compact special Lagrangian m-fold in a fixed Calabi–Yau
   m-fold (M, J, g, Ω). Let $\mathcal{M}_N$ be the moduli space of *special Lagrangian
   deformations* of N, that is, the connected component of the set of special
   Lagrangian m-folds containing N. What can we say about $\mathcal{M}_N$? For
   instance, is it a smooth manifold, and of what dimension?
2. Let $\{(M, J_t, g_t, \Omega_t) : t \in (-\epsilon, \epsilon)\}$ be a smooth 1-parameter family of
   Calabi–Yau m-folds. Suppose N_0 is an SL m-fold in (M, J_0, g_0, Ω_0). Under what conditions can we extend N_0 to a smooth family of special
   Lagrangian m-folds N_t in (M, J_t, g_t, Ω_t) for $t \in (-\epsilon, \epsilon)$?
3. In general the moduli space $\mathcal{M}_N$ in Question 1 will be noncompact. Can
   we enlarge $\mathcal{M}_N$ to a compact space $\overline{\mathcal{M}}_N$ by adding a 'boundary' consisting of *singular* special Lagrangian m-folds? If so, what is the nature of
   the singularities that develop?

Briefly, these questions concern the *deformations* of special Lagrangian
m-folds, *obstructions* to their existence, and their *singularities* respectively.
The local answers to Questions 1 and 2 are well understood, and we shall
discuss them in this section. Question 3 is the subject of §11–§12.

10.1   SL m-folds in Calabi–Yau m-folds

Here is the definition.

Definition 10.1 Let (M, J, g, Ω) be a Calabi–Yau m-fold. Then $\operatorname{Re} \Omega$ is a
calibration on (M, g). An oriented real m-submanifold N in M is called a
special Lagrangian submanifold (SL m-fold) if it is calibrated with respect
to $\operatorname{Re} \Omega$.

From Proposition 8.3 we deduce an *alternative definition* of SL m-folds.
It is often more useful than Definition 10.1.

Proposition 10.2 Let (M, J, g, Ω) be a Calabi–Yau m-fold, with Kähler
form ω, and L a real m-dimensional submanifold in M. Then N admits
an orientation making it into an SL m-fold in M if and only if $\omega|_N \equiv 0$
and $\operatorname{Im} \Omega|_N \equiv 0$.

Regard N as an immersed submanifold, with immersion $\iota : N \to M$.
Then $[\omega|_N]$ and $[\operatorname{Im} \Omega|_N]$ are unchanged under continuous variations of the
immersion ι. Thus, $[\omega|_N] = [\operatorname{Im} \Omega|_N] = 0$ is a necessary condition not just for
N to be special Lagrangian, but also for any isotopic submanifold N' in M
to be special Lagrangian. This proves:

Corollary 10.3 *Let (M, J, g, Ω) be a Calabi-Yau m-fold, and N a compact real m-submanifold in M. Then a necessary condition for N to be isotopic to a special Lagrangian submanifold N' in M is that $[\omega|_N] = 0$ in $H^2(N, \mathbb{R})$ and $[\mathrm{Im}\,\Omega|_N] = 0$ in $H^m(N, \mathbb{R})$.*

This gives a simple, necessary topological condition for an isotopy class of m-submanifolds in a Calabi-Yau m-fold to contain a special Lagrangian submanifold.

10.2   Deformations of Compact Special Lagrangian m-folds

The deformation theory of compact special Lagrangian manifolds was studied by McLean, who proved the following result [155, Th. 3.6].

Theorem 10.4 *Let (M, J, g, Ω) be a Calabi-Yau m-fold, and N a compact special Lagrangian m-fold in M. Then the moduli space $\mathcal{M}_N$ of special Lagrangian deformations of N is a smooth manifold of dimension $b^1(N)$, the first Betti number of N.*

Sketch proof. Suppose for simplicity that N is an embedded submanifold. There is a natural orthogonal decomposition $TM|_N = TN \oplus \nu$, where $\nu \to N$ is the *normal bundle* of N in M. As N is Lagrangian, the complex structure $J : TM \to TM$ gives an isomorphism $J : \nu \to TN$. But the metric g gives an isomorphism $TN \cong T^*N$. Composing these two gives an isomorphism $\nu \cong T^*N$.

Let T be a small *tubular neighbourhood* of N in M. Then we can identify T with a neighbourhood of the zero section in ν. Using the isomorphism $\nu \cong T^*N$, we have an identification between T and a neighbourhood of the zero section in T^*N. This can be chosen to identify the Kähler form ω on T with the natural symplectic structure on T^*N. Let $\pi : T \to N$ be the obvious projection.

Under this identification, submanifolds N' in $T \subset M$ which are C^1 close to N are identified with the graphs of small smooth sections α of T^*N. That is, submanifolds N' of M close to N are identified with 1-*forms* α on N. We need to know: which 1-forms α are identified with *special Lagrangian* submanifolds N'?

Well, N' is special Lagrangian if $\omega|_{N'} \equiv \mathrm{Im}\,\Omega|_{N'} \equiv 0$. Now $\pi|_{N'} : N' \to N$ is a diffeomorphism, so we can push $\omega|_{N'}$ and $\mathrm{Im}\,\Omega|_{N'}$ down to N, and regard them as functions of α. Calculation shows that

$$\pi_*(\omega|_{N'}) = \mathrm{d}\alpha \quad \text{and} \quad \pi_*(\mathrm{Im}\,\Omega|_{N'}) = F(\alpha, \nabla\alpha),$$

where F is a nonlinear function of its arguments. Thus, the moduli space $\mathcal{M}_N$ is locally isomorphic to the set of small 1-forms α on N such that $\mathrm{d}\alpha \equiv 0$ and $F(\alpha, \nabla\alpha) \equiv 0$.

Now it turns out that F satisfies $F(\alpha, \nabla\alpha) \approx d(*\alpha)$ when α is small. Therefore $\mathcal{M}_N$ is locally approximately isomorphic to the vector space of 1-forms α with $d\alpha = d(*\alpha) = 0$. But by Hodge theory, this is isomorphic to the de Rham cohomology group $H^1(N, \mathbb{R})$, and is a manifold with dimension $b^1(N)$.

To carry out this last step rigorously requires some technical machinery: one must work with certain *Banach spaces* of sections of T^*N, $\Lambda^2 T^*N$ and $\Lambda^m T^*N$, use *elliptic regularity results* to prove that $\alpha \mapsto (d\alpha, F(\alpha, \nabla\alpha))$ has *closed image* in these Banach spaces, and then use the *Implicit Function Theorem for Banach spaces* to show that the kernel of the map is what we expect. $\square$

10.3   Obstructions to the Existence of Compact SL m-folds

Next we address Question 2 above. Let $\{(M, J_t, g_t, \Omega_t) : t \in (-\epsilon, \epsilon)\}$ be a smooth 1-parameter family of Calabi–Yau m-folds. Suppose N_0 is a special Lagrangian m-fold of (M, J_0, g_0, Ω_0). When can we extend N_0 to a smooth family of special Lagrangian m-folds N_t in (M, J_t, g_t, Ω_t) for $t \in (-\epsilon, \epsilon)$?

By Corollary 10.3, a necessary condition is that $[\omega_t|_{N_0}] = [\operatorname{Im}\Omega_t|_{N_0}] = 0$ for all t. Our next result shows that locally, this is also a *sufficient* condition.

Theorem 10.5 *Let* $\{(M, J_t, g_t, \Omega_t) : t \in (-\epsilon, \epsilon)\}$ *be a smooth 1-parameter family of Calabi–Yau m-folds, with Kähler forms ω_t. Let N_0 be a compact SL m-fold in (M, J_0, g_0, Ω_0), and suppose that $[\omega_t|_{N_0}] = 0$ in $H^2(N_0, \mathbb{R})$ and $[\operatorname{Im}\Omega_t|_{N_0}] = 0$ in $H^m(N_0, \mathbb{R})$ for all $t \in (-\epsilon, \epsilon)$. Then N_0 extends to a smooth 1-parameter family $\{N_t : t \in (-\delta, \delta)\}$, where $0 < \delta \leqslant \epsilon$ and N_t is a compact SL m-fold in (M, J_t, g_t, Ω_t).*

This can be proved using similar techniques to Theorem 10.4, though McLean did not prove it. Note that the condition $[\operatorname{Im}\Omega_t|_{N_0}] = 0$ for all t can be satisfied by choosing the phases of the Ω_t appropriately, and if the image of $H_2(N, \mathbb{Z})$ in $H_2(M, \mathbb{R})$ is zero, then the condition $[\omega|_N] = 0$ holds automatically.

Thus, the obstructions $[\omega_t|_{N_0}] = [\operatorname{Im}\Omega_t|_{N_0}] = 0$ in Theorem 10.5 are actually fairly mild restrictions, and special Lagrangian m-folds should be thought of as pretty stable under small deformations of the Calabi–Yau structure.

10.4   Natural Coordinates on the Moduli Space $\mathcal{M}_N$

Let N be a compact SL m-fold in a Calabi–Yau m-fold (M, J, g, Ω). Theorem 10.4 shows that the moduli space $\mathcal{M}_N$ has dimension $b^1(N)$. By Poincaré duality $b^1(N) = b^{m-1}(N)$. Thus $\mathcal{M}_N$ has the same dimension as the de Rham cohomology groups $H^1(M, \mathbb{R})$ and $H^{m-1}(M, \mathbb{R})$.

We shall construct natural local diffeomorphisms $\Phi : \mathcal{M}_N \to H^1(N, \mathbb{R})$ and $\Psi : \mathcal{M}_N \to H^{m-1}(N, \mathbb{R})$. These induce two natural *affine structures* on

$\mathcal{M}_N$, and can be thought of as two *natural coordinate systems* on $\mathcal{M}_N$. The material of this section can be found in Hitchin [96, §4].

Here is how to define Φ, Ψ. Let U be a connected and simply-connected open neighbourhood of N in $\mathcal{M}_N$. We will construct smooth maps $\Phi : U \to H^1(N, \mathbb{R})$ and $\Psi : U \to H^{m-1}(N, \mathbb{R})$ with $\Phi(N) = \Psi(N) = 0$, which are local diffeomorphisms.

Let $N' \in U$. Then as U is connected, there exists a smooth path $\gamma :$ $[0, 1] \to U$ with $\gamma(0) = N$ and $\gamma(1) = N'$, and as U is simply-connected, γ is unique up to isotopy. Now γ parametrizes a family of submanifolds of M diffeomorphic to N, which we can lift to a smooth map $\Gamma : N \times [0, 1] \to M$ with $\Gamma(N \times \{t\}) = \gamma(t)$.

Consider the 2-form $\Gamma^*(\omega)$ on $N \times [0, 1]$. As each fibre $\gamma(t)$ is Lagrangian, we have $\Gamma^*(\omega)|_{N \times \{t\}} \equiv 0$ for each $t \in [0, 1]$. Therefore we may write $\Gamma^*(\omega) = \alpha_t \wedge dt$, where α_t is a closed 1-form on N for $t \in [0, 1]$. Define $\Phi(N') = \left[\int_0^1 \alpha_t \, dt \right] \in H^1(N, \mathbb{R})$. That is, we integrate the 1-forms α_t with respect to t to get a closed 1-form $\int_0^1 \alpha_t \, dt$, and then take its cohomology class.

Similarly, write $\Gamma^*(\operatorname{Im} \Omega) = \beta_t \wedge dt$, where β_t is a closed $(m-1)$-form on N for $t \in [0, 1]$, and define $\Psi(N') = \left[\int_0^1 \beta_t \, dt \right] \in H^{m-1}(N, \mathbb{R})$. Then Φ and Ψ are independent of choices made in the construction (exercise). We need to restrict to a simply-connected subset U of $\mathcal{M}_N$ so that γ is unique up to isotopy. Alternatively, one can define Φ and Ψ on the universal cover $\tilde{\mathcal{M}}_N$ of $\mathcal{M}_N$.

10.5 SL m-folds in Almost Calabi–Yau Manifolds

Next we explain a generalization of special Lagrangian geometry to the class of *almost Calabi–Yau manifolds*.

Definition 10.6 Let $m \geqslant 2$. An *almost Calabi–Yau m-fold*, or *ACY m-fold* for short, is a quadruple (M, J, g, Ω) such that (M, J, g) is a compact m-dimensional Kähler manifold, and Ω is a non-vanishing holomorphic $(m, 0)$-form on M.

The difference between this and Definition 4.3 is that we do not require Ω and the Kähler form ω of g and Ω to satisfy equation (11), and hence g need not be Ricci-flat, nor have holonomy $SU(m)$. Here is the appropriate definition of special Lagrangian m-folds in ACY m-folds.

Definition 10.7 Let (M, J, g, Ω) be an ACY m-fold with Kähler form ω, and N a real m-dimensional submanifold of M. We call N a *special Lagrangian submanifold*, or *SL m-fold* for short, if $\omega|_N \equiv \operatorname{Im} \Omega|_N \equiv 0$. It easily follows that $\operatorname{Re} \Omega|_N$ is a nonvanishing m-form on N. Thus N is orientable, with a unique orientation in which $\operatorname{Re} \Omega|_N$ is positive.

By Proposition 10.2, if (M, J, g, Ω) is Calabi–Yau rather than almost Calabi–Yau, then N is special Lagrangian in the sense of Definition 4.3.

Thus, this is a genuine extension of the idea of special Lagrangian submanifold. Many of the good properties of special Lagrangian submanifolds in Calabi–Yau manifolds also apply in almost Calabi–Yau manifolds. In particular:

Theorem 10.8 *Corollary 10.3 and Theorems 10.4 and 10.5 also hold in almost Calabi–Yau manifolds rather than Calabi–Yau manifolds.*

This is because the proofs of these results only really depend on the conditions $\omega|_N \equiv \operatorname{Im} \Omega|_N \equiv 0$, and the pointwise connection (11) between ω and Ω is not important.

Let (M, J, g, Ω) be an ACY m-fold. In general, SL m-folds in M are neither calibrated nor minimal with respect to g. However, let $f : M \to (0, \infty)$ be the unique smooth function such that $f^{2m}\omega^m/m! = (-1)^{m(m-1)/2}(i/2)^m \Omega \wedge \bar{\Omega}$, and define $\tilde{g}$ to be the conformally equivalent metric $f^2 g$ on M. Then $\operatorname{Re} \Omega$ is a calibration on the Riemannian manifold $(M, \tilde{g})$, and SL m-folds N in (M, J, g, Ω) are calibrated with respect to it, so that they are minimal with respect to $\tilde{g}$.

The idea of extending special Lagrangian geometry to almost Calabi–Yau manifolds appears in the work of Goldstein [69, §3.1], Bryant [31, §1], who uses the term 'special Kähler' instead of 'almost Calabi–Yau', and the author [114].

One important reason for considering SL m-folds in almost Calabi–Yau rather than Calabi–Yau m-folds is that they have much stronger *genericness properties*. There are many situations in geometry in which one uses a genericity assumption to control singular behaviour.

For instance, pseudo-holomorphic curves in an arbitrary almost complex manifold may have bad singularities, but the possible singularities in a generic almost complex manifold are much simpler. In the same way, it is reasonable to hope that in a *generic* Calabi–Yau m-fold, compact SL m-folds may have better singular behaviour than in an arbitrary Calabi–Yau m-fold.

But because Calabi–Yau manifolds come in only finite-dimensional families, choosing a generic Calabi–Yau structure is a fairly weak assumption, and probably will not help very much. However, almost Calabi–Yau manifolds come in *infinite-dimensional* families, so choosing a generic almost Calabi–Yau structure is a much more powerful thing to do, and will probably simplify the singular behaviour of compact SL m-folds considerably. We will return to this idea in §11.

10.6   Coassociative 4-folds in 7-manifolds with Holonomy G_2

Let (M, g) be a Riemannian 7-manifold with holonomy G_2. Then as in §6.1 there is a natural 4-form $*\varphi$ on M, of the form (13) at each point. As in §8.5 it is a calibration, and the corresponding calibrated submanifolds are called *coassociative 4-folds*.

By Proposition 8.5, a 4-fold N in M is coassociative if and only if $\varphi|_N \equiv 0$. Thus, coassociative 4-folds may be defined by the vanishing of a closed form, in the same way as SL m-folds are. This gives coassociative 4-folds similar properties to SL m-folds. Here is the analogue of Theorem 10.4, proved by McLean [155, Th. 4.5].

Theorem 10.9 *Let (M, φ, g) be a G_2-manifold, and N a compact coasso-ciative 4-fold in M. Then the moduli space of coassociative 4-folds isotopic to N in M is a smooth manifold of dimension $b_+^2(N)$.*

Briefly, the theorem holds because the normal bundle ν of N in M is naturally isomorphic to the bundle $\Lambda_+^2 N$ of *self-dual 2-forms* on N. Nearby submanifolds N' correspond to small sections α of $\Lambda_+^2 N$, and to leading order N' is coassociative if and only α is closed. So the tangent space $T_N \mathcal{M}_N$ at N to the moduli space $\mathcal{M}_N$ of coassociative 4-folds is the vector space of closed self-dual 2-forms on N, which has dimension $b_+^2(N)$.

There are also analogues for coassociative 4-folds of Theorem 10.5 and §10.4, which we will not give.

10.7   Associative 3-folds and Cayley 4-folds

Let (M, g) be a Riemannian 7-manifold with holonomy G_2. Then as in §6.1 there is a natural 3-form φ on M, of the form (13) at each point. As in §8.5 it is a calibration, and the corresponding calibrated submanifolds are called *associative 3-folds*.

Similarly, let (M, g) be a Riemannian 8-manifold with holonomy Spin(7). Then as in §6.2 there is a natural 4-form Ω on M, of the form (14) at each point. As in §8.6 it is a calibration, and the corresponding calibrated submanifolds are called *Cayley 4-folds*.

Associative 3-folds and Cayley 4-folds cannot be defined in terms of the vanishing of closed forms, and this gives their deformation and obstruction theory a different character to the special Lagrangian and coassociative cases. Here is how the theories work, drawn mostly from McLean [155, §5–§6].

Let N be a compact associative 3-fold or Cayley 4-fold in a 7- or 8-manifold M. Then there are vector bundles $E, F \to N$ with $E \cong \nu$, the normal bundle of N in M, and a first-order elliptic operator $D_N : C^\infty(E) \to C^\infty(F)$ on N. The *kernel* Ker D_N is the set of *infinitesimal deformations* of N as an associative 3-fold or Cayley 4-fold. The *cokernel* Coker D_N is the *obstruction space* for these deformations.

Both are finite-dimensional vector spaces, and

$$\dim \operatorname{Ker} D_N - \dim \operatorname{Coker} D_N = \operatorname{ind}(D_N),$$

the *index* of D_N. It is a topological invariant, given in terms of characteristic classes by the *Atiyah–Singer Index Theorem*. In the associative case we have

$E \cong F$, and D_N is anti-self-adjoint, so that $\mathrm{Ker}(D_N) \cong \mathrm{Coker}(D_N)$ and $\mathrm{ind}(D_N) = 0$ automatically. In the Cayley case we have

$$\mathrm{ind}(D_N) = \tau(N) - \tfrac{1}{2}\chi(N) - \tfrac{1}{2}[N] \cdot [N],$$

where $\tau(N)$ is the signature, $\chi(N)$ the Euler characteristic and $[N] \cdot [N]$ the self-intersection of N.

In a *generic* situation we expect $\mathrm{Coker}\, D_N = 0$, and then deformations of N will be unobstructed, so that the moduli space $\mathcal{M}_N$ of associative or Cayley deformations of N will locally be a smooth manifold of dimension $\mathrm{ind}(D_N)$. However, in nongeneric situations the obstruction space may be nonzero, and then we cannot predict the dimension of the moduli space.

This general structure is found in the deformation theory of many other important mathematical objects — for instance, pseudo-holomorphic curves in almost complex manifolds, and instantons and Seiberg–Witten solutions on 4-manifolds. In each case, we can only predict the dimension of the moduli space under a *genericity assumption* which forces the obstructions to vanish.

However, special Lagrangian and coassociative submanifolds do not follow this pattern. Instead, there are *no obstructions*, and the dimension of the moduli space is *always* given by a topological formula. This should be regarded as a minor mathematical miracle.

10.8   Exercises

10.1 Show that the maps Φ, Ψ between special Lagrangian moduli space $\mathcal{M}_N$ and $H^1(N, \mathbb{R})$, $H^{m-1}(N, \mathbb{R})$ defined in §10.4 are well-defined and independent of choices.

Prove also that Φ and Ψ are *local diffeomorphisms*, that is, that $\mathrm{d}\Phi|_{N'}$ and $\mathrm{d}\Psi|_{N'}$ are isomorphisms from $T_{N'}\mathcal{M}_N$ to $H^1(N, \mathbb{R})$, $H^{m-1}(N, \mathbb{R})$ for each $N' \in U$.

10.2 Putting together the maps Φ, Ψ of Exercise 10.1 gives a map $\Phi \times \Psi : U \to H^1(N, \mathbb{R}) \times H^{m-1}(N, \mathbb{R})$. Now $H^1(N, \mathbb{R})$ and $H^{m-1}(N, \mathbb{R})$ are dual by Poincaré duality, so $H^1(N, \mathbb{R}) \times H^{m-1}(N, \mathbb{R})$ has a natural *symplectic structure*. Show that the image of U is a *Lagrangian submanifold* in $H^1(N, \mathbb{R}) \times H^{m-1}(N, \mathbb{R})$.

Hint: From the proof of McLean's theorem in §10.2, the tangent space $T_N\mathcal{M}_N$ is isomorphic to the vector space of 1-forms α with $\mathrm{d}\alpha = \mathrm{d}(*\alpha) = 0$. Then $\mathrm{d}\Phi|_N : T_N\mathcal{M} \to H^1(M, \mathbb{R})$ takes $\alpha \mapsto [\alpha]$, and $\mathrm{d}\Psi|_N : T_N\mathcal{M} \to H^{m-1}(M, \mathbb{R})$ takes $\alpha \mapsto [*\alpha]$. Use the fact that for 1-forms α, β on an oriented Riemannian manifold we have $\alpha \wedge (*\beta) = \beta \wedge (*\alpha)$.

11   Singularities of Special Lagrangian m-folds

Now we move on to Question 3 of §10, and discuss the *singularities* of special Lagrangian m-folds. We can divide it into two sub-questions:

3(a) What kinds of singularities are possible in singular special Lagrangian m-folds, and what do they look like?

3(b) How can singular SL m-folds arise as limits of nonsingular SL m-folds, and what does the limiting behaviour look like near the singularities?

The basic premise of the author's approach to special Lagrangian singularities is that singularities of SL m-folds in Calabi–Yau m-folds should look locally like singularities of SL m-folds in $\mathbb{C}^m$, to the first few orders of approximation. That is, if M is a Calabi–Yau m-fold and N an SL m-fold in M with a singular point at $x \in M$, then near x, M resembles $\mathbb{C}^m = T_x M$, and N resembles an SL m-fold L in $\mathbb{C}^m$ with a singular point at 0. We call L a *local model* for N near x.

Therefore, to understand singularities of SL m-folds in Calabi–Yau manifolds, we begin by studying singularities of SL m-folds in $\mathbb{C}^m$, first by constructing as many examples as we can, and then by aiming for some kind of rough classification of the most common kinds of special Lagrangian singularities, at least in low dimensions such as $m = 3$.

11.1   Cones, and Asymptotically Conical SL m-folds

In Examples 9.4–9.6, the singular SL 3-folds we constructed were cones. Here a closed SL m-fold C in $\mathbb{C}^m$ is called a *cone* if $C = tC$ for all $t > 0$, where $tC = \{tx : x \in C\}$. Note that 0 is always a singular point of C, unless C is a special Lagrangian plane $\mathbb{R}^m$ in $\mathbb{C}^m$. The simplest kind of SL cones (from the point of view of singular behaviour) are those in which 0 is the only singular point. Then $\Sigma = C \cap S^{2m-1}$ is a nonsingular, compact, minimal, Legendrian $(m-1)$-submanifold in the unit sphere S^{2m-1} in $\mathbb{C}^m$.

In one sense, *all* singularities of SL m-folds are modelled on special Lagrangian cones, to highest order. It follows from [93, §II.5] that if M is a Calabi–Yau m-fold and N an SL m-fold in M with a singular point at x, then N has a *tangent cone* at x in the sense of Geometric Measure Theory, which is a special Lagrangian cone C in $\mathbb{C}^m = T_x M$. (It is not yet known whether the tangent cone need be unique.)

If C has multiplicity one and 0 is its only singular point, then N really does look like C near x. However, things become more complicated if the singularities of C are not isolated (for instance, if C is the union of two SL 3-planes $\mathbb{R}^3$ in $\mathbb{C}^3$ intersecting in a line) or if the multiplicity of C is greater than 1 (this happens in Example 9.7, as the tangent cone is a double $\mathbb{R}^3$). In these cases, the tangent cone captures only the simplest part of the singular behaviour, and we have to work harder to find out what is going on.

Now suppose for simplicity that we are interested in SL m-folds with singularities modelled on a multiplicity 1 SL cone C in $\mathbb{C}^m$ with an isolated singular point. To answer question 3(b) above, and understand how such singularities arise as limits of nonsingular SL m-folds, we need to study *asymptotically conical (AC)* SL m-folds L in $\mathbb{C}^m$ asymptotic to C.

We shall be interested in two classes of asymptotically conical SL m-folds L, *weakly AC*, which converge to C like $o(r)$, and *strongly AC*, which converge to C like $O(r^{-1})$. Here is a more precise definition.

Definition 11.1 Let C be a special Lagrangian cone in $\mathbb{C}^m$ with isolated singularity at 0, and let $\Sigma = C \cap S^{2m-1}$, so that Σ is a compact, nonsingular $(m-1)$-manifold. Define the *number of ends of C at infinity* to be the number of connected components of Σ. Let h be the metric on Σ induced by the metric g on $\mathbb{C}^m$, and r the radius function on $\mathbb{C}^m$. Define $\iota : \Sigma \times (0, \infty) \to \mathbb{C}^m$ by $\iota(\sigma, r) = r\sigma$. Then the image of ι is $C \setminus \{0\}$, and $\iota^*(g) = r^2 h + dr^2$ is the cone metric on $C \setminus \{0\}$.

Let L be a closed, nonsingular SL m-fold in $\mathbb{C}^m$. We call L *weakly asymptotically conical (weakly AC)* with cone C if there exists a compact subset $K \subset L$ and a diffeomorphism $\phi : \Sigma \times (R, \infty) \to L \setminus K$ for some $R > 0$, such that $|\phi - \iota| = o(r)$ and $\left|\nabla^k(\phi - \iota)\right| = o(r^{1-k})$ as $r \to \infty$ for $k = 1, 2, \ldots$, where ∇ is the Levi-Civita connection of the cone metric $\iota^*(g)$, and $|.|$ is computed using $\iota^*(g)$.

Similarly, we call L *strongly asymptotically conical (strongly AC)* with asymptotic cone C if $|\phi - \iota| = O(r^{-1})$ and $\left|\nabla^k(\phi - \iota)\right| = O(r^{-1-k})$ as $r \to \infty$ for $k = 1, 2, \ldots$, using the same notation.

These two asymptotic conditions are useful for different purposes. If L is a weakly AC SL m-fold then tL converges to C as $t \to 0_+$. Thus, weakly AC SL m-folds provide models for how singularities modelled on cones C can arise as limits of nonsingular SL m-folds. The weakly AC condition is in practice the weakest asymptotic condition which ensures this; if the $o(r)$ condition were made any weaker then the asymptotic cone C at infinity might not be unique or well-defined.

On the other hand, explicit constructions tend to produce strongly AC SL m-folds, and they appear to be the easiest class to prove results about. For example, one can show:

Proposition 11.2 *Suppose C is an SL cone in $\mathbb{C}^m$ with an isolated singular point at 0, invariant under a connected Lie subgroup G of $\mathrm{SU}(m)$. Then any strongly AC SL m-fold L in $\mathbb{C}^m$ with cone C is also G-invariant.*

This should be a help in classifying strongly AC SL m-folds with cones with a lot of symmetry. For instance, using $\mathrm{U}(1)^2$ symmetry one can show that the only strongly AC SL 3-folds in $\mathbb{C}^3$ with cone L_0 from Example 9.4 are the SL 3-folds L_a from Example 9.4 for $a > 0$, and two other families obtained from the L_a by cyclic permutations of coordinates (z_1, z_2, z_3).

11.2   Moduli Spaces of AC SL m-folds

Next we discuss moduli space problems for AC SL m-folds. I shall state our problems as conjectures, because the proofs are not yet complete. One should be able to prove analogues of Theorem 10.4 for AC SL m-folds. Here is the appropriate result for strongly AC SL m-folds.

Conjecture 11.3 Let L be a strongly AC SL m-fold in $\mathbb{C}^m$, with cone C, and let k be the number of ends of C at infinity. Then the moduli space $\mathcal{M}_L^s$ of strongly AC SL m-folds in $\mathbb{C}^m$ with cone C is near L a smooth manifold of dimension $b^1(L) + k - 1$.

Before generalizing this to the weak case, here is a definition.

Definition 11.4 Let C be a special Lagrangian cone in $\mathbb{C}^m$ with an isolated singularity at 0, and let $\Sigma = C \cap S^{2m-1}$. Regard Σ as a compact Riemannian manifold, with metric induced from the round metric on S^{2m-1}. Let $\Delta = \mathrm{d}^*\mathrm{d}$ be the Laplacian on functions on Σ. Define the *Legendrian index* l-ind(C) to be the number of eigenvalues of Δ in $(0, 2m)$, counted with multiplicity.

We call this the *Legendrian index* since it is the index of the area functional at Σ under Legendrian variations of the submanifold Σ in S^{2m-1}. It is not difficult to show that the restriction of any real linear function on $\mathbb{C}^m$ to Σ is an eigenfunction of Δ with eigenvalue $m - 1$. These contribute m to l-ind(C) for each connected component of Σ which is a round unit sphere S^{m-1}, and $2m$ to l-ind(C) for each other connected component of Σ. This gives a useful lower bound for l-ind(C). In particular, l-ind$(C) \geqslant 2m$.

Conjecture 11.5 Let L be a weakly AC SL m-fold in $\mathbb{C}^m$, with cone C, and let k be the number of ends of C at infinity. Then the moduli space $\mathcal{M}_L^w$ of weakly AC SL m-folds in $\mathbb{C}^m$ with cone C is near L a smooth manifold of dimension $b^1(L) + k - 1 + \text{l-ind}(C)$.

Here are some remarks on these conjectures:

- The author's student, Stephen Marshall, is working on proofs of Conjectures 11.3 and 11.5, which we hope to be able to publish soon. Some related results have recently been proved by Tommaso Pacini [171].
- If L is a weakly AC SL m-fold in $\mathbb{C}^m$, then any translation of L is also weakly AC, with the same cone. Since C has an isolated singularity by assumption, it cannot have translation symmetries. Hence L also has no translation symmetries, so the translations of L are all distinct, and $\mathcal{M}_L^w$ has dimension at least $2m$. The inequality l-ind$(C) \geqslant 2m$ above ensures this.
- The dimension of $\mathcal{M}_L^s$ in Conjecture 11.3 is purely topological, as in Theorem 10.4, which is another indication that strongly AC is in many ways the nicest asymptotic condition to work with. But the dimension of $\mathcal{M}_L^w$ in Conjecture 11.5 has an analytic component, the eigenvalue count in l-ind(C).

- It is an interesting question whether moduli spaces of weakly AC SL m-folds always contain a strongly AC SL m-fold.

11.3 SL Singularities in Generic Almost Calabi–Yau m-folds

We move on to discuss the singular behaviour of compact SL m-folds in Calabi–Yau m-folds. For simplicity we shall restrict our attention to a class of SL cones with no nontrivial deformations.

Definition 11.6 Let C be a special Lagrangian cone in $\mathbb{C}^m$ with an isolated singularity at 0 and k ends at infinity, and set $\Sigma = C \cap \mathcal{S}^{2m-1}$. Let $\Sigma_1, \dots, \Sigma_k$ be the connected components of Σ. Regard each Σ_j as a compact Riemannian manifold, and let $\Delta_j = \mathrm{d}^*\mathrm{d}$ be the Laplacian on functions on Σ_j. Let G_j be the Lie subgroup of $\mathrm{SU}(m)$ preserving Σ_j, and V_j the eigenspace of Δ_j with eigenvalue $2m$. We call the SL cone C *rigid* if $\dim V_j = \dim \mathrm{SU}(m) - \dim G_j$ for each $j = 1, \dots, k$.

Here is how to understand this definition. We can regard C as the union of one-ended SL cones $C_1, \dots, C_k$ intersecting at 0, where $C_j \setminus \{0\}$ is naturally identified with $\Sigma_j \times (0, \infty)$. The cone metric on C_j is $g_j = r^2 h_j + \mathrm{d}r^2$, where h_j is the metric on Σ_j. Suppose f_j is an eigenfunction of Δ_j on Σ_j with eigenvalue $2m$. Then $r^2 f_j$ is *harmonic* on C_j.

Hence, $\mathrm{d}(r^2 f_j)$ is a closed, coclosed 1-form on C_j which is linear in r. By the Principle in §8.4, the basis of the proof of Theorem 10.4, such 1-forms correspond to small deformations of C_j as an SL cone in $\mathbb{C}^m$. Therefore, we can interpret V_j as the space of *infinitesimal deformations* of C_j as a special Lagrangian cone.

Clearly, one way to deform C_j as a special Lagrangian cone is to apply elements of $\mathrm{SU}(m)$. This gives a family $\mathrm{SU}(m)/G_j$ of deformations of C_j, with dimension $\dim \mathrm{SU}(m) - \dim G_j$, so that $\dim V_j \geqslant \dim \mathrm{SU}(m) - \dim G_j$. (The corresponding functions in V_j are moment maps of $\mathfrak{su}(m)$ vector fields.) We call C *rigid* if equality holds for all j, that is, if all infinitesimal deformations of C come from applying motions in $\mathfrak{su}(m)$ to the component cones $C_1, \dots, C_k$.

Not all SL cones in $\mathbb{C}^m$ are rigid. One can show using integrable systems that there exist families of SL T^2-cones C in $\mathbb{C}^3$ up to $\mathrm{SU}(3)$ equivalence, of arbitrarily high dimension. If the dimension of the family is greater than $\dim \mathrm{SU}(3) - \dim G_1$, where G_1 is the Lie subgroup of $\mathrm{SU}(3)$ preserving C, then C is not rigid.

Now we can give a first approximation to the kinds of results the author expects to hold for singular SL m-folds in (almost) Calabi–Yau m-folds.

Conjecture 11.7 Let C be a rigid special Lagrangian cone in $\mathbb{C}^m$ with an isolated singularity at 0 and k ends at infinity, and L a weakly AC SL m-fold in $\mathbb{C}^m$ with cone C. Let (M, J, g, Ω) be a generic almost Calabi–Yau m-fold, and $\mathcal{M}$ a connected moduli space of compact nonsingular SL m-folds N in M.

Suppose that at the boundary of $\mathcal{M}$ there is a moduli space $\mathcal{M}_C$ of compact, singular SL m-folds with one isolated singular point modelled on the cone C, which arise as limits of SL m-folds in $\mathcal{M}$ by collapsing weakly AC SL m-folds with the topology of L. Then

$$\dim \mathcal{M} = \dim \mathcal{M}_C + b^1(L) + k - 1 + \text{l-ind}(C) - 2m. \qquad (25)$$

Here are some remarks on the conjecture:

- I have an outline proof of this conjecture which works when $m < 6$. The analytic difficulties increase with dimension; I am not sure whether the conjecture holds in high dimensions.
- Similar results should hold for non-rigid singularities, but the dimension formulae will be more complicated.
- Closely associated to this result is an analogue of Theorem 10.4 for SL m-folds with isolated conical singularities of a given kind, under an appropriate genericity assumption.
- Here is one way to arrive at equation (25). Assuming Conjecture 11.5, the moduli space $\mathcal{M}_L^w$ of weakly AC SL m-folds containing L has dimension $b^1(L) + k - 1 + \text{l-ind}(C)$. Now translations in $\mathbb{C}^m$ act freely on $\mathcal{M}_L^w$, so the family of weakly AC SL m-folds up to translations has dimension $b^1(L) + k - 1 + \text{l-ind}(C) - 2m$.
  The idea is that each singular SL m-fold N_0 in $\mathcal{M}_C$ can be 'resolved' to give a nonsingular SL m-fold N by gluing in any 'sufficiently small' weakly AC SL m-fold L', up to translation. Thus, desingularizing should add $b^1(L) + k - 1 + \text{l-ind}(C) - 2m$ degrees of freedom, which is how we get equation (25).
- However, (25) may not give the right answer in every case. One can imagine situations in which there are *cohomological obstructions* to gluing in AC SL m-folds L'. For instance, it might be necessary that the symplectic area of a disc in $\mathbb{C}^m$ with boundary in L' be zero to make $N = L' \# N_0$ Lagrangian. These could reduce the number of degrees of freedom in desingularizing N_0, and then (25) would require correction. An example of this is considered in [121, §4.4].

Now we can introduce the final important idea in this section. Suppose we have a suitably generic (almost) Calabi–Yau m-fold M and a compact, singular SL m-fold N_0 in M, which is the limit of a family of compact nonsingular SL m-folds N in M.

We (loosely) define the *index* of the singularities of N_0 to be the codimension of the family of singular SL m-folds with singularities like those of N_0 in the family of nonsingular SL m-folds N. Thus, in the situation of Conjecture 11.7, the index of the singularities is $b^1(L) + k - 1 + \text{l-ind}(C) - 2m$.

More generally, one can work not just with a fixed generic almost Calabi–Yau m-fold, but with a *generic family* of almost Calabi–Yau m-folds. So, for instance, if we have a generic k-dimensional family of almost Calabi–Yau m-folds M, and in each M we have an l-dimensional family of SL m-folds, then in the total $(k+l)$-dimensional family of SL m-folds we are guaranteed to meet singularities of index at most $k+l$.

Now singularities with *small index* are the most commonly occurring, and so arguably the most interesting kinds of singularity. Also, for various problems it will only be necessary to know about singularities with index up to a certain value.

For example, in [121] the author proposed to define an invariant of almost Calabi–Yau 3-folds by counting special Lagrangian homology 3-spheres (which occur in 0-dimensional moduli spaces) in a given homology class, with a certain topological weight. This invariant will only be interesting if it is essentially conserved under deformations of the underlying almost Calabi–Yau 3-fold. During such a deformation, nonsingular SL 3-folds can develop singularities and disappear, or new ones appear, which might change the invariant.

To prove the invariant is conserved, we need to show that it is unchanged along generic 1-parameter families of Calabi–Yau 3-folds. The only kinds of singularities of SL homology 3-spheres that arise in such families will have index 1. Thus, to resolve the conjectures in [121], we only have to know about index 1 singularities of SL 3-folds in almost Calabi–Yau 3-folds.

Another problem in which the index of singularities will be important is the *SYZ Conjecture*, to be discussed in §12. This has to do with dual 3-dimensional families $\mathscr{M}_X$, $\mathscr{M}_{\hat{X}}$ of SL 3-tori in (almost) Calabi–Yau 3-folds $X, \hat{X}$. If $X, \hat{X}$ are generic then the only kinds of singularities that can occur at the boundaries of $\mathscr{M}_X, \mathscr{M}_{\hat{X}}$ are of index 1, 2 or 3. So, to study the SYZ Conjecture in the generic case, we only have to know about singularities of SL 3-folds with index 1, 2, 3 (and possibly 4).

It would be an interesting and useful project to find examples of, and eventually to classify, special Lagrangian singularities with small index, at least in dimension 3. For instance, consider rigid SL cones C in $\mathbb{C}^3$ as in Conjecture 11.7, of index 1. Then $b^1(L) + k - 1 + \text{l-ind}(C) - 6 = 1$, and $\text{l-ind}(C) \geqslant 6$, so $b^1(L) + k \leqslant 2$. But $k \geqslant 1$ and $b^1(L) \geqslant \frac{1}{2}b^1(\Sigma)$, so $b^1(\Sigma) \leqslant 2$. As Σ is oriented, one can show that either $k = 1$, $\text{l-ind}(C) = 6$ and Σ is a torus T^2, or $k = 2$, $\text{l-ind}(C) = 6$, Σ is 2 copies of S^2, and C is the union of two SL 3-planes in $\mathbb{C}^3$ intersecting only at 0.

The eigenvalue count $\text{l-ind}(C)$ implies an upper bound for the area of Σ. Hopefully, one can then use integrable systems results as in §9.4 to pin down the possibilities for C. The author guesses that the T^2-cone L_0 of Example 9.4, and perhaps also the T^2-cone $L_0^{1,2}$ of Example 9.6, are the only examples of index 1 SL T^2-cones in $\mathbb{C}^3$ up to SU(3) isomorphisms.

11.4 Exercises

11.1 Prove Proposition 11.2.

12   The SYZ Conjecture, and SL Fibrations

Mirror Symmetry is a mysterious relationship between pairs of Calabi–Yau 3-folds $X, \hat{X}$, arising from a branch of physics known as *String Theory*, and leading to some very strange and exciting conjectures about Calabi–Yau 3-folds, many of which have been proved in special cases.

The *SYZ Conjecture* is an attempt to explain Mirror Symmetry in terms of dual 'fibrations' $f : X \to B$ and $\hat{f} : \hat{X} \to B$ of $X, \hat{X}$ by special Lagrangian 3-folds, including singular fibres. We give brief introductions to String Theory, Mirror Symmetry, and the SYZ Conjecture, and then a short survey of the state of mathematical research into the SYZ Conjecture, biased in favour of the author's own interests.

12.1   String Theory and Mirror Symmetry

String Theory is a branch of high-energy theoretical physics in which particles are modelled not as points but as 1-dimensional objects – 'strings' – propagating in some background space-time M. String theorists aim to construct a *quantum theory* of the string's motion. The process of quantization is extremely complicated, and fraught with mathematical difficulties that are as yet still poorly understood.

The most popular version of String Theory requires the universe to be 10-dimensional for this quantization process to work. Therefore, String Theorists suppose that the space we live in looks locally like $M = \mathbb{R}^4 \times X$, where $\mathbb{R}^4$ is Minkowski space, and X is a compact Riemannian 6-manifold with radius of order 10^{-33}cm, the Planck length. Since the Planck length is so small, space then appears to macroscopic observers to be 4-dimensional.

Because of supersymmetry, X has to be a *Calabi–Yau 3-fold*. Therefore String Theorists are very interested in Calabi–Yau 3-folds. They believe that each Calabi–Yau 3-fold X has a quantization, which is a *Super Conformal Field Theory* (SCFT), a complicated mathematical object. Invariants of X such as the Dolbeault groups $H^{p,q}(X)$ and the number of holomorphic curves in X translate to properties of the SCFT.

However, two entirely different Calabi–Yau 3-folds X and $\hat{X}$ may have the *same* SCFT. In this case, there are powerful relationships between the invariants of X and of $\hat{X}$ that translate to properties of the SCFT. This is the idea behind *Mirror Symmetry* of Calabi–Yau 3-folds.

It turns out that there is a very simple automorphism of the structure of a SCFT — changing the sign of a U(1)-action — which does *not* correspond to a classical automorphism of Calabi–Yau 3-folds. We say that X and $\hat{X}$ are *mirror* Calabi–Yau 3-folds if their SCFT's are related by this automorphism. Then one can argue using String Theory that

$$H^{1,1}(X) \cong H^{2,1}(\hat{X}) \quad \text{and} \quad H^{2,1}(X) \cong H^{1,1}(\hat{X}).$$

Effectively, the mirror transform exchanges even- and odd-dimensional cohomology. This is a very surprising result!

More involved String Theory arguments show that, in effect, the Mirror Transform exchanges things related to the complex structure of X with things related to the symplectic structure of $\hat{X}$, and vice versa. Also, a generating function for the number of holomorphic rational curves in X is exchanged with a simple invariant to do with variation of complex structure on $\hat{X}$, and so on.

Because the quantization process is poorly understood and not at all rigorous — it involves non-convergent path-integrals over horrible infinite-dimensional spaces — String Theory generates only conjectures about Mirror Symmetry, not proofs. However, many of these conjectures have been verified in particular cases, as we shall see in Part II.

12.2   Mathematical Interpretations of Mirror Symmetry

In the beginning (the 1980's), Mirror Symmetry seemed mathematically completely mysterious. But there are now two complementary conjectural theories, due to Kontsevich and Strominger–Yau–Zaslow, which explain Mirror Symmetry in a fairly mathematical way. Probably both are true, at some level.

The first proposal was due to Kontsevich [135] in 1994. This says that for mirror Calabi–Yau 3-folds X and $\hat{X}$, the derived category $D^b(X)$ of coherent sheaves on X is equivalent to the derived category $D^b(\mathrm{Fuk}(\hat{X}))$ of the Fukaya category of $\hat{X}$, and vice versa. Basically, $D^b(X)$ has to do with X as a complex manifold, and $D^b(\mathrm{Fuk}(\hat{X}))$ with $\hat{X}$ as a symplectic manifold, and its Lagrangian submanifolds. We shall not discuss this here.

The second proposal, due to Strominger, Yau and Zaslow [184] in 1996, is known as the *SYZ Conjecture*. Here is an attempt to state it.

The SYZ Conjecture *Suppose X and $\hat{X}$ are mirror Calabi–Yau 3-folds. Then (under some additional conditions) there should exist a compact topological 3-manifold B and surjective, continuous maps $f : X \to B$ and $\hat{f} : \hat{X} \to B$, such that*

(i) *There exists a dense open set $B_0 \subset B$, such that for each $b \in B_0$, the fibres $f^{-1}(b)$ and $\hat{f}^{-1}(b)$ are nonsingular special Lagrangian 3-tori T^3 in X and $\hat{X}$. Furthermore, $f^{-1}(b)$ and $\hat{f}^{-1}(b)$ are in some sense dual to one another.*

(ii) *For each $b \in \Delta = B \setminus B_0$, the fibres $f^{-1}(b)$ and $\hat{f}^{-1}(b)$ are expected to be singular special Lagrangian 3-folds in X and $\hat{X}$.*

For another point of view on the SYZ Conjecture, see §19. We call f and $\hat{f}$ *special Lagrangian fibrations*, and the set of singular fibres Δ is called the *discriminant*. In part (i), the nonsingular fibres of f and $\hat{f}$ are supposed to be *dual tori*. What does this mean?

On the topological level, we can define duality between two tori $T, \hat{T}$ to be a choice of isomorphism $H^1(T, \mathbb{Z}) \cong H_1(\hat{T}, \mathbb{Z})$. We can also define duality between tori equipped with flat Riemannian metrics. Write $T = V/\Lambda$, where V is a Euclidean vector space and Λ a *lattice* in V. Then the dual torus $\hat{T}$ is defined to be V^*/Λ^*, where V^* is the dual vector space and Λ^* the dual lattice. However, there is no notion of duality between non-flat metrics on dual tori.

Strominger, Yau and Zaslow argue only that their conjecture holds when $X, \hat{X}$ are close to the 'large complex structure limit'. In this case, the diameters of the fibres $f^{-1}(b), \hat{f}^{-1}(b)$ are expected to be small compared to the diameter of the base space B, and away from singularities of $f, \hat{f}$, the metrics on the nonsingular fibres are expected to be approximately flat.

So, part (i) of the SYZ Conjecture says that for $b \in B \setminus B_0$, $f^{-1}(b)$ is approximately a flat Riemannian 3-torus, and $\hat{f}^{-1}(b)$ is approximately the dual flat Riemannian torus. Really, the SYZ Conjecture makes most sense as a statement about the limiting behaviour of *families* of mirror Calabi–Yau 3-folds $X_t, \hat{X}_t$ which approach the 'large complex structure limit' as $t \to 0$.

12.3 The Symplectic Topological Approach to SYZ

The most successful approach to the SYZ Conjecture so far could be described as *symplectic topological*. In this approach, we mostly forget about complex structures, and treat $X, \hat{X}$ just as *symplectic manifolds*. We mostly forget about the 'special' condition, and treat $f, \hat{f}$ just as *Lagrangian fibrations*. We also impose the condition that B is a *smooth* 3-manifold and $f : X \to B$ and $\hat{f} : \hat{X} \to B$ are *smooth maps*. (It is not clear that $f, \hat{f}$ can in fact be smooth at every point, though).

Under these simplifying assumptions, Gross [82,84,86,85], Ruan [175,177], and others have built up a beautiful, detailed picture of how dual SYZ fibrations work at the global topological level, in particular for examples such as the quintic and its mirror, and for Calabi–Yau 3-folds constructed as hypersurfaces in toric 4-folds, using combinatorial data. This is described in part in §19.

12.4 Local Geometric Approach, and SL Singularities

There is also another approach to the SYZ Conjecture, begun by the author in [114,119], and making use of the ideas and philosophy set out in §11. We could describe it as a *local geometric* approach.

In it we try to take the special Lagrangian condition seriously from the outset, and our focus is on the local behaviour of special Lagrangian submanifolds, and especially their singularities, rather than on global topological questions. Also, we are interested in what fibrations of *generic* (almost) Calabi–Yau 3-folds might look like.

One of the first-fruits of this approach has been the understanding that for *generic* (almost) Calabi–Yau 3-folds X, special Lagrangian fibrations $f : X \to B$ will not be smooth maps, but only piecewise smooth. Furthermore, their behaviour at the singular set is rather different to the smooth Lagrangian fibrations discussed in §12.3.

For smooth special Lagrangian fibrations $f : X \to B$, the discriminant Δ is of codimension 2 in B, and the typical singular fibre is singular along an S^1. But in a generic special Lagrangian fibration $f : X \to B$ the discriminant Δ is of codimension 1 in B, and the typical singular fibre is singular at finitely many points.

One can also show that if $X, \hat{X}$ are a mirror pair of generic (almost) Calabi–Yau 3-folds and $f : X \to B$ and $\hat{f} : \hat{X} \to B$ are dual special Lagrangian fibrations, then in general the discriminants Δ of f and $\hat{\Delta}$ of $\hat{f}$ cannot coincide in B, because they have different topological properties in the neighbourhood of a certain kind of codimension 3 singular fibre.

This contradicts part (ii) of the SYZ Conjecture, as we have stated it in §12.2. In the author's view, these calculations support the idea that the SYZ Conjecture in its present form should be viewed primarily as a limiting statement, about what happens at the 'large complex structure limit', rather than as simply being about pairs of Calabi–Yau 3-folds. A similar conclusion is reached by Mark Gross in [85, §5].

12.5   U(1)-invariant SL Fibrations in $\mathbb{C}^3$

We finish by describing work of the author in [119, §8] and [114], which aims to describe what the singularities of SL fibrations of *generic* (almost) Calabi–Yau 3-folds look like, providing they exist. This proceeds by first studying SL fibrations of subsets of $\mathbb{C}^3$ invariant under the U(1)-action (19), using the ideas of §9.5. For a brief survey of the main results, see [120].

Then we argue (without a complete proof, as yet) that the kinds of singularities we see in codimension 1 and 2 in generic U(1)-invariant SL fibrations in $\mathbb{C}^3$, also occur in codimension 1 and 2 in SL fibrations of generic (almost) Calabi–Yau 3-folds, without any assumption of U(1)-invariance.

Following [119, Def. 8.1], we use the results of §9.5 to construct a family of SL 3-folds N_α in $\mathbb{C}^3$, depending on boundary data $\Phi(\alpha)$.

Definition 12.1 Let S be a strictly convex domain in $\mathbb{R}^2$ invariant under $(x, y) \mapsto (x, -y)$, let U be an open set in $\mathbb{R}^3$, and $\alpha \in (0, 1)$. Suppose $\Phi : U \to C^{3,\alpha}(\partial S)$ is a continuous map such that if $(a, b, c) \neq (a, b', c')$ in U then $\Phi(a, b, c) - \Phi(a, b', c')$ has exactly one local maximum and one local minimum in ∂S.

For $\alpha = (a, b, c) \in U$, let $f_\alpha \in C^{3,\alpha}(S)$ or $C^1(S)$ be the unique (weak) solution of (23) with $f_\alpha|_{\partial S} = \Phi(\alpha)$, which exists by Theorem 9.3. Define $u_\alpha = \frac{\partial f_\alpha}{\partial y}$ and $v_\alpha = \frac{\partial f_\alpha}{\partial x}$. Then (u_α, v_α) is a solution of (22) in $C^{2,\alpha}(S)$ if

$a \neq 0$, and a weak solution of (21) in $C^0(S)$ if $a = 0$. Also u_α, v_α depend continuously on $\alpha \in U$ in $C^0(S)$, by Theorem 9.3.

For each $\alpha = (a, b, c)$ in U, define N_α in $\mathbb{C}^3$ by

$$N_\alpha = \{(z_1, z_2, z_3) \in \mathbb{C}^3 : z_1 z_2 = v_\alpha(x, y) + iy, \quad z_3 = x + iu_\alpha(x, y),$$
$$|z_1|^2 - |z_2|^2 = 2a, \quad (x, y) \in S^\circ \}. \tag{26}$$

Then N_α is a noncompact SL 3-fold without boundary in $\mathbb{C}^3$, which is non-singular if $a \neq 0$, by Proposition 9.1.

In [119, Th. 8.2] we show that the N_α are the fibres of an *SL fibration*.

Theorem 12.2 *In the situation of Definition 12.1, if $\alpha \neq \alpha'$ in U then $N_\alpha \cap N_{\alpha'} = \emptyset$. There exists an open set $V \subset \mathbb{C}^3$ and a continuous, surjective map $F : V \to U$ such that $F^{-1}(\alpha) = N_\alpha$ for all $\alpha \in U$. Thus, F is a special Lagrangian fibration of $V \subset \mathbb{C}^3$, which may include singular fibres.*

It is easy to produce families Φ satisfying Definition 12.1. For example [119, Ex. 8.3], given any $\phi \in C^{3,\alpha}(\partial S)$ we may define $U = \mathbb{R}^3$ and $\Phi : \mathbb{R}^3 \to C^{3,\alpha}(\partial S)$ by $\Phi(a, b, c) = \phi + bx + cy$. So this construction produces very large families of U(1)-invariant SL fibrations, including singular fibres, which can have any multiplicity and type.

Here is a simple, explicit example. Define $F : \mathbb{C}^3 \to \mathbb{R} \times \mathbb{C}$ by

$$F(z_1, z_2, z_3) = (a, b), \quad \text{where} \quad 2a = |z_1|^2 - |z_2|^2$$
$$\text{and} \quad b = \begin{cases} z_3, & a = z_1 = z_2 = 0, \\ z_3 + \bar{z}_1 \bar{z}_2 / |z_1|, & a \geqslant 0, \ z_1 \neq 0, \\ z_3 + \bar{z}_1 \bar{z}_2 / |z_2|, & a < 0. \end{cases} \tag{27}$$

This is a piecewise-smooth SL fibration of $\mathbb{C}^3$. It is not smooth on $|z_1| = |z_2|$.

The fibres $F^{-1}(a, b)$ are T^2-cones singular at $(0, 0, b)$ when $a = 0$, and nonsingular $S^1 \times \mathbb{R}^2$ when $a \neq 0$. They are isomorphic to the SL 3-folds of Example 9.4 under transformations of $\mathbb{C}^3$, but they are assembled to make a fibration in a novel way.

As a goes from positive to negative the fibres undergo a surgery, a Dehn twist on S^1. The reason why the fibration is only piecewise-smooth, rather than smooth, is really this topological transition, rather than the singularities themselves. The fibration is not differentiable at every point of a singular fibre, rather than just at singular points, and this is because we are jumping from one moduli space of SL 3-folds to another at the singular fibres.

I conjecture that F is the local model for codimension one singularities of SL fibrations of generic almost Calabi–Yau 3-folds. The justification for this is that the T^2-cone singularities have 'index one' in the sense of §11.3, and so should occur in codimension one in families of SL 3-folds in generic almost Calabi–Yau 3-folds. Since they occur in codimension one in this family, the singular behaviour should be stable under small perturbations of the underlying almost Calabi–Yau structure.

I also have a U(1)-invariant model for codimension two singularities, described in [114], in which two of the codimension one T^2-cones come together and cancel out. I conjecture that it too is a typical codimension two singular behaviour in SL fibrations of generic almost Calabi–Yau 3-folds. I do not expect codimension three singularities in generic SL fibrations to be locally U(1)-invariant, and so this approach will not help.

12.6   Exercises

12.1 Let $f : \mathbb{C}^3 \to \mathbb{R} \times \mathbb{C}$ be as in equation (27).

    (a) Show that f is continuous, surjective, and piecewise smooth.

    (b) Show that $f^{-1}(a,b)$ is a (possibly singular) special Lagrangian 3-fold for all $(a,b) \in \mathbb{R} \times \mathbb{C}$.

    (c) Identify the singular fibres and describe their singularities. Describe the topology of the singular and the nonsingular fibres.

The idea of a 'special Lagrangian fibration' $f : X \to B$ is in some ways a rather unnatural one. One of the problems is that the map f doesn't satisfy a particularly nice equation, locally; the level sets of f do, but the 'coordinates' on B are determined globally rather than locally. To understand the problems with special Lagrangian fibrations, try the following (rather difficult) exercise.

12.2 Let X be a Calabi–Yau 3-fold, N a compact SL 3-fold in X diffeomorphic to T^3, $\mathcal{M}_N$ the family of special Lagrangian deformations of N, and $\overline{\mathcal{M}}_N$ be $\mathcal{M}_N$ together with the singular SL 3-folds occurring as limits of elements of $\mathcal{M}_N$.

In good cases, SYZ hope that $\overline{\mathcal{M}}_N$ is the family of level sets of an SL fibration $f : X \to B$, where B is homeomorphic to $\overline{\mathcal{M}}_N$. How many different ways can you think of for this not to happen? (There are at least two mechanisms not involving singular fibres, and others which do).

Part II

Calabi–Yau Manifolds and Mirror Symmetry

Mark Gross

13   Introduction

This part grew out of a post-graduate level course taught at the University of Warwick in the winter of 1998. The goal was to introduce the students to some of the basic geometry of Calabi–Yau manifolds and lead into mirror symmetry. It began with a discussion of holonomy, Ricci curvature, and Yau's proof of the Calabi conjecture, which is not included here, and then moved on to an introduction to the different tools needed to understand the mirror symmetry conjecture. It ended with a detailed working out of the example of the quintic. Because of the diverse nature of the students following the course, with varied backgrounds in both algebraic and differential geometry, I always tried to present ideas with as little technical machinery as possible. I believe this resulted in a text quite different than some of the already existing excellent sources about mirror symmetry, and the student who wishes to pursue a career in the subject is strongly urged to consult more advanced texts, especially [197] and the very thorough [44], after reading this one.

We begin in §14 with some of the "classical" results about Calabi–Yau manifolds, pre-dating the "quantum" era of mirror symmetry. The most fundamental result, underlying essentially every further important idea in the theory of Calabi–Yau manifolds, is the Bogomolov–Tian–Todorov unobstructedness theorem, which tells us that a Calabi–Yau manifold has a smooth (unobstructed) universal deformation space. We will give an essentially complete proof of this result. Out of the Bogomolov–Tian–Todorov theorem comes the important results of Wilson on the structure and deformation invariance of the Kähler cone of Calabi–Yau 3-folds; this will be given briefer treatment. We will end the classical discussion with a brief survey of the enormous range of examples of Calabi–Yau 3-folds currently known.

Sections 15 and 16 begin the study of the two structures equated by mirror symmetry. Put in a very crude way, mirror symmetry is a conjecture that there exist *mirror pairs* of Calabi–Yau manifolds, i.e. Calabi–Yau manifolds X and $\check{X}$ which have a certain relationship. As this relationship takes quite a bit of work to even write down, we will have to cover a fair amount of material before we can even state these relationships. The basic idea is that there are certain moduli spaces, the *Kähler* and *complex* moduli spaces of X, which, roughly put, are manifolds whose points represent Kähler structures and complex structures on X, respectively. Let's denote these by $\mathcal{M}_{Kah}(X)$ and $\mathcal{M}_{cx}(X)$. Again, quite roughly, we should think of X and $\check{X}$ as being a mirror pair if there is an isomorphism between $\mathcal{M}_{Kah}(X)$ and $\mathcal{M}_{cx}(\check{X})$ that identifies certain functions we can define on the two moduli spaces. The functions on the Kähler moduli space are functions that involve concepts such as the number of rational curves on X, while the functions on complex moduli are things that, in theory, we can calculate in a straightforward manner. Thus the calculation on complex moduli of $\check{X}$ should yield information about rational curves on X. These are the basic ideas we will be developing.

Section 15 will be devoted to the structures on Kähler moduli space, which arise from Gromov–Witten invariants and quantum cohomology, while §16 will be devoted to the more classical notions of variation and degeneration of Hodge structures.

In §17 we will equate the structures arising on these two sides for mirror pairs of Calabi–Yau manifolds, and state the general mirror conjecture. Section 18 carries out the extended example of the quintic, following the foundational paper of Candelas, de la Ossa, Green and Parkes [36], to compute the predictions for the numbers of rational curves on the quintic using the structures we have set up previously.

Finally, in §19, we will discuss recent progress on new approaches to mirror symmetry, notably the Strominger–Yau–Zaslow conjecture.

The more expert reader will notice there are many fundamental topics that we have not covered. We have not gone into great detail on key issues surrounding J-holomorphic curves, such as Gromov compactness; nor have we delved into the many rich structures associated with quantum cohomology, such as Frobenius manifolds. We have not touched on the fundamental Batyrev or Batyrev–Borisov constructions for mirrors of families of hypersurfaces or complete intersections in toric varieties, nor on the Gelfand-Kapronov–Zelevinsky equations giving the Picard–Fuchs equations for such families. We ignore the Givental and Lian–Liu–Yau approach to verifying mirror symmetry predictions. These are all covered well in many other sources. Here we are motivated by the philosophy of taking the shortest path through the material with a goal of explaining the basic underlying ideas of mirror symmetry with a focus on the geometry of the Calabi–Yau manifolds themselves, rather than on their ambient spaces. However any reader who wishes to learn more about these topics is encouraged to continue reading other texts after this one.

I would like to thank M. Kreuzer for giving permission to reproduce Figure 2 from [138], and supplying the necessary computer files to do so.

14   The Classical Geometry of Calabi–Yau Manifolds

The basic object of study in this part will be Calabi–Yau manifolds. We will give a slightly different definition for Part II than was given in Part I:

Definition 14.1 A *Calabi–Yau manifold* is a compact complex Kähler manifold X with trivial canonical bundle $\omega_X \cong \mathcal{O}_X$.

From §5, we know that a Calabi–Yau manifold will always carry Ricci-flat metrics, but we won't include the data of metric or nowhere zero holomorphic n-form in this part. We will often focus on the 3-fold case, or insist that $H^1(X, \mathcal{O}_X) = 0$ (e.g. X is simply connected). See §14.2 for some discussion of this latter condition.

We will now study some of the fundamental "classical" (pre-mirror symmetry) results about Calabi–Yau manifolds.

14.1 Complex Structures and the Bogomolov–Tian–Todorov Theorem

Let X be a compact complex manifold. The manifold X has an underlying differentiable structure, but given this fixed underlying structure there may be many different complex structures on X. In particular, there might be a range of complex structures on X varying in a continuous manner. Ideally, we might study this by fixing a differentiable manifold X and considering the *Teichmüller space* of X, namely

$$\mathrm{Teich}(X) = \{\text{complex structures on } X\}/\sim_0,$$

where two complex structures J and J' on X satisfy $J \sim_0 J'$ if there exists a diffeomorphism $\varphi : X \to X$ isotopic to the identity such that $\varphi^* J' = J$. Meanwhile the moduli space of complex structures on X is

$$\mathcal{M}_{cx}(X) = \{\text{complex structures on } X\}/\sim,$$

where $J \sim J'$ if there exists a diffeomorphism $\varphi : X \to X$ such that $\varphi^* J' = J$. Note that

$$\mathcal{M}_{cx}(X) = \mathrm{Teich}(X)/(\mathrm{Diff}(X)/\mathrm{Diff}_0(X)),$$

where $\mathrm{Diff}_0(X)$ is the connected component of the identity of the group of diffeomorphisms, $\mathrm{Diff}(X)$.

Now in general $\mathcal{M}_{cx}(X)$ may not have particularly good properties, and there may not be much sensible one can say about it. Ideally, one would like $\mathcal{M}_{cx}(X)$ to be a nice complex manifold, say a quasi-projective variety, but this will almost always not be the case. As a result, we tend not to deal with the complex moduli space of X directly, but engage in a more local study of deformations of a specific complex structure. We will essentially be studying small neighbourhoods of points in $\mathrm{Teich}(X)$.

We can proceed as follows: Fix a given complex structure on X.

Definition 14.2 A *deformation* of X consists of a smooth proper morphism $\mathcal{X} \to S$, where $\mathcal{X}$ and S are connected complex spaces, and an isomorphism $X \cong \mathcal{X}_0$, where $0 \in S$ is a distinguished point. We call $\mathcal{X} \to S$ a *family of complex manifolds*.

One should keep in mind that S is not necessarily a manifold, and can be singular, reducible, or non-reduced, (e.g. $S = \mathrm{Spec}\,\mathbb{C}[x]/(x^2)$). Often, we only care about the germ of S at 0. We can then define:

Definition 14.3 A deformation $\mathcal{X} \to (S, 0)$ of X is called *universal* if any other deformation $\mathcal{X}' \to (S', 0')$ is isomorphic to the pull-back under a uniquely determined morphism $\varphi : S' \to S$ with $\varphi(0') = 0$.

The universal family is unique up to unique isomorphism. It will be denoted by $\mathcal{X} \to \mathrm{Def}(X)$, where $(\mathrm{Def}(X), 0)$ is again considered as the germ of a complex space.

The main result of the deformation theory of complex manifolds is the following existence theorem, which we will only state for manifolds without holomorphic vector fields.

Theorem 14.4 (Kuranishi) *If X is a compact complex manifold without global holomorphic vector fields, i.e. $H^0(X, \mathcal{T}_X) = 0$ where $\mathcal{T}_X$ is the holomorphic tangent bundle, then a universal deformation of X exists. Moreover, the universal deformation is universal for any of its fibres.*

We wish to understand how we build small, or infinitesimal, deformations of X. We begin with the following standard observation:

Theorem 14.5 *Let $f : \mathcal{X} \to U$ be a family of complex manifolds with U contractible. Then there is a diffeomorphism $\mathcal{X} \cong X \times U$, where X is diffeomorphic to any fibre of f, making the diagram*

$$
\begin{array}{ccc}
\mathcal{X} & \longrightarrow & X \times U \\
\downarrow & & \downarrow \\
U & \xrightarrow{=} & U
\end{array}
$$

commute, where the second vertical arrow is projection.

Thus given a deformation $\mathcal{X} \to S$ over contractible S (say a germ), we can view this as giving a family of complex structures on the fixed differentiable manifold X, and thus can view $\mathrm{Def}(X)$ as a germ of Teichmüller space.

Now given the fixed complex structure of X, thinking for a moment of $\mathcal{T}_X$ as the real C^∞ tangent bundle, we obtain decompositions on the complexified tangent and cotangent bundles

$$
\mathcal{T}_X \otimes_{\mathbb{R}} \mathbb{C} = \mathcal{T}_X^{1,0} \oplus \mathcal{T}_X^{0,1}
$$
$$
\mathcal{T}_X^* \otimes_{\mathbb{R}} \mathbb{C} = \Omega_X^{1,0} \oplus \Omega_X^{0,1}
$$

into tangent vectors and forms of type $(1,0)$ and $(0,1)$. Let $\pi^{1,0} : \mathcal{T}_X^* \otimes_{\mathbb{R}} \mathbb{C} \to \Omega_X^{1,0}$ and $\pi^{0,1} : \mathcal{T}_X^* \otimes_{\mathbb{R}} \mathbb{C} \to \Omega_X^{0,1}$ denote the two projections.

Now, for a nearby complex structure on X, which we denote by X', we get another decomposition

$$
\mathcal{T}_X^* \otimes_{\mathbb{R}} \mathbb{C} = \Omega_{X'}^{1,0} \oplus \Omega_{X'}^{0,1}.
$$

We can take the word "nearby" to mean that the projection

$$
\pi^{1,0}|_{\Omega_{X'}^{1,0}} : \Omega_{X'}^{1,0} \to \Omega_X^{1,0}
$$

is an isomorphism. In other words, as the complex structure on X varies continuously, the space of $(1,0)$-forms varies in a continuous way, so for nearby complex structures this projection will be an isomorphism. We then can define a map $s : \Omega_X^{1,0} \to \Omega_X^{0,1}$ as

$$
s = -\pi^{0,1} \circ \left(\pi^{1,0}|_{\Omega_{X'}^{1,0}}\right)^{-1},
$$

or equivalently we can view s as a C^∞ section of the vector bundle $T_X^{1,0} \otimes \Omega_X^{0,1}$.

Conversely, given a "small" element $s \in \Gamma(X, T_X^{1,0} \otimes \Omega_X^{0,1})$, this element determines an almost complex structure on X as follows: We take the bundle $\Omega_{X'}^{1,0}$ to be the graph of $-s : \Omega_X^{1,0} \to \Omega_X^{0,1}$ as a subbundle of $T_X^* \otimes_{\mathbb{R}} \mathbb{C}$. The bundle $\Omega_{X'}^{0,1}$ is taken to be the complex conjugate of $\Omega_{X'}^{1,0}$, and it is then easy to see that if s is sufficiently close to zero then we obtain a splitting $T_X^* \otimes_{\mathbb{R}} \mathbb{C} = \Omega_{X'}^{1,0} \oplus \Omega_{X'}^{0,1}$. This determines an almost complex structure J by determining the $\pm i$ eigenspaces of J. Of course, this almost complex structure is not necessarily integrable, and we want to determine the integrability condition. We will write this locally first.

If $z_1, \ldots, z_n$ are local holomorphic coordinates on X then we can write

$$s = \sum_{i,j} s_{ij} \frac{\partial}{\partial z_i} \otimes d\bar{z}_j,$$

and then

$$dz_i - s(dz_i) = dz_i - \sum_j s_{ij} d\bar{z}_j$$

form a local basis of $(1,0)$-forms on X'.

Theorem 14.6 *The almost complex structure induced by s is integrable if and only if*

$$\sum_i \left(\frac{\partial s_{lk}}{\partial z_i} s_{ij} - \frac{\partial s_{lj}}{\partial z_i} s_{ik} \right) = \frac{\partial s_{lj}}{\partial \bar{z}_k} - \frac{\partial s_{lk}}{\partial \bar{z}_j}$$

for all j, k and l.

Proof. Put

$$\theta_i = dz_i - \sum_j s_{ij} d\bar{z}_j.$$

$\theta_1, \ldots, \theta_n$ span the space of $(1,0)$-forms with respect to the almost complex structure induced by s. To show that the almost complex structure induced by $\theta_1, \ldots, \theta_n$ is integrable we need to show that $d\theta_i$ is of type $(2,0) + (1,1)$ for all i. (See [41], page 15, for this integrability condition). Since $\theta_1, \ldots, \theta_n, d\bar{z}_1, \ldots, d\bar{z}_n$ form a basis for the space of one-forms, we can write

$$d\theta_l = \sum_{i,j} \frac{1}{2} A_{ij}^l \theta_i \wedge \theta_j + \sum_{ij} B_{ij}^l \theta_i \wedge d\bar{z}_j + \sum_{i,j} \frac{1}{2} C_{ij}^l d\bar{z}_i \wedge d\bar{z}_j.$$

The almost complex structure is integrable if and only if $C_{ij}^l = 0$ for all i, j and l since $\theta_i \wedge \theta_j$ and $\theta_i \wedge d\bar{z}_j$ span the space of forms of type $(2,0) + (1,1)$. Here A^l and C^l are skew-symmetric matrices. Note

$$d\theta_l = -\sum_{i,j} \frac{\partial s_{lj}}{\partial z_i} dz_i \wedge d\bar{z}_j - \sum_{i,j} \frac{\partial s_{lj}}{\partial \bar{z}_i} d\bar{z}_i \wedge d\bar{z}_j.$$

Since $d\theta_l$ contains no $dz_i \wedge dz_j$ terms, we must have $A_{ij}^l = 0$, and therefore $B_{ij}^l = -\partial s_{lj}/\partial z_i$. Then the almost complex structure is integrable if and only if

$$d\theta_l = \sum_{i,j} B_{ij}^l \theta_i \wedge d\bar{z}_j$$

$$= -\sum_{i,j} \frac{\partial s_{lj}}{\partial z_i} dz_i \wedge d\bar{z}_j + \sum_{i,j,k} \frac{\partial s_{lj}}{\partial z_i} s_{ik} d\bar{z}_k \wedge d\bar{z}_j,$$

which holds if and only

$$\sum_i \left(\frac{\partial s_{lk}}{\partial z_i} s_{ij} - \frac{\partial s_{lj}}{\partial z_i} s_{ik} \right) = \frac{\partial s_{lj}}{\partial \bar{z}_k} - \frac{\partial s_{lk}}{\partial \bar{z}_j}.$$

This is the desired condition.   □

The condition of Theorem 14.6 can be formulated as follows: Consider the complex $(\mathcal{T}_X^{1,0} \otimes \Omega_X^{0,\bullet}, \bar{\partial})$, i.e. the complex

$$\mathcal{T}_X^{1,0} = \mathcal{T}_X^{1,0} \otimes \Omega_X^{0,0} \xrightarrow{\bar{\partial}} \mathcal{T}_X^{1,0} \otimes \Omega_X^{0,1} \xrightarrow{\bar{\partial}} \mathcal{T}_X^{1,0} \otimes \Omega_X^{0,2} \xrightarrow{\bar{\partial}} \cdots.$$

Here $\bar{\partial}$ acts on the second factor:

$$\bar{\partial}\left(\frac{\partial}{\partial z_i} \otimes \alpha \right) = \frac{\partial}{\partial z_i} \otimes \bar{\partial}\alpha.$$

This complex can be equipped with some extra structure, known as a differential graded Lie algebra (DGLA) structure. Here $\bar{\partial}$ is the differential, and we can define a Lie bracket

$$[\cdot,\cdot] : (\mathcal{T}_X^{1,0} \otimes \Omega_X^{0,p}) \times (\mathcal{T}_X^{1,0} \otimes \Omega_X^{0,q}) \to \mathcal{T}_X^{1,0} \otimes \Omega_X^{0,p+q}$$

by

$$[\alpha \otimes d\bar{z}_I, \alpha' \otimes d\bar{z}_J] = [\alpha, \alpha'] \otimes d\bar{z}_I \wedge d\bar{z}_J,$$

where on the right hand side $[,]$ is the usual Lie bracket of vector fields and I and J are multi-indices. The bracket is then extended bilinearly. This satisfies, for $\alpha \in \mathcal{T}_X^{1,0} \otimes \Omega_X^{0,i}$, $\beta \in \mathcal{T}_X^{1,0} \otimes \Omega_X^{0,j}$, $\gamma \in \mathcal{T}_X^{1,0} \otimes \Omega_X^{0,k}$, (graded) skew-commutativity

$$[\alpha, \beta] + (-1)^{ij}[\beta, \alpha] = 0,$$

and the (graded) Jacobi identity

$$(-1)^{ki}[\alpha, [\beta, \gamma]] + (-1)^{ij}[\beta, [\gamma, \alpha]] + (-1)^{jk}[\gamma, [\alpha, \beta]] = 0.$$

Furthermore,

$$\bar{\partial}[\alpha, \beta] = [\bar{\partial}\alpha, \beta] + (-1)^i[\alpha, \bar{\partial}\beta].$$

It is these three equalities which make $(\mathcal{T}_X^{1,0} \otimes \Omega_X^{0,\bullet}, [,], \bar{\partial})$ into a DGLA. We can then restate Theorem 14.6:

Corollary 14.7 *If $s \in \Gamma(X, T_X^{1,0} \otimes \Omega_X^{0,1})$ induces an almost complex structure on X, then it is integrable if and only if*

$$\bar{\partial} s + \frac{1}{2}[s, s] = 0.$$

Remark 14.8 Another way to think of the complex structure determined by $s \in \Gamma(X, T_X^{1,0} \otimes \Omega_X^{0,1})$ is to describe the holomorphic functions on X in this new complex structure. A function f is holomorphic if df is of pure type $(1,0)$; this means we must have the equality

$$\sum_{i=1}^{n} \frac{\partial f}{\partial z_i} dz_i + \sum_{i=1}^{n} \frac{\partial f}{\partial \bar{z}_i} d\bar{z}_i = \sum_{i=1}^{n} \frac{\partial f}{\partial z_i} \left(dz_i - \sum_{j=1}^{n} s_{ij} d\bar{z}_j \right),$$

i.e.

$$\frac{\partial f}{\partial \bar{z}_j} = -\sum_{i=1}^{n} s_{ij} \frac{\partial f}{\partial z_i}.$$

This can be written as

$$(\bar{\partial} + s \circ \partial)f = 0.$$

Now consider a germ of a curve in $\mathrm{Def}(X)$ through $0 \in \mathrm{Def}(X)$, which we can think of as a continuous family $s(t) \in \Gamma(X, T_X^{1,0} \otimes \Omega_X^{0,1})$ with $s(0) = 0$. Here $t \in \mathbb{C}$ is a complex parameter and, for t close to zero, $s(t)$ induces an almost complex structure on X; the equation

$$\bar{\partial} s(t) + \frac{1}{2}[s(t), s(t)] = 0 \tag{28}$$

is equivalent to integrability of these almost complex structures. If (28) holds, then $s(t)$ determines a family of complex structures on X, i.e. a curve in $\mathrm{Def}(X)$. Of course, we need to divide out by diffeomorphisms of X isotopic to the identity. This can be done by specifying a family $v(t) \in \Gamma(X, T_X^{1,0})$ of vector fields, and integrating these to obtain a family of diffeomorphisms $\phi_t : X \to X$ with ϕ_0 the identity.

Let us use this to determine the tangent space of $\mathrm{Def}(X)$ at a point corresponding to the given complex structure on X. Take a curve $s(t)$ of complex structures as above with $s(0) = 0$, and expand $s(t)$ in a power series $s(t) = \sum_{n=1}^{\infty} s_n t^n$ with $s_i \in \Gamma(X, T_X^{1,0} \otimes \Omega_X^{0,1})$. We want to understand the possible values s_1 can take. To see this, we just consider equation (28) modulo t^2, and we get the linear equation

$$\bar{\partial} s_1 = 0.$$

However, we also need to divide out by the action of a family of diffeomorphisms $\phi_t : X \to X$, ϕ_0 the identity. To first order, locally, ϕ_t can be described by

$$z_i \mapsto z_i + t f_i(z, \bar{z}) + \cdots,$$

and only this first order expansion affects s_1. More precisely, $s(t)$ gives $(1,0)$-forms of the form

$$dz_i - \sum_j s_{1,ij} t \, d\bar{z}_j + \text{higher order terms},$$

which under ϕ_t is transformed to

$$dz_i + t df_i - \sum_j s_{1,ij} t \, d\bar{z}_j + \text{higher order terms}$$

$$= dz_i + t\left(\sum_j \left(-s_{1,ij} + \frac{\partial f_i}{\partial \bar{z}_j}\right) d\bar{z}_j\right) + t\sum_j \frac{\partial f_i}{\partial z_j} dz_j + \text{higher order terms}.$$

These $(1,0)$-forms span the same space, to first order, as

$$dz_i - t\left(\sum_j \left(s_{1,ij} - \frac{\partial f_i}{\partial \bar{z}_j}\right) d\bar{z}_j\right) + \text{higher order terms}.$$

Now differentiating ϕ_t at $t = 0$ gives the vector field $v = f_i \frac{\partial}{\partial z_i}$. Thus the space of possible s_1's modulo the action of diffeomorphisms is

$$\frac{\{s \in \Gamma(X, T_X^{1,0} \otimes \Omega_X^{0,1}) | \bar{\partial} s = 0\}}{\bar{\partial}\Gamma(X, T_X^{1,0})} = H^1(X, T_X),$$

where T_X is the holomorphic tangent bundle on X.

This discussion may be a bit misleading, as given an $s_1 \in H^1(X, T_X)$, there may not be a solution $s(t) = \sum_{n=1}^{\infty} s_n t^n$ of (28); in other words, we have looked at infinitesimal deformations, but we don't know if they can be extended to general deformations. We will try to solve (28) order by order, but we will first formalize the above situation.

Let A be an Artin local $\mathbb{C}$-algebra, i.e. A has a unique maximal ideal $\mathbf{m}_A$ and is a finite dimensional $\mathbb{C}$-vector space. We can tensor the complex $\Gamma(X, T_X^{1,0} \otimes \Omega_X^{0,\bullet})$ with $\mathbf{m}_A$, and then try to solve (28) over A, i.e. set

$$D_X(A) = \frac{\{s \in \Gamma(X, T_X^{1,0} \otimes \Omega_X^{0,1}) \otimes \mathbf{m}_A | \bar{\partial} s + \frac{1}{2}[s,s] = 0\}}{\sim},$$

where equivalence is given by an action of $\Gamma(X, T_X^{1,0}) \otimes \mathbf{m}_A$ on $\Gamma(X, T_X^{1,0} \otimes \Omega_X^{0,1}) \otimes \mathbf{m}_A$. Elements of $\Gamma(X, T_X^{1,0}) \otimes \mathbf{m}_A$ act by "infinitesimal" diffeomorphisms, as shown above when $A = \mathbb{C}[t]/(t^2)$ (in which case $D_X(A) = H^1(X, T_X)$). The actual action is somewhat complicated, and we do not give it here, but see [66], [67] for details. Note D_X is a covariant functor: given a map $\varphi : A \to B$, we obtain a map $D_X(\varphi) : D_X(A) \to D_X(B)$.

Remark 14.9 Given an element $s \in D_X(A)$, we can think of this as defining an analytic space X_A defined over Spec A. The underlying topological space of X_A is X, and the sheaf $\mathcal{O}_{X_A}$ of holomorphic functions on X_A is the subsheaf of $C^\infty(X) \otimes A$ of functions f satisfying $(\bar{\partial} + s \circ \partial)f = 0$, as in Remark 14.8. The sheaf of holomorphic functions $\mathcal{O}_{X_A}$ is then a sheaf of A-algebras, which gives a morphism $X_A \to$ Spec A. This can be viewed as an infinitesimal family of complex structures on X.

The integrability equation $\bar{\partial}s + \frac{1}{2}[s, s] = 0$ then implies that locally there exists $f_1, \dots, f_n$ ($n = \dim X$) with $(\bar{\partial} + s \circ \partial)f_i = 0$, such that $f_1, \dots, f_n$ yields an isomorphism of an open set on X_A with $U \times$ Spec A, where $U \subseteq \mathbb{C}^n$ is some open set.

We will usually write X_A/A, meaning we are considering X_A along with the structure map $X_A \to$ Spec A.

We can then define the relative holomorphic cotangent bundle $\Omega^1_{X_A/A}$ as the locally free sheaf of $\mathcal{O}_{X_A}$-modules locally generated by $df_1, \dots, df_n$, where $f_1, \dots, f_n$ are as above. We also obtain $\Omega^p_{X_A/A} = \bigwedge^p \Omega^1_{X_A/A}$, and the sheaf of $\mathcal{O}_{X_A}$-linear homomorphisms from $\Omega^1_{X_A/A}$ to $\mathcal{O}_{X_A}$ is $\mathcal{T}_{X_A/A}$.

Alternatively, we could have defined

$$D_X(A) = \{\text{deformations } X_A \to \text{Spec } A \text{ of } X\}/\cong.$$

Now the key result for us is:

Theorem 14.10 (The Bogomolov–Tian–Todorov unobstructedness theorem) *Let X be a compact Calabi–Yau manifold with $H^0(X, \mathcal{T}_X) = 0$. Then* $\mathrm{Def}(X)$ *is a germ of a smooth manifold with tangent space $H^1(X, \mathcal{T}_X)$.*

Remark 14.11 When X is a Calabi–Yau manifold there are some useful identifications. If $\dim X = n$, we have the pairing given by the wedge product

$$\Omega^1_X \times \Omega^{n-1}_X \to \Omega^n_X = \omega_X \cong \mathcal{O}_X.$$

Here Ω^i_X denotes the bundle of holomorphic i-forms. Note that the last isomorphism depends on the choice of a non-zero section of Ω^n_X. This pairing is always perfect, so we see that $\Omega^{n-1}_X \cong (\Omega^1_X)^\vee \cong \mathcal{T}_X$. Thus, we see the important fact that, given a choice of non-zero section of ω_X for a Calabi–Yau manifold, there is a canonical isomorphism $\Omega^{n-1}_X \cong \mathcal{T}_X$. In particular, this says that

$$H^0(X, \mathcal{T}_X) \cong H^0(X, \Omega^{n-1}_X) \cong \overline{H^{n-1}(X, \mathcal{O}_X)} \cong \overline{H^1(X, \mathcal{O}_X)^\vee}$$

by Serre duality. Hence the non-existence of global holomorphic vector fields on X is equivalent to $H^1(X, \mathcal{O}_X) = 0$. Also,

$$H^1(X, \mathcal{T}_X) \cong H^1(X, \Omega^{n-1}_X) \cong H^{n-1,1}(X)$$

and

$$H^2(X, \mathcal{T}_X) \cong H^{n-1,2}(X).$$

Proof of BTT. We will follow Kawamata's proof [130] of this theorem. The basic idea of all proofs (see [189], [190], [172]) is as follows: Set $A_n = \mathbb{C}[t]/(t^{n+1})$. An element $s_1 \in D_X(A_1)$ can be viewed as a tangent vector to $\mathrm{Def}(X)$, and the difficulty is there may not be a curve through $0 \in \mathrm{Def}(X)$ with tangent direction s_1. If there is such a curve for every tangent vector s_1, then $\mathrm{Def}(X)$ is smooth at 0 (and hence in a neighbourhood of 0). So the idea is to start with $s \in D_X(A_1)$ arbitrary, and try to lift this order by order to $D_X(A_2)$, $D_X(A_3)$, etc. In the end we get a *formal* germ of a curve through $0 \in \mathrm{Def}(X)$, and this is sufficient to prove non-singularity. Thus it will be enough to show $D_X(A_{n+1}) \to D_X(A_n)$ is surjective for all n.

Set

$$B_n = \mathbb{C}[x,y]/(x^{n+1}, y^2)$$

with maps $\alpha_n : A_{n+1} \to A_n$ the projection, $\beta_n : B_n \to A_n$ given by $x \mapsto t$, $y \mapsto 0$, $\xi_n : B_n \to B_{n-1}$ the projection, and $\epsilon_n : A_n \to B_{n-1}$ given by $t \mapsto x + y$. Denote an element $s \in D_X(A_n)$ by X_n/A_n as in Remark 14.9, and set

$$T^1(X_n/A_n) = \{u \in D_X(B_n)|D_X(\beta_n)(u) = s\}.$$

This can be thought of as

$$\{u \in \Gamma(X, T_X^{1,0} \otimes \Omega_X^{0,1}) \otimes A_n|s + yu \in D_X(B_n)\}/ \sim$$

where $u_1 \sim u_2$ if $s+yu_1 = s+yu_2$ in $D_X(B_n)$. As such, this is an A_n-module: if $u_1, u_2 \in T^1(X_n/A_n)$, $f \in A_n$, then

$$\bar{\partial}(s + y(fu_1 + u_2)) + \frac{1}{2}[s + y(fu_1 + u_2), s + y(fu_1 + u_2)] = 0 \quad \text{in } B_n,$$

so $fu_1 + u_2 \in T^1(X_n/A_n)$. Similarly, if X_{n-1}/A_{n-1} denotes $D_X(\alpha_{n-1})(s)$, we obtain a space $T^1(X_{n-1}/A_{n-1})$ and a map induced by $D_X(\xi_n)$

$$\phi_n : T^1(X_n/A_n) \to T^1(X_{n-1}/A_{n-1}).$$

Claim: (Ran's T^1 lifting criterion.) If ϕ_n is surjective for all X_n/A_n then $D(\alpha_n) : D_X(A_{n+1}) \to D_X(A_n)$ is surjective.

Proof. Suppose that $s = \sum_{i=1}^n s_i t^i \in D_X(A_n)$ defines X_n/A_n. We wish to lift this to $\tilde{s} \in D_X(A_{n+1})$. Now

$$D_X(\epsilon_n)(s) = \sum_{i=1}^{n-1} s_i x^i + \sum_{i=1}^n i s_i x^{i-1} y \in T^1(X_{n-1}/A_{n-1}).$$

By assumption of surjectivity of ϕ_n, this lifts to

$$\sum_{i=1}^n s_i x^i + \sum_{i=1}^n i s_i x^{i-1} y + s' x^n y \in T^1(X_n/A_n) \subseteq D_X(B_n).$$

This must satisfy the equation (28), i.e. focusing on the $x^n y$ term in (28) we must have

$$
\begin{aligned}
0 &= \bar{\partial}s' + \frac{1}{2}\left(\sum_{i=1}^{n}[s_i,(n-i+1)s_{n-i+1}] + [is_i, s_{n-i+1}]\right) \\
&= \bar{\partial}s' + \frac{1}{2}\sum_{i=1}^{n}(n+1)[s_i, s_{n-i+1}].
\end{aligned}
\tag{29}
$$

Then put

$$
\tilde{s} = \sum_{i=1}^{n} s_i t^i + \frac{s'}{n+1}t^{n+1}.
$$

Looking at the t^{n+1} term, equation (28) holds for $\tilde{s}$ if

$$
\bar{\partial}s' + \frac{1}{2}\sum_{i=1}^{n}(n+1)[s_i, s_{n-i+1}] = 0.
$$

But this is the same as (29). Thus $\tilde{s}$ is a lifting of s. $\qquad\square$

Claim: For X a Calabi-Yau manifold, ϕ_n is always surjective.

Proof. This is the only part of the proof where we need to use the fact that X is a Calabi-Yau manifold. We will be a bit sketchy here. Using local holomorphic coordinates on X_A as in Remark 14.9, it is easy to see, as in the calculation of $D_X(A_1)$, that $T^1(X_n/A_n) \cong H^1(X, \mathcal{T}_{X_n/A_n})$. The isomorphism $\mathcal{T}_{X_n/A_n} \cong \Omega^{d-1}_{X_n/A_n}$ (where $d = \dim X$) holds as in Remark 14.11. (To see this, one needs a global non-vanishing section of $\Omega^d_{X_n/A_n} = \omega_{X_n/A_n}$. But this is a deformation of the trivial bundle $\mathcal{O}_X \cong \omega_X$, and since $H^1(X, \mathcal{O}_X) = 0$, it must remain trivial.) Thus surjectivity of $T^1(X_n/A_n) \to T^1(X_{n-1}/A_{n-1})$ is the same as surjectivity of

$$
H^1(X, \Omega^{d-1}_{X_n/A_n}) \to H^1(X, \Omega^{d-1}_{X_{n-1}/A_{n-1}}).
$$

However, this is a standard result from Hodge theory. This says essentially that the Hodge numbers of Kähler manifolds are invariant under deformation or, more precisely, the Hodge bundles behave well under base-change. See [47], Theorem 5.5 for the precise relevant statement. $\qquad\square$

This completes the proof of the BTT theorem. $\qquad\square$

Remark 14.12 Why is BTT surprising? We have managed, in the proof of BTT, to lift elements of $D_X(A_n)$ to $D_X(A_{n+1})$. In general, this should not be possible, but to understand this, we need to explain obstruction theory.

Let $\phi : B \to A$ be a *small extension* of A, i.e. a surjective map of local Artin $\mathbb{C}$-algebras with kernel J such that $m_B J = 0$. Let $s_A \in D_X(A)$, which we can think of as represented by an element $s_A \in \Gamma(X, \mathcal{T}_X^{1,0} \otimes \Omega_X^{0,1}) \otimes \mathbf{m}_A$ satisfying (28). We lift this element arbitrarily to an element $s_B \in \Gamma(X, \mathcal{T}_X^{1,0} \otimes \Omega_X^{0,1}) \otimes \mathbf{m}_B$, and try to find an element $s_J \in \Gamma(X, \mathcal{T}_X^{1,0} \otimes \Omega_X^{0,1}) \otimes J$ that satisfies the equation

$$\bar{\partial}(s_B + s_J) + \frac{1}{2}[s_B + s_J, s_B + s_J] = 0.$$

But since $\mathbf{m}_B J = 0$, $[s_B, s_J] = [s_J, s_J] = 0$, so we in fact have a linear equation for s_J, namely

$$\bar{\partial}s_J + (\bar{\partial}s_B + \frac{1}{2}[s_B, s_B]) = 0.$$

This can be solved if and only if

$$\bar{\partial}s_B + \frac{1}{2}[s_B, s_B] \in \Gamma(X, \mathcal{T}_X^{1,0} \otimes \Omega_X^{0,2}) \otimes J$$

represents zero in $H^2(X, \mathcal{T}_X) \otimes J$. Furthermore, $\bar{\partial}s_B + \frac{1}{2}[s_B, s_B]$ defines a well-defined element of $H^2(X, \mathcal{T}_X) \otimes J$, independent of the lifting of s_B. We denote this element by $o(s_A)$. Thus in general, we have an "exact sequence"

$$D_X(B) \longrightarrow D_X(A) \overset{o}{\longrightarrow} H^2(X, \mathcal{T}_X) \otimes J,$$

i.e. an element $s_A \in D_X(A)$ lifts to $D_X(B)$ if and only if $o(s_A) = 0$. As a result, $H^2(X, \mathcal{T}_X)$ is called the *obstruction space* for X.

If $H^2(X, \mathcal{T}_X) = 0$ then all deformations always lift, and therefore $\mathrm{Def}(X)$ is smooth. A surprise is that for a Calabi–Yau manifold, $H^2(X, \mathcal{T}_X)$ is in general non-zero by Remark 14.11: for example, for a Calabi–Yau 3-fold, the obstruction space is dual to $H^{1,1}(X)$, which is always non-zero if X is Kähler, and often big. Experience with deformation theory says it is extremely rare that there is any general unobstructedness result when the obstruction space is non-zero.

We end this section with a discussion of the Kodaira–Spencer map. Given a family $f : \mathcal{X} \to S$ of compact complex manifolds with a fibre $X = \mathcal{X}_s$, we get a map ψ from an open neighbourhood U of s to $\mathrm{Def}(X)$. We obtain a map

$$\psi_* : \mathcal{T}_{S,s} \to \mathcal{T}_{\mathrm{Def}(X),X} \cong H^1(X, \mathcal{T}_X)$$

called the *Kodaira–Spencer map*.

This map can be described explicitly. Look at the normal bundle exact sequence for $X = \mathcal{X}_s \subseteq \mathcal{X}$,

$$0 \to \mathcal{T}_X \to \mathcal{T}_{\mathcal{X}}|_X \to \mathcal{N}_{X/\mathcal{X}} \to 0.$$

Taking cohomology, this gives a long exact sequence, a part of which is

$$H^0(X, \mathcal{T}_{\mathcal{X}}|_X) \to H^0(X, \mathcal{N}_{X/\mathcal{X}}) \to H^1(X, \mathcal{T}_X).$$

Now the normal bundle $\mathcal{N}_{X/\mathcal{X}}$ is the pull-back of the normal bundle of $s \in \mathcal{S}$, which is just the vector space $\mathcal{T}_{S,s}$, so $H^0(X, \mathcal{N}_{X/\mathcal{X}}) = \mathcal{T}_{S,s}$. Thus we obtain a map

$$\mathcal{T}_{S,s} \to H^1(X, \mathcal{T}_X),$$

which is in fact the Kodaira-Spencer map. (See Exercise 14.4.)

There is another way to view the Kodaira-Spencer map for Calabi-Yau manifolds. Given a family of n dimensional Calabi-Yau manifolds $\mathcal{X} \to \mathcal{S}$, then locally for the base we can choose a holomorphic n-form on $\mathcal{X}$ that restricts to a non-vanishing holomorphic n-form on each fibre. Think of this as providing a family of holomorphic n-forms; since we can locally trivialise the fibration $\mathcal{X} \to \mathcal{S}$, we will denote this by a family of n-forms $\Omega(s)$ on a fibre $\mathcal{X}_{s_0} = X$. The cohomology class $[\Omega(s)] \in H^n(X, \mathbb{C})$ varies with s, but does not depend on the particular choice of trivialization. Also, given a tangent vector $\partial/\partial s \in \mathcal{T}_{S,s_0}$, we can then compute $\partial\Omega(s)/\partial s$ at s_0, which is a form on $\mathcal{X}_s$ of type $(n, 0) + (n - 1, 1)$. (See Theorem 16.4 for a proof). Because $\partial\Omega(s)/\partial s$ is d-closed, the $(n - 1, 1)$ part is necessarily $\bar{\partial}$-closed and hence determines a class in $H^{n-1,1}(X)$. Thus we have defined a map

$$\mathcal{T}_{S,s} \to H^{n-1,1}(X).$$

This map is easily seen to be independent of the trivialization. But what about dependence of this map on the choice of family of holomorphic n-forms? If $f(s)$ is a nowhere-vanishing holomorphic function on $\mathcal{S}$, then $\tilde{\Omega}(s) = f(s)\Omega(s)$ is another holomorphic family of holomorphic n-forms. How does the derivative change? We have

$$\partial\tilde{\Omega}(s)/\partial s = \partial(f(s)\Omega(s))/\partial s$$
$$= (\partial f(s)/\partial s)\Omega(s) + f(s)(\partial\Omega(s)/\partial s).$$

If $f(s_0) = 1$, then $\partial\Omega'(s)/\partial s - \partial\Omega(s)/\partial s$ at s_0 is a form of type $(n, 0)$, and so the $(n - 1, 1)$ part of $\partial\Omega(s)/\partial s$ does not depend on the particular choice of family of n-forms but only on $\Omega(s_0)$. Thus having fixed $\Omega(s_0)$, we get a well-defined map

$$\mathcal{T}_{S,s} \to H^{n-1,1}(X).$$

Now the identification $\mathcal{T}_X \xrightarrow{\cong} \Omega_X^{n-1}$ of Remark 14.11 depends on a choice of a holomorphic n-form on X, and given the choice $\Omega(s_0)$, we get a map $\mathcal{T}_{S,s} \to H^1(X, \mathcal{T}_X)$. One can show this coincides with the Kodaira-Spencer map (see Exercise 14.4).

14.2 The Structure of the Kähler Cone

Let X be a compact Kähler manifold. We set $H^{1,1}(X, \mathbb{R}) := H^2(X, \mathbb{R}) \cap H^{1,1}(X, \mathbb{C})$. Every element of $H^{1,1}(X, \mathbb{R})$ is represented by a real closed $(1, 1)$-form.

Definition 14.13 The Kähler cone of X is the cone in $H^{1,1}(X, \mathbb{R})$ given by

$$\mathcal{K}_X = \{c \in H^{1,1}(X, \mathbb{R}) | c \text{ is represented by a Kähler form}\}.$$

A basic fact:

Proposition 14.14 $\mathcal{K}_X$ *is an open subset of* $H^{1,1}(X, \mathbb{R})$.

Proof. Let α be a real closed $(1, 1)$ form. Then locally we can write

$$\alpha = i \sum_{j,k} \alpha_{jk} \mathrm{d}z_j \wedge \mathrm{d}\bar{z}_k,$$

with (α_{jk}) a Hermitian matrix. Now if ω is a Kähler form, then

$$\omega = i \sum_{j,k} h_{jk} \mathrm{d}z_j \wedge \mathrm{d}\bar{z}_k$$

with h_{jk} positive definite Hermitian. For sufficiently small ϵ, $h_{jk} + \epsilon \alpha_{jk}$ is still positive definite Hermitian at each point of X (using the fact that X is compact). Thus we can deform ω to another Kähler form $\omega + \epsilon \alpha$ in any direction $\alpha \in H^{1,1}(X, \mathbb{R})$, and so $\mathcal{K}_X$ is open. $\qquad\square$

If X is simply connected then $H^1(X, \mathcal{O}_X) = 0$. If also $H^2(X, \mathcal{O}_X) = 0$ then we are in a particularly nice situation, since then $H^{0,2}(X, \mathbb{C}) = H^{2,0}(X, \mathbb{C}) = 0$, and by the Hodge decomposition, $H^2(X, \mathbb{C}) = H^{1,1}(X, \mathbb{C})$ (and the same equality holds for $\mathbb{C}$ replaced by $\mathbb{R}$). The exponential exact sequence

$$0 \longrightarrow \mathbb{Z} \longrightarrow \mathcal{O}_X \overset{\exp(2\pi i \cdot)}{\longrightarrow} \mathcal{O}_X^* \longrightarrow 0$$

in general yields an exact sequence

$$H^1(X, \mathcal{O}_X) \longrightarrow \mathrm{Pic}(X) \overset{c_1}{\longrightarrow} H^2(X, \mathbb{Z}) \longrightarrow H^2(X, \mathcal{O}_X),$$

and so if $H^i(X, \mathcal{O}_X) = 0$ for $i = 1, 2$, c_1 yields an isomorphism between $\mathrm{Pic}(X)$ and $H^2(X, \mathbb{Z})$. Thus the Picard group is independent of the complex structure on X and is a topological invariant. In particular, if $\mathcal{X} \to \Delta$ is a deformation of complex structures, then the Picard group does not vary on the fibres.

How restrictive are these conditions for manifolds with $c_1(X) = 0$? The *Bogomolov decomposition theorem* gives a crude classification of manifolds with $c_1 = 0$. It is proved in Theorem 5.4 using Riemannian holonomy and the Calabi Conjecture.

Theorem 14.15 *Let X be a compact Kähler manifold with $c_1(X) = 0$. Then X has a finite unramified cover Y with*

$$Y \cong Z \times \prod_i S_i \times \prod_i C_i,$$

where

1. Z is a complex torus (i.e. of the form $\mathbb{C}^n/\Lambda$ for $\Lambda \cong \mathbb{Z}^{2n}$ a lattice in $\mathbb{C}^n$).
2. Each S_i is a simply connected holomorphic symplectic manifold with $\dim H^2(S_i, \mathcal{O}_{S_i}) = 1$.
3. Each C_i is a simply connected Calabi–Yau manifold with $H^2(C_i, \mathcal{O}_{C_i}) = 0$.

Proof. See Theorem 5.4, or [16], Exposé XVI.   □

Complex tori, though very interesting objects in their own right, hold little interest from our point of view, other than to provide some elementary examples of mirror symmetry. Holomorphic symplectic manifolds are covered in §21, so here we will focus on the third case. Thus the assumption that $H^1(X, \mathcal{O}_X) = H^2(X, \mathcal{O}_X) = 0$ is a natural one. In any event, we will focus in the rest of this section on the 3-fold case. In this case $H^1(X, \mathcal{O}_X) \cong H^2(X, \omega_X)^\vee$ by Serre duality. If X is Calabi–Yau, then $\omega_X \cong \mathcal{O}_X$, so $H^1(X, \mathcal{O}_X) \cong H^2(X, \mathcal{O}_X)^\vee$; thus if one is zero, so is the other.

So we will now let X be a Calabi–Yau 3-fold with $H^1(X, \mathcal{O}_X) = 0$, and we will summarize results due to P.M.H. Wilson on the structure of $\mathcal{K}_X$. We need to restrict to dimension three here because many results in birational geometry that we need are not known in higher dimensions. However, some of the results stated probably do extend to higher dimensions.

Definition 14.16 If X is a compact 3-fold, the *cubic intersection form* on $H^2(X, \mathbb{R})$ is given, for $D_1, D_2, D_3 \in H^2(X, \mathbb{R})$, by

$$D_1 \cdot D_2 \cdot D_3 = \int_X D_1 \wedge D_2 \wedge D_3.$$

This is an integral cubic form on $H^2(X, \mathbb{Z})$. We also set W to be the cone

$$W = \{D \in H^2(X, \mathbb{R}) | D^3 = 0\}.$$

We then have:

Theorem 14.17 *Let X be a projective Calabi–Yau 3-fold. If U is an open neighbourhood of W in $H^2(X, \mathbb{R})$ (in the Euclidean topology), then the closure of the Kähler cone $\overline{\mathcal{K}}_X$ is rational polyhedral outside of U. Furthermore, for each codimension 1 face σ of $\overline{\mathcal{K}}_X$ not contained in W there is a birational contraction morphism $f : X \to Y$ to a normal 3-fold Y such that $f^*(\mathrm{Pic}(Y)) \otimes \mathbb{R}$ is the span of σ.*

Proof. See [204],[205].   □

One way of paraphrasing this is saying that $\overline{\mathcal{K}}_X$ is *locally rational polyhedral* away from W. Things that can prevent $\overline{\mathcal{K}}_X$ itself from being rational polyhedral include the possibility that faces of $\overline{\mathcal{K}}_X$ accumulate towards an accumulation point on W, or $\overline{\mathcal{K}}_X$ might include a curvy bit contained in W.

Birational contractions $f : X \to Y$ induced by codimension one faces of $\overline{\mathcal{K}}_X$ are *primitive*, i.e. cannot be factored in the algebraic category. Wilson gives a coarse clasification of such contractions, and it is a good idea to keep this classification in mind in order to have a feeling of what kind of phenomena can happen.

Type I contractions are contractions in which a finite number of curves are contracted to points. These curves will then all be $\mathbb{P}^1$'s (not necessarily disjoint).

Type II contractions contract an irreducible surface to a point. These surfaces will always be (possibly singular) del Pezzo surfaces, i.e. surfaces whose anti-canonical bundle is ample.

Type III contractions contract a surface E to a non-singular curve C, and the fibres of the map $E \to C$ are conics (either non-singular conics, unions of two lines, or a doubled line).

Using this coarse classification, Wilson showed the following key theorem:

Theorem 14.18 *Let X be a projective Calabi–Yau 3-fold, $\mathcal{X} \to \Delta$ a family of Calabi–Yau manifolds over a disk Δ with $\mathcal{X}_0 \cong X$. Then there is an open neighbourhood of $0 \in \Delta$ such that $\overline{\mathcal{K}}_{\mathcal{X}_0} = \overline{\mathcal{K}}_{\mathcal{X}_t}$ for all $t \in U$ unless X contains a quasi-ruled elliptic surface, i.e. a surface $E \subseteq X$ such that either*

1. *E is a $\mathbb{P}^1$-bundle over an elliptic curve, or*
2. *there is a map $E \to C$, where C is an elliptic curve, such that each fibre is a union of two $\mathbb{P}^1$'s meeting at a point.*

Proof. (Sketch). Wilson proved this result by performing a case-by-case analysis of the possible primitive contractions on X. Each codimension one face of $\overline{\mathcal{K}}_X$ not contained in W yields a primitive contraction, and if one can prove that the contraction persists under deformation, then that face also persists. Type I and II contractions always deform to type I and II contractions. Type III contractions always deform to type III contractions if $g(C) = 0$; deform to type III or I contractions if $g(C) > 1$; and may disappear entirely only if $g(C) = 1$. A more detailed analysis leads to the above two cases. See [205] for details. □

14.3   Examples of Calabi–Yau Manifolds

We will survey some of the broad range of examples of Calabi–Yau manifolds. Most of the exploration of examples has been driven by string theorists searching for Calabi–Yau compactifications of space-time (see §12.1). There are essentially two key tools.

The first tool is adjunction, which tells us that if X is a variety and $D \subseteq X$ is a divisor (i.e. a hypersurface locally defined by a single equation) then

$$\omega_D \cong \mathcal{N}_D \otimes \omega_X|_D$$

where $\mathcal{N}_D$ is the normal bundle of D in X (see for example [77, p. 147]).

The second is the notion of a crepant resolution of singularities. Let X be a singular variety. Suppose that X is normal, so that in particular the singular locus of X is codimension $\geqslant 2$ in X. Let $i : X_{ns} \hookrightarrow X$ be the inclusion of the non-singular part of X in X. Then the canonical bundle $\omega_{X_{ns}}$ is defined, and we can put $\omega_X := i_* \omega_{X_{ns}}$, where i_* denotes the push-forward of sheaves. In other words, we define the space of holomorphic n-forms on an open subset $U \subseteq X$ to be the space of holomorphic n-forms on $U \cap X_{ns}$. Now ω_X is not necessarily a line bundle, but when it is, we say X is *Gorenstein*. As the adjunction formula still holds even when D is singular, any hypersurface in a non-singular variety is Gorenstein.

If X is Gorenstein, we say a resolution of singularities $f : Y \to X$ (i.e. a birational morphism from a non-singular variety Y to X) is *crepant* if $f^* \omega_X \cong \omega_Y$. Under certain conditions such a resolution can be found. If $\omega_X \cong \mathcal{O}_X$, then of course $\omega_Y \cong \mathcal{O}_Y$.

Let's see these two techniques in action.

Example 14.19 (0) Let f be a homogeneous quintic polynomial in variables $x_0, \ldots, x_4$. Then $f = 0$ defines a hypersurface X in $\mathbb{P}^4$. Since

$$\omega_{\mathbb{P}^4} \cong \mathcal{O}_{\mathbb{P}^4}(-5),$$

$\omega_X \cong \mathcal{O}_X$ by adjunction. If X is non-singular, i.e. f and $\partial f / \partial x_0, \ldots, \partial f / \partial x_4$ do not vanish simultaneously, then X is a non-singular Calabi-Yau manifold.

(I). Take a $\mathbb{P}^2 \subseteq \mathbb{P}^4$, say given by $x_0 = x_1 = 0$. Take any quintic containing the $\mathbb{P}^2$, i.e. any quintic X of the form

$$x_0 f_4 + x_1 g_4,$$

where f_4 and g_4 are general degree 4 polynomials. One checks easily that X is singular at 16 points given by the equation

$$x_0 = x_1 = f_4 = g_4 = 0.$$

If one blows up the $\mathbb{P}^2$ inside of $\mathbb{P}^4$, one obtains as the proper transform of X a non-singular 3-fold Y such that $f : Y \to X$ contracts 16 projective lines, one over each singular point of Y. Furthermore, $\omega_Y = \mathcal{O}_Y$. This follows because the exceptional locus of f is codimension 2, so there are no divisors supported on this exceptional locus.

(II). Take a quintic X with a triple point, e.g.

$$x_0^2(x_1^3 + x_2^3 + x_3^3 + x_4^3) + x_0 f_4 + f_5 = 0,$$

where f_4 and f_5 are generic polynomials of degree 4 and 5 in $x_1, \ldots, x_4$. Convince yourself that this is only singular at the point $(1, 0, 0, 0, 0)$. To resolve the singularities blow up $\mathbb{P}^4$ at $(1, 0, 0, 0, 0)$ to obtain the proper transform Y of X. Over $(1, 0, 0, 0, 0)$ sits an exceptional surface whose equation is $x_1^3 + x_2^3 + x_3^3 + x_4^3 = 0$ in $\mathbb{P}^3$, a classic example of a del Pezzo surface. One can show this is a crepant resolution (see Exercise 14.3).

(III). Take a general quintic X singular along the $\mathbb{P}^1$ given by $x_0 = x_1 = x_2 = 0$. An easy way to do this is to take a quintic in the ideal $(x_0, x_1, x_2)^2$. The blow-up of a general such X along the line $x_0 = x_1 = x_2 = 0$ yields a non-singular Calabi–Yau 3-fold X. One can check that the exceptional divisor E is a conic bundle over the $\mathbb{P}^1$. (See Exercise 14.3).

Note examples I-III are examples of the three types of contractions listed in §14.2.

Example 14.20 It is worth keeping in mind the Hodge diamond for a Calabi–Yau 3-fold X. Recall that $H^{3,0}(X) \cong H^0(X, \omega_X)$, which is 1-dimensional, and recall also the symmetry that if $h^{p,q} = \dim H^{p,q}(X)$ then $h^{p,q} = h^{n-p,n-q}$ ($n = \dim X = 3$) and $h^{p,q} = h^{q,p}$. We end up with the following Hodge diamond:

$$
\begin{array}{ccccccc}
& & & 1 & & & \\
& & 0 & & 0 & & \\
& 0 & & h^{1,1} & & 0 & \\
1 & & h^{2,1} & & h^{2,1} & & 1 \\
& 0 & & h^{1,1} & & 0 & \\
& & 0 & & 0 & & \\
& & & 1 & & &
\end{array}
$$

Now $h^{1,1} = \dim H^2(X, \mathbb{C})$, while $h^{2,1}$, as we saw in §14.1, is the dimension of the complex moduli of X.

In all these examples, we can make a naive dimension count to compute the Hodge numbers. This method does not always work, but in the examples above they do.

Consider first the case of the non-singular quintic. The Lefschetz hyperplane theorem ([77, p. 156]) tells us that $H^2(X, \mathbb{C}) \cong H^2(\mathbb{P}^4, \mathbb{C}) \cong \mathbb{C}$, so $h^{1,1} = 1$. Now we can compute $h^{1,2}$ by computing the dimension of the space of quintic hypersurfaces. The vector space of homogeneous quintic polynomials in five variables is 126-dimensional. However, proportional polynomials give the same hypersurface, and two polynomials related by an element of PGL(5), the automorphism group of $\mathbb{P}^4$, yield isomorphic hypersurfaces. This yields a moduli space of dimension $126 - 1 - \dim \mathrm{PGL}(5) = 101$. This is only a heuristic, as one would need both to check that a quintic can't be deformed to something not isomorphic to a quintic, and there are no unexpected isomorphisms between different quintic hypersurfaces. This calculation can be verified, however (see Exercise 14.1).

In Example 14.19, (I)-(III), in each case one can argue intuitively that the blow-up introduces an additional divisor, and that therefore $H^2(Y, \mathbb{C})$

is two-dimensional. Performing the same dimension count on the parameter space, one obtains $h^{1,2} = 86, 90, 86$ respectively in the three examples. This can again be verified, but the reader should be warned this reasoning is far more tenuous. (See Exercise 14.3.)

One can imagine playing a similar game to the above examples, taking more degenerate singular quintics containing more surfaces, or more triple points, or singular along more curves. One often gets Calabi–Yau 3-folds in this manner, and the number of topologically distinct Calabi–Yau 3-folds which are obtained as resolutions of singular quintics is presumably enormous.

Example 14.21 A rich field of examples explored thoroughly in [99] and references therein are complete intersections in products of projective spaces. If $Y = \mathbb{P}^{n_1} \times \cdots \times \mathbb{P}^{n_q}$, and (d_{ij}) a $p = -3 + \sum_{i=1}^{q} n_i$ by q matrix of non-negative integers with no rows being identically zero, let f_i be a general multi-homogeneous polynomial on Y of multi-degree $(d_{i1}, \ldots, d_{iq})$. As long as $\sum_{i=1}^{p} d_{ij} = n_j + 1$ for each j, adjunction tells us that $f_1 = \cdots = f_p = 0$ defines a variety X with $\omega_X \cong \mathcal{O}_X$, which for general choice of the f_i's is non-singular of dimension three. Additional conditions may be required to ensure that X is non-empty and connected.

The simplest such examples are with $q = 1$, and these are only the quintic in $\mathbb{P}^4$, complete intersections of type $(2, 4)$ and $(3, 3)$ in $\mathbb{P}^5$ (i.e. the intersection of a quadric and quartic or two cubics), type $(2, 2, 3)$ in $\mathbb{P}^6$, and $(2, 2, 2, 2)$ in $\mathbb{P}^7$. Any other choice of degrees will necessarily contain a 1, and hence be contained in a smaller dimensional projective space.

All of these examples have topological Euler characteristic < 0, with $h^{1,1}$ tending to be relatively small and $h^{1,2}$ tending to be relatively large. For example, for the complete intersections in $\mathbb{P}^n$ listed above, we always have $h^{1,1} = 1$ by the Lefschetz hyperplane theorem, and $h^{1,2} = 101, 89, 73, 73,$ and 65 in the order listed.

As an historical note, the first string theory paper to propose Calabi–Yau compactifications of space-time was [37]. The only simply connected examples known to the authors at that time were the above five complete intersections in projective space and one additional example with $\chi = 72$ constructed as a crepant resolution of a quotient of an abelian 3-fold. This motivated the search for new Calabi–Yau manifolds, since physical considerations at that time demanded a Calabi–Yau 3-fold with $\chi = \pm 6$. It was perhaps hoped there would be few of these; in the end, huge numbers of such examples were found.

Example 14.22 *Hypersurfaces in weighted projective space.* Historically these were perhaps the most important examples in the initial discovery of mirror symmetry. A weighted projective space with weights $(a_0, \ldots, a_n)$ is

$$W\mathbb{P}^n(a_0, \ldots, a_n) := (\mathbb{C}^{n+1} \setminus \{0\}) / \sim,$$

where the equivalence relation is

$$(x_0, \ldots, x_n) \sim (\lambda^{a_0} x_0, \ldots, \lambda^{a_n} x_n)$$

whenever $\lambda \in \mathbb{C} \backslash \{0\}$. Ordinary projective space is $W\mathbb{P}^n(1,\ldots,1)$. While most weighted projective spaces are singular, they still enjoy many properties that ordinary projective spaces enjoy (see [51]). In particular, for f a weighted homogeneous polynomial in $n + 1$ variables $x_0,\ldots,x_n$ with $\deg x_i = a_i$, $f = 0$ defines a hypersurface X in $W\mathbb{P}^n(a_0,\ldots a_n)$. If $\deg f = \sum a_i + 1$, then $\omega_X \cong \mathcal{O}_X$. However, since weighted projective spaces in general have singularities, we need to have a crepant resolution $f : Y \to X$ to get an example of a Calabi–Yau manifold. In [38] the authors provided a list of 5-tuples $(a_0,\ldots,a_4)$ for which a three-dimensional hypersurface in $W\mathbb{P}^4(a_0,\ldots,a_4)$ has a crepant Calabi–Yau resolution. In addition, they computed the Hodge numbers of these 3-folds, producing a list of about 6000 examples and finding a wide range of Hodge numbers. The surprising result was a symmetry in these Hodge numbers. Frequently, though not always, these Calabi–Yau 3-folds came in pairs in which $h^{1,1}$ and $h^{1,2}$ were exchanged between the two members of each pair. This was very surprising given that there had been very few previous examples of Calabi–Yau 3-folds with $h^{1,1} > h^{1,2}$, i.e. positive topological Euler characteristic. It was these examples, along with the simultaneous construction of the mirror quintic via physical arguments by Greene and Plesser [74] which jump-started mirror symmetry. See [73] for a dramatized account of this discovery.

Example 14.23 An even broader class of examples which has full symmetry was introduced by Batyrev in [5]. This paper lay the mathematical foundations for much of the future research in mirror symmetry, though we will actually say very little about this here as it involves introducing a great deal of toric geometry. However, the most elementary aspects of the construction can be summarized.

Let $M = \mathbb{Z}^n$ be a lattice, $M_\mathbb{R} \cong M \otimes_\mathbb{Z} \mathbb{R}$, and $N = \text{Hom}_\mathbb{Z}(M, \mathbb{Z})$, $N_\mathbb{R} = N \otimes_\mathbb{Z} \mathbb{R}$. Let $\Delta \subseteq M_\mathbb{R}$ be a convex polytope with vertices in M. The polar polytope $\Delta^* \subseteq N_\mathbb{R}$ is defined by

$$\Delta^* = \{y \in N_\mathbb{R} | \langle y, x \rangle \geqslant -1 \quad \forall x \in \Delta\}.$$

We say Δ is *reflexive* if 0 is in the interior of Δ and the affine hyperplane spanned by any $n - 1$-dimensional face of Δ is given by the affine linear equation $\langle y, \cdot \rangle = 1$ for some $y \in N$.

Any element $m = (m_1,\ldots,m_n) \in M$ defines a Laurent polynomial

$$z^m := z_1^{m_1} \cdot \cdots \cdot z_n^{m_n} \in \mathbb{C}[z_1, z_1^{-1},\ldots,z_n, z_n^{-1}]$$

which can be viewed as a function on $(\mathbb{C}^*)^n$. Then one can use the set of Laurent monomials $\{z^m | m \in \Delta \cap M\}$ to define a morphism $f : (\mathbb{C}^*)^n \to \mathbb{P}^{P-1}$, where $P = \#\Delta \cap M$. We define $\mathbb{P}_\Delta$ to be the normalization of the closure of the image of this map. This variety is known as a *projective toric variety*.

Batyrev proved that if Δ is a reflexive polytope in dimension $\leqslant 4$, then a general linear section of $\mathbb{P}_\Delta$ is a variety X with a crepant Calabi-Yau resolution Y. He also proved in the three-dimensional case that if $\check{Y}$ is obtained in the same way from Δ^*, then $h^{1,1}(Y) = h^{1,2}(\check{Y})$ and $h^{1,2}(Y) = h^{1,1}(\check{Y})$. Thus at a combinatorial level, mirror symmetry is elegantly simple.

This construction was later generalised to complete intersections in $\mathbb{P}_\Delta$ by Batyrev and Borisov [8], giving the largest range of known examples of mirror symmetric pairs of Calabi-Yau manifolds (though there are also now other families of examples, such as complete intersections in flag varieties, see [9] and [10]).

How many examples does this produce? In [138], Kreuzer and Skarke produced a complete list of four-dimensional reflexive polytopes, producing $473,800,776$ examples, including $30,108$ distinct pairs of Hodge numbers. As the Hodge numbers are a crude topological invariant, this places a lower bound on the number of topologically distinct Calabi-Yau 3-folds. Kreuzer and Skarke plotted a scatter plot of all pairs of Hodge numbers they found, and this plot is shown in Figure 2, reproduced from [138]. The intriguing nature of the curve bounding the inhabited region of Hodge numbers from above has never been explained. It is an interesting question as to whether there are some bounds that the Hodge numbers of Calabi-Yau 3-folds must always satisfy which explains this curve.

If there is any moral to be drawn from this range of examples, it is that there are a vast but finite list of known families of Calabi-Yau 3-folds. There is to date no method of constructing an infinite number of topologically distinct Calabi-Yau 3-folds. It is an open question as to whether there are a finite number of diffeomorphism types of Calabi-Yau 3-folds. See [80], [81], [83], [173] for further discussion of this problem.

14.4 Exercises

14.1 Verify that if X is a non-singular quintic in $\mathbb{P}^4$, then $h^{1,2}(X) = 101$. This can be done as follows: We know $h^{1,1}(X) = 1$ by the Lefschetz hyperplane theorem. Also, $\chi(X)$, the topological Euler characteristic of X, satisfies $\chi(X) = 2(h^{1,1}(X) - h^{1,2}(X))$, so we want to show $\chi(X) = -200$. But $\chi(X) = c_3(\mathcal{T}_X)$ by the Gauss-Bonnet theorem. Compute this third Chern class via the exact sequences

$$0 \to \mathcal{T}_X \to \mathcal{T}_{\mathbb{P}^4}|_X \to \mathcal{N}_{X/\mathbb{P}^4} \to 0$$

and

$$0 \to \mathcal{O}_{\mathbb{P}^4} \to \mathcal{O}_{\mathbb{P}^4}(1)^{\oplus 5} \to \mathcal{T}_{\mathbb{P}^4} \to 0.$$

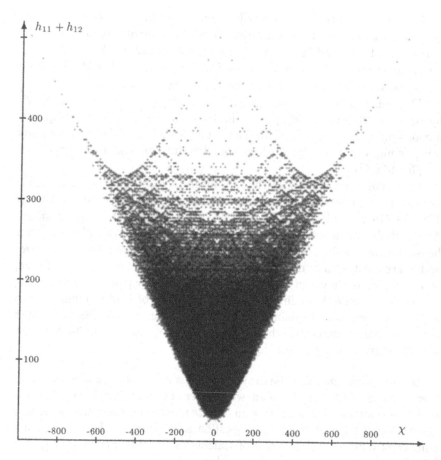

Figure 2. $h_{11} + h_{12}$ vs. Euler number $\chi = 2(h_{11} - h_{12})$ of Calabi–Yau 3-folds.

14.2(a) Consider $\mathbb{P}^1 \times \mathbb{P}^4$ with coordinates u, v on $\mathbb{P}^1$ and $x_0, \ldots, x_4$ on $\mathbb{P}^4$. Using the notation of Example 14.19, (I), let Y be the subvariety of $\mathbb{P}^1 \times \mathbb{P}^4$ defined by the equations

$$u f_4 + v g_4 = 0, u x_1 - v x_0 = 0.$$

Show that for generic f_4 and g_4, Y is smooth. Let $p : Y \to \mathbb{P}^4$ be the projection onto the second factor. Show that $p(Y) = X$, where X is defined by $x_0 f_4 + x_1 g_4 = 0$. If X_{ns} is the non-singular part of X, show that $p^{-1}(X_{ns}) \to X_{ns}$ is an isomorphism. Finally, show that if $x \in X$ is a singular point, then $p^{-1}(x) = \mathbb{P}^1$.

(b) Using the Lefschetz hyperplane theorem, conclude $h^{1,1}(Y) = 2$. Using the same techniques as in Exercise 14.1, show $h^{1,2}(Y) = 86$.

14.3(a) Perform the blowups in Example 14.19, (II) and (III), and check that the proper transforms of the quintics given are non-singular. Recall the blow-up formula which states that if $f : X \to Y$ is a blowup of a smooth subvariety of codimension d then $K_X = f^*K_Y + (d-1)E$, where E is the exceptional divisor. Using this and adjunction, show that these resolutions of Example 14.19, (II) and (III) are indeed crepant.

  (b) If Y is the blowup as in (a), show that $h^{1,2}(Y) = 90$ or 86 in cases (II) and (III) respectively.

14.4 In §14.1, two descriptions of the Kodaira–Spencer map were given. Show both these descriptions really do yield the Kodaira–Spencer map as follows:

  (a) In the first description, let U_i be an open covering of a neighbourhood of X in $\mathcal{X}$ such that there exists holomorphic liftings v_i of a $\partial/\partial s \in T_{S,s}$ to $T_{\mathcal{X}}|_{U_i \cap X}$. Then the image of $\partial/\partial s$ in $H^1(X, T_X)$ is represented by a Čech 1-cocycle $(v_i - v_j, U_i \cap U_j \cap X)$. Understand how this 1-cocycle can be identified with a Dolbeault representative of $H^1(X, T_X)$, i.e. a $\bar{\partial}$-closed element of $\Gamma(X, T_X^{1,0} \otimes \Omega_X^{0,1})$, and then show this represents the right element of $T_{\mathrm{Def}(X)}$.

  (b) In the Calabi–Yau manifold case, show that the map defined in §14.1 from $T_{S,s}$ to $H^{n-1,1}(X)$ coincides with the Kodaira–Spencer map after a choice of a holomorphic n-form on X.

15 Kähler Moduli and Gromov–Witten Invariants

15.1 Complexified Kähler Moduli Space

While we have not yet defined a mirror pair of Calabi–Yau 3-folds, one feature of a mirror pair X and $\check{X}$ is the interchange of Hodge numbers, i.e. $h^{1,1}(X) = h^{1,2}(\check{X})$ and $h^{1,2}(X) = h^{1,1}(\check{X})$. However, mirror symmetry should involve more than just this exchange. In the early physics literature, it was suggested that if X and $\check{X}$ were a mirror pair of Calabi–Yau manifolds, there should be, very roughly speaking, natural isomorphisms between the complex moduli space of $\check{X}$ and some other moduli space of X. We have studied the local structure of complex moduli of $\check{X}$ in §14.1, and learned this space is of dimension $h^{1,2}(\check{X}) = h^{1,1}(X)$. The corresponding moduli space of X is called the Kähler moduli space of X. Initially this was believed to be the space of Ricci-flat metrics on X. Now by Theorem 5.2 we know that each Kähler class of X is represented by a unique Ricci-flat metric. The set of Kähler classes on X is the Kähler cone $\mathcal{K}_X$, which is an open cone in $H^{1,1}(X, \mathbb{R})$, and hence is a real manifold of dimension $h^{1,1}(X)$. Since complex moduli space is in particular a complex manifold, this idea of an isomorphism needed some correction.

It turns out there is another bit of data in the moduli space of physical theories, called the B-field, and this extra bit of data solves this problem. The B-field is a class in $H^2(X, \mathbb{R})$ defined up to an integral cohomology class. Precisely,

Definition 15.1 If X is a Calabi–Yau 3-fold with $H^1(X, \mathcal{O}_X) = 0$, the *complexified Kähler moduli space* of X is

$$\mathcal{M}_{Kah}(X) = (H^2(X, \mathbb{R}) + i\mathcal{K}_X)/H^2(X, \mathbb{Z}).$$

Elements of this space are usually written $B + i\omega$, and B is referred to as the B-field, and is only well-defined up to shifts by elements of $H^2(X, \mathbb{Z})$.

For example, if $h^{1,1}(X) = 1$, then the Kähler cone $\mathcal{K}_X$ is just an open ray, and

$$\mathcal{M}_{Kah}(X) = (\mathbb{R} + i\mathbb{R}_{>0})/\mathbb{Z} = \mathcal{H}/\mathbb{Z}$$

where the action of $\mathbb{Z}$ on the upper half-plane is $z \mapsto z + 1$. The quotient can be identified with the punctured unit disk

$$\Delta^* = \{q \in \mathbb{C} | q \neq 0, |q| < 1\}$$

via $q = e^{2\pi i z}$.

If $h^{1,1}(X) > 1$, we sometimes follow a modified procedure. A *framing* of X is a choice of basis $e_1, \ldots, e_n$ for $H^2(X, \mathbb{Z})$, with $e_1, \ldots, e_n \in \overline{\mathcal{K}}_X$. This yields a cone

$$\Sigma = \{\alpha \in H^2(X, \mathbb{R}) | \alpha = \sum_i t_i e_i, t_i > 0\}.$$

We then set

$$\mathcal{M}_{Kah, \Sigma}(X) = (H^2(X, \mathbb{R}) + i\Sigma)/H^2(X, \mathbb{Z}).$$

This is isomorphic to, for $n = h^{1,1}(X)$,

$$(\Delta^*)^n = \{(q_1, \ldots, q_n) \in \mathbb{C}^n | \prod_i q_i \neq 0, |q_i| < 1\}$$

via

$$\sum_i z_i e_i \mapsto (e^{2\pi i z_1}, \ldots, e^{2\pi i z_n}).$$

The reader should not come to the conclusion that complex moduli space of $\check{X}$ is really isomorphic to $\mathcal{M}_{Kah, \Sigma}(X)$; the isomorphisms will only be local in a neighbourhood of some boundary point of complex moduli. This will be made precise in §17.

15.2 Pseudo-holomorphic Curves

Our goal now is to define certain additional structures on the Kähler moduli spaces we have defined above. These structures involve enumerating rational curves. More precisely, let X be a Calabi–Yau 3-fold, and consider all maps $f : \mathbb{P}^1 \to X$ which are generically 1-1 and whose images represent a given homology class $\eta \in H_2(X, \mathbb{Z})$. Consider two such maps f_1, f_2 equivalent if they differ by an automorphism $\psi : \mathbb{P}^1 \to \mathbb{P}^1$, i.e. $f_1 \circ \psi = f_2$. We would like to count the number of equivalence classes of these maps, but this yields some difficulties. The quintic $\sum x_i^5 = 0$ in $\mathbb{P}^4$ actually has an infinite number of lines, so we can't get a finite answer to the above question. The solution in this case is that you can deform X to a generic quintic where there are exactly 2875 lines, and now there is a finite answer. So the first strategy would be to deform the complex structure on X to a generic complex structure, and then count. This cannot always work, however. Consider the Calabi–Yau 3-fold Y we obtained by blowing up a triple point on a quintic (Example 14.19, (II)). Wilson's results mentioned in the sketch of the proof of Theorem 14.18 tells us that if we deform the complex structure on Y, the cubic surface $E \subseteq Y$, which is the exceptional locus of the blow-up of the singular quintic we began with, must also deform with Y. Thus any complex deformation of Y still contains a cubic surface, and cubic surfaces always contain an infinite number of rational curves (because a cubic surface is isomorphic to $\mathbb{P}^2$ blown up in 6 points, for example). So deforming to a generic complex structure is not enough, and we need to go further, actually deforming the complex structure on Y to a generic *almost complex structure*.

This brings us to the notion of a J-holomorphic curve. For many more details, see McDuff and Salamon's book [153].

Definition 15.2 Let (X, J) be an almost complex manifold and (Σ, j) be a Riemann surface, J and j the almost complex structures on X and Σ respectively. A map $u : \Sigma \to X$ is called *J-holomorphic* if u_* satisfies

$$J \circ u_* = u_* \circ j.$$

It is convenient to rewrite this condition as follows. For a map $u : \Sigma \to X$, put

$$\bar{\partial}_J(u) = \frac{1}{2}(u_* + J \circ u_* \circ j).$$

This is clearly zero if and only if u is J-holomorphic. This is in fact the complex anti-linear part of u_*, the complex linear part being $\frac{1}{2}(u_* - J \circ u_* \circ j)$. Indeed,

$$\frac{1}{2}(u_* + J \circ u_* \circ j)(j(v)) = \frac{1}{2}(u_* \circ j - J \circ u_*)(v)$$

$$= -J \circ (\frac{1}{2}(J \circ u_* \circ j + u_*))(v).$$

So we can think of $\bar{\partial}_J(u)$ as a u^*T_X-valued $(0,1)$ form, i.e. an element of $\Gamma(\Sigma, \Omega_\Sigma^{0,1} \otimes u^*T_X)$.

Example 15.3 Let $z = x+iy$ be a local complex coordinate on Σ, and suppose that J is the standard almost complex structure on $\mathbb{C}^n = \mathbb{R}^{2n}$ with coordinates $z_i = x_i + iy_i$. As usual, $J(\partial/\partial x_i) = \partial/\partial y_i$ and $J(\partial/\partial y_i) = -\partial/\partial x_i$. Write $u = f + ig$ with $f = (f_1, \ldots, f_n)$, $g = (g_1, \ldots, g_n)$. Then

$$(u_* \circ j)(\partial/\partial x) = u_*(\partial/\partial y)$$

$$= \sum_i \left(\frac{\partial f_i}{\partial y} \frac{\partial}{\partial x_i} + \frac{\partial g_i}{\partial y} \frac{\partial}{\partial y_i} \right)$$

while

$$J \circ u_*(\partial/\partial x) = J \left(\sum_i \left(\frac{\partial f_i}{\partial x} \frac{\partial}{\partial x_i} + \frac{\partial g_i}{\partial x} \frac{\partial}{\partial y_i} \right) \right)$$

$$= \sum_i \left(\frac{\partial f_i}{\partial x} \frac{\partial}{\partial y_i} - \frac{\partial g_i}{\partial x} \frac{\partial}{\partial x_i} \right)$$

Equating these two expressions yields

$$\frac{\partial f_i}{\partial y} = -\frac{\partial g_i}{\partial x}$$

and

$$\frac{\partial g_i}{\partial y} = \frac{\partial f_i}{\partial x}.$$

These are of course the familiar Cauchy–Riemann equations.

Now let $Map(\Sigma, X)$ be the set of all smooth maps $u : \Sigma \to X$. This is an infinite dimensional manifold. We define a vector bundle $\mathcal{E}$ on $Map(\Sigma, X)$ whose fibre at u is

$$\mathcal{E}_u = \Gamma(\Sigma, \Omega_\Sigma^{0,1} \otimes u^*T_X),$$

(again an infinite dimensional space), and then we can think of $\bar{\partial}_J$ as being a section of this vector bundle.

We say that a map $u : \Sigma \to X$ is *somewhere injective*, or *simple*, if there is a point $z \in \Sigma$ such that u_* is injective at z and $u^{-1}(u(z)) = \{z\}$. Fix a homology class $\eta \in H_2(X, \mathbb{Z})$. Let $Map(\Sigma, X, \eta)$ be the subset of $Map(\Sigma, X)$ consisting of simple maps whose images represent the homology class η. You should think of η as specifying the degree of the curve. For example, if X is the quintic 3-fold, then $H_2(X, \mathbb{Z}) = \mathbb{Z}$ and the homology class is determined by the degree of the image. We'll write $\mathcal{X} = Map(\Sigma, X, \eta)$ for convenience. We still have the vector bundle $\mathcal{E}$ on $\mathcal{X}$ with

$$\mathcal{E}_u = \Gamma(\Sigma, \Omega^{0,1}_{\Sigma} \otimes u^* \mathcal{T}_X)$$

and section $\bar{\partial}_J \in \Gamma(\mathcal{X}, \mathcal{E})$.

Let $M(\eta, J, \Sigma)$ denote the zero-locus of $\bar{\partial}_J$. This is precisely the moduli space of simple J-holomorphic maps representing the homology class η. One of the key questions for us is: when is $M(\eta, J, \Sigma)$ a manifold? Can we compute its dimension?

To address this question, we define the notion of a *regular* J-holomorphic curve. Thinking of $\bar{\partial}_J$ as a map $\bar{\partial}_J : \mathcal{X} \to \mathcal{E}$, then at a J-holomorphic curve $u \in \mathcal{X}$, we have

$$(\bar{\partial}_J)_* : \mathcal{T}_{\mathcal{X},u} \to \mathcal{T}_{\mathcal{E},(u,0)} = \mathcal{T}_{\mathcal{X},u} \oplus \mathcal{E}_u.$$

Let π be the projection from $\mathcal{T}_{\mathcal{E},(u,0)}$ onto $\mathcal{E}_u$, and put $D_u = \pi \circ (\bar{\partial}_J)_*$. We say u is *regular* if D_u is surjective.

This is essentially saying that the section $\bar{\partial}_J$ and the zero section of $\mathcal{E}$ meet transversally at the point $(u, 0)$. If they meet transversally, then we expect the intersection to be a manifold of some predictable dimension.

We would now like to consider the behaviour of $M(\eta, J, \Sigma)$ as J varies, so we need a suitable class of almost complex structures. One way to proceed is to fix a real symplectic form ω on X, say the Kähler form of a Kähler metric. An almost complex structure J is *compatible* with ω if $\omega(v, Jv) > 0$ for all non-zero $v \in \mathcal{T}_X$ and $\omega(Jv, Jw) = \omega(v, w)$ for all $v, w \in \mathcal{T}_X$. This permits one to define a Riemannian metric by $g(v, w) = \omega(v, Jw)$. We denote by $\mathcal{J}(\omega)$ the set of almost complex structures compatible with ω.

We set

$$\mathcal{J}_{reg}(\eta, \omega, \Sigma) = \{ J \in \mathcal{J}(\omega) | u \text{ is regular for all } u \in M(\eta, J, \Sigma). \}$$

The first key theorem is:

Theorem 15.4 *1. If $J \in \mathcal{J}_{reg}(\eta, \omega, \Sigma)$, then the space $M(\eta, J, \Sigma)$ is a smooth manifold of real dimension*

$$n(2 - 2g) + 2c_1(X) \cdot \eta$$

Furthermore, $M(\eta, J, \Sigma)$ carries a natural orientation.

2. The set $\mathcal{J}_{reg}(\eta, \omega, \Sigma)$ has second category in $\mathcal{J}(\omega)$, i.e. contains the intersection of countably many dense open subsets of $\mathcal{J}(\omega)$.

Proof. See [153, §3].   □

Remark 15.5 Theorem 15.4 includes the statement that $M(\eta, J, \Sigma)$ comes with a natural orientation. It is not hard to see that if J is in fact integrable, then $M(\eta, J, \Sigma)$ comes with a natural almost complex structure, and hence a natural orientation. If J is not integrable, one has to work harder to construct this orientation, and one byproduct is that one might obtain, for Calabi–Yau 3-folds, moduli in which some of the points count negatively. This is a subtle point, but it is worth noting it can happen.

There is a very useful criterion in practice for telling if a J-holomorphic curve is regular. First recall that by a theorem of Grothendieck (see for example [91], V, Ex. 2.6), any holomorphic vector bundle on $\mathbb{P}^1$ decomposes as a direct sum of line bundles. Recall also that any line bundle on $\mathbb{P}^1$ is determined by its first Chern class, and we write $\mathcal{O}_{\mathbb{P}^1}(a)$ to denote the line bundle with first Chern class a.

Theorem 15.6 *If J is an integrable almost complex structure on X, and $u : \mathbb{P}^1 \to X$ a J-holomorphic curve, then u is regular if in the decomposition $u^* T_X = \bigoplus_i \mathcal{O}_{\mathbb{P}^1}(a_i)$, we have $a_i \geqslant -1$ for all i.*

Proof. See [153], Lemma 3.5.1. □

Example 15.7 Let X be a Calabi–Yau 3-fold. Then $n = 3$, $c_1(X) = 0$, and for $g = 0$, we then get a moduli space $M(\eta, J, \mathbb{P}^1)$ of real dimension 6. On the other hand, the group of automorphisms of $\mathbb{P}^1$ is PGL(2), which is also of real dimension six. Given any J-holomorphic map $u : \mathbb{P}^1 \to X$ and automorphism ψ of $(\mathbb{P}^1, j)$, $u \circ \psi$ is another J-holomorphic curve. This gives an action of PGL(2) on $M(\eta, J, \mathbb{P}^1)$, and thus in this case we would expect $M(\eta, J, \mathbb{P}^1)/\text{PGL}(2)$ to be finite.

To apply the regularity criterion above in this case, note that if u is an embedding, we have an exact sequence

$$0 \to T_{\mathbb{P}^1} \to u^* T_X \to N_{\mathbb{P}^1} \to 0$$

and $T_{\mathbb{P}^1} = \mathcal{O}_{\mathbb{P}^1}(2)$, $N_{\mathbb{P}^1} = \mathcal{O}_{\mathbb{P}^1}(a) \oplus \mathcal{O}_{\mathbb{P}^1}(b)$. Since $c_1(u^* T_X) = c_1(T_{\mathbb{P}^1}) + c_1(N_{\mathbb{P}^1}) = 2 + a + b$. Since $c_1(T_X) = 0$, we must have $2 + a + b = 0$, so $a + b = -2$. Thus we can apply the regularity criterion if $a = b = -1$.

In the 3-fold case, it then makes sense to talk about

$$\#M(\eta, J, \mathbb{P}^1)/\text{PGL}(2),$$

counting points with their orientation, so there might be negative contributions to this number.

We also need to know that the moduli space $M(\eta, J, \Sigma)$ is essentially independent of J. So we might consider two choices $J_0, J_1 \in \mathcal{J}(\omega)$. A *smooth homotopy* of almost complex structures between J_0 and J_1 is just a smooth map $[0,1] \to \mathcal{J}(\omega)$, with $\lambda \mapsto J_\lambda$ almost complex structures interpolating between J_0 and J_1. Of course, even if $J_0, J_1 \in \mathcal{J}_{reg}(\eta, \omega, \Sigma)$ there may be no homotopy with $J_\lambda \in \mathcal{J}_{reg}(\eta, \omega, \Sigma)$ for all λ, so we might expect something to go wrong along the way. In a sense this doesn't happen. Let $\mathcal{J}(J_0, J_1)$ be the set of all smooth homotopies between J_0 and J_1.

Theorem 15.8 *In the above situation there exists a dense set*

$$\mathcal{J}_{reg}(\eta, J_0, J_1) \subseteq \mathcal{J}(J_0, J_1)$$

such that for every $\{J_\lambda\} \in \mathcal{J}_{reg}(\eta, J_0, J_1)$, $\bigcup_{\lambda=0}^{1} M(\eta, J_\lambda, \Sigma)$ *is a smooth manifold of dimension* $n(2-2g) + 2c_1(\eta) + 1$. *This manifold carries a natural orientation.*

Proof. See [153, §3]. $\qquad\qquad\qquad\qquad\qquad\qquad\qquad\qquad\qquad\qquad\qquad$ □

In other words, the one-parameter family of moduli spaces $M(\eta, J_\lambda, \Sigma)$ forms a manifold fibering over $[0, 1]$; even though some of the fibres are singular, the total space is not. This gives an oriented cobordism between $M(\eta, J_0, \Sigma)$ and $M(\eta, J_1, \Sigma)$. If we know that the total space is compact, then this would imply certain things are left invariant. For example, in the Calabi–Yau 3-fold case, this would imply

$$\#M(\eta, J_0, \mathbb{P}^1)/\mathrm{PGL}(2) = \#M(\eta, J_1, \mathbb{P}^1)/\mathrm{PGL}(2)$$

counting the number of points with sign.

However, in general, compactness fails for a very simple reason. A sequence of J-holomorphic curves can converge to a union of curves. For example, consider in $\mathbb{P}^2$ the family of curves $xz - ty^2 = 0$ parametrised by t as $t \to 0$. These curves can be parametrised by $f_t : \mathbb{P}^1 \to \mathbb{P}^2$ given by $f_t(u, v) = (tu^2, uv, v^2)$, or alternatively by $g_t(u, v) = (u^2, uv, tv^2)$. Note while f_t and g_t have the same image for $t \neq 0$, a conic, their images for $t = 0$ are two distinct lines. One should view the union of these two lines as the correct limit. This phenomenon is usually called *bubbling*. Thus to compactify, we should allow unions of rational curves.

The Gromov compactness theorem roughly tells us that if we include such unions of rational curves, then we do get compact moduli spaces. See [153] for details. This gives compactifications

$$\overline{M(\eta, J, \mathbb{P}^1)/\mathrm{PGL}(2)}.$$

In the Calabi–Yau 3-fold case, one then sees there is a well-defined invariant

$$\#\overline{M(\eta, J, \mathbb{P}^1)/\mathrm{PGL}(2)},$$

for $J \in \mathcal{J}_{reg}(\eta, \omega, \mathbb{P}^1)$ with ω the Kähler form of a Kähler metric, and for general $J \in \mathcal{J}_{reg}(\eta, \omega, \mathbb{P}^1)$,

$$\overline{M(\eta, J, \mathbb{P}^1)/\mathrm{PGL}(2)} = M(\eta, J, \mathbb{P}^1)/\mathrm{PGL}(2).$$

In this case we only need the compactifications to ensure that the number is actually independent of J. As a result, we omit the details about this compactification, and encourage the reader to consult [153] or other sources on this important topic.

15.3   Gromov–Witten Invariants and the $(1,1)$ Yukawa Coupling

Finally we can define Gromov–Witten invariants for Calabi–Yau 3-folds. Suppose $D_1, D_2, D_3 \in H^2(X, \mathbb{C})$. We define the Gromov–Witten invariant

$$\Phi_\eta(D_1, D_2, D_3) = (D_1 \cdot \eta)(D_2 \cdot \eta)(D_3 \cdot \eta)\overline{\#M(\eta, J, \mathbb{P}^1)/\mathrm{PGL}(2)}.$$

This is a bit ad hoc. There are lots of other Gromov–Witten invariants in more general situations, and one should in fact think of the above number as follows. If the D_i are integral classes and Z_1, Z_2, Z_3 are four dimensional submanifolds in X, which are Poincaré dual to the classes D_1, D_2 and D_3 respectively, then intuitively we can think of this number as

$$\#\{u \in M(\eta, J, \mathbb{P}^1) | u(0) \in Z_1, u(1) \in Z_2, u(\infty) \in Z_3\}.$$

Of course, the first definition doesn't require the D_i to be integral classes.

We now put all the Gromov–Witten invariants for different η in one generating series. We can define a "quantum deformation" of the usual cubic intersection form on $H^2(X, \mathbb{C})$ formally by

$$\langle D_1, D_2, D_3 \rangle := D_1 \cdot D_2 \cdot D_3 + \sum_{0 \neq \eta \in H_2(X, \mathbb{Z})} \left(\Phi_\eta(D_1, D_2, D_3) \sum_{m=1}^{\infty} q^{m\eta} \right)$$

$$= D_1 \cdot D_2 \cdot D_3 + \sum_{0 \neq \eta \in H_2(X, \mathbb{Z})} \left(\Phi_\eta(D_1, D_2, D_3) \frac{q^\eta}{1 - q^\eta} \right)$$

Here the dot represents the usual cup product on $H^2(X, \mathbb{C})$, but we have to make sense of the remainder of this expression.

For any semigroup S, define the *formal semigroup ring with coefficients in a ring R* to be

$$R[[q; S]] = \left\{ \sum_{\eta \in S} a_\eta q^\eta \, | \, a_\eta \in R \right\}.$$

To multiply, we put

$$\left(\sum_{\eta_1 \in S} a_{\eta_1} q^{\eta_1} \right) \cdot \left(\sum_{\eta_2 \in S} b_{\eta_1} q^{\eta_1} \right) = \sum_{\eta \in S} \left(\sum_{\substack{(\eta_1, \eta_2) \in S \times S \\ \eta_1 + \eta_2 = \eta}} a_{\eta_1} b_{\eta_2} \right) q^\eta.$$

In order for this to make sense, S must satisfy the *finite partition property*, i.e. for each $\eta \in S$, there only exists a finite number of η_1 and η_2 in S with $\eta_1 + \eta_2 = \eta$.

We define the *integral Mori semigroup*

$$\overline{NE}(X, \mathbb{Z}) := \{\eta \in H_2(X, \mathbb{Z}) | \omega \cdot \eta \geqslant 0 \text{ for all } \omega \in \overline{\mathcal{K}}_X\}.$$

This is essentially the dual cone to $\overline{\mathcal{K}}_X$ intersected with the integral lattice $H_2(X, \mathbb{Z})$, though the latter group might have some torsion which contributes to the group. It satisfies the finite partition property, and furthermore, any class $\eta \in H_2(X, \mathbb{Z})$ represented by any holomorphic curve whatsoever must be in $\overline{NE}(X, \mathbb{Z})$. So we can think of $\langle D_1, D_2, D_3 \rangle$ as being an element of

$$\mathbb{C}[[q, \overline{NE}(X, \mathbb{Z})]].$$

There is another, more concrete way we can think of these cubic forms if $H_2(X, \mathbb{Z})$ is torsion free. Suppose we have chosen a framing for X, i.e. a basis $e_1, \ldots, e_n \in \overline{\mathcal{K}}_X$ for $H^2(X, \mathbb{Z})$. This gives coordinates $q_1, \ldots, q_n$ on $\mathcal{M}_{Kah, \Sigma}(X)$, where, for $z = \sum_j z_j e_j$, $q_j = e^{2\pi i z_j}$. We can then write

$$\langle D_1, D_2, D_3 \rangle = D_1 \cdot D_2 \cdot D_3 + \sum_{0 \neq \eta \in H_2(X, \mathbb{Z})} \Phi_\eta(D_1, D_2, D_3) \frac{\prod_i q_i^{e_i \cdot \eta}}{1 - \prod_i q_i^{e_i \cdot \eta}}.$$

Since $e_i \in \overline{\mathcal{K}}_X$, $e_i \cdot \eta \geq 0$, so we can think of this as living in the formal power series ring $\mathbb{C}[[q_1, \ldots, q_n]]$. But we would also like to think of it as a function on $\mathcal{M}_{Kah, \Sigma}(X)$. We can't quite do this because we don't know if this power series has non-zero radius of convergence (an important unsolved question, in fact). Nevertheless, we should think of this cubic form as really being a cubic form defined on the tangent space $H^2(X, \mathbb{C})$ at each point of $\mathcal{M}_{Kah, \Sigma}(X)$.

To summarize, what we have is, modulo this convergence issue:

Definition 15.9 Given a framing Σ of a Calabi–Yau 3-fold X with $H_2(X, \mathbb{Z})$ torsion-free, the $(1,1)$-Yukawa coupling of X is the cubic form on the tangent bundle to $\mathcal{M}_{Kah, \Sigma}(X)$ defined as follows. Given a point in $\mathcal{M}_{Kah, \Sigma}(X)$ with coordinate

$$q = (q_1, \ldots, q_n),$$

the tangent space to $\mathcal{M}_{Kah, \Sigma}(X)$ at q is canonically isomorphic to $H^2(X, \mathbb{C})$. We define the $(1,1)$-Yukawa coupling of $D_1, D_2, D_3 \in H^2(X, \mathbb{C})$ by

$$\langle D_1, D_2, D_3 \rangle = D_1 \cdot D_2 \cdot D_3 + \sum_{0 \neq \eta \in H_2(X, \mathbb{Z})} \Phi_\eta(D_1, D_2, D_3) \frac{\prod_i q_i^{e_i \cdot \eta}}{1 - \prod_i q_i^{e_i \cdot \eta}}.$$

Of course, we don't know whether or not this series converges, so it should properly be regarded as something which makes sense only in a formal neighbourhood of a boundary point of $\mathcal{M}_{Kah, \Sigma}(X)$; but we shall ignore this technicality.

The goal of mirror symmetry will be to find similar structure on the complex moduli space of the mirror, and match this data up.

15.4 Exercises

15.1 Check that the integral Mori semigroup satisfies the finite partition property.

15.2 Show that if $u : \Sigma \to X$ is a J-holomorphic curve, and ω is a symplectic form on X compatible with J, then the area of $u(\Sigma)$ with respect to the induced metric on X is $\int_\Sigma u^*\omega$.

15.3 Let $X \subseteq \mathbb{P}^3$ be a non-singular cubic surface. One way to calculate the number of lines contained in X is as follows. Let $\mathrm{Gr}(2,4)$ denote the Grassmannian of two-dimensional subspaces of a four-dimensional complex vector space; this parametrizes lines in $\mathbb{P}^3$. There is an *incidence correspondence* $I \subseteq \mathbb{P}^3 \times \mathrm{Gr}(2,4)$ defined by

$$I = \{(x,l)\,|\,x \in l\}.$$

Let $p_1 : I \to \mathbb{P}^3$ and $p_2 : I \to \mathrm{Gr}(2,4)$ be the projections. Note p_2 is a $\mathbb{P}^1$-bundle.

(a) Show that $\mathcal{E} = p_{2*}p_1^*\mathcal{O}_{\mathbb{P}^3}(3)$ is a rank four vector bundle on $\mathrm{Gr}(2,4)$. [Hint: use [91], III Theorem 12.11.]

(b) Show that if $s \in \Gamma(\mathbb{P}^3, \mathcal{O}_{\mathbb{P}^3}(3))$ is a section such that $s = 0$ is the surface X, then p_1^*s defines a section of $p_1^*\mathcal{O}_{\mathbb{P}^3}(3)$ that vanishes along a fibre of p_2 if and only if X contains the corresponding line. Show that the corresponding section $p_{2*}p_1^*s$ of $\mathcal{E}$ vanishes along the set

$$\{l \in \mathrm{Gr}(2,4)\,|\,l \subseteq X\}.$$

(c) Since $\dim \mathrm{Gr}(2,4) = 4$, a general section of $\mathcal{E}$ is expected to vanish on a zero-dimensional variety with degree $c_4(\mathcal{E})$. Show $c_4(\mathcal{E}) = 27$. [Hint: Use the Grothendieck–Riemann–Roch theorem, see [91], Appendix A, Theorem 5.3. If you have never done this sort of thing before, [63] may be useful.] Note this does not prove the number of lines in X is 27; it just shows the number, if finite, is 27 if counted with suitable multiplicities. It is true that any non-singular cubic surface in $\mathbb{P}^3$ contains 27 distinct lines.

15.4 Show the quintic $x_0^5 + \cdots + x_4^5 = 0$ in $\mathbb{P}^4$ contains an infinite number of lines.

Show, using the same techniques as in Exercise 15.3, that the expected number of lines on a quintic 3-fold is 2875.

16 Variation and Degeneration of Hodge Structures

Here we will discuss fairly standard material on Hodge structures. For the more elementary facts about Hodge theory one can consult [77]. For the presentation of the subtler aspects of the theory we follow [76]. The reader can consult the latter book for many more details on variations of Hodge structures and mixed Hodge structures.

16.1   Variation of Hodge Structures and the Yukawa Coupling

Recall that for a given compact Kähler manifold X, one has the Hodge decomposition, i.e. an identification

$$H^n(X, \mathbb{C}) = \bigoplus_{p+q=n} H^{p,q}(X).$$

The groups $H^{p,q}(X)$ satisfy

$$\overline{H^{p,q}(X)} = H^{q,p}(X).$$

More abstractly:

Definition 16.1 A *Hodge structure of weight n* is a lattice $H_{\mathbb{Z}}$ of finite rank together with a decomposition

$$H_{\mathbb{Z}} \otimes_{\mathbb{Z}} \mathbb{C} = \bigoplus_{p+q=n} H^{p,q}$$

of complex subspaces with

$$\overline{H^{p,q}} = H^{q,p}.$$

Example 16.2 Recalling the Hodge diamond for a Calabi–Yau 3-fold X with $H^1(X, \mathcal{O}_X) = 0$ is

$$
\begin{array}{ccccccc}
 & & & 1 & & & \\
 & & 0 & & 0 & & \\
 & 0 & & h^{1,1} & & 0 & \\
1 & & h^{2,1} & & h^{2,1} & & 1 \\
 & 0 & & h^{1,1} & & 0 & \\
 & & 0 & & 0 & & \\
 & & & 1 & & &
\end{array}
$$

we see that the only interesting cohomology group from the point of view of Hodge theory is $H^3(X, \mathbb{Z})$, which carries an interesting Hodge structure of weight 3.

It is often convenient, instead of considering the Hodge decomposition, to consider the Hodge filtration. For a weight n Hodge structure, put

$$F^p = H^{n,0} \oplus \cdots \oplus H^{p,n-p}$$

so that

$$H = F^0 \supseteq F^1 \supseteq \cdots \supseteq F^n.$$

Note the additional relations that

$$H = F^p \oplus \overline{F^{n-p+1}}$$

and

$$H^{p,q} = F^p \cap \overline{F^q}.$$

(Hodge theory always involves a lot of annoying indices.) We can thus define a Hodge structure of weight n as a filtration on $H = H_{\mathbb{Z}} \otimes \mathbb{C}$,

$$H = F^0 \supseteq \cdots \supseteq F^n$$

satisfying the condition $H = F^p \oplus \overline{F^{n-p+1}}$. Such a filtration is called a *Hodge filtration*.

We will now define an (abstract) variation of Hodge structures. The idea of a variation of Hodge structures is to put together a whole family of Hodge structures. This family comes with a lot of data. But to motivate it, we first give the main example, and then the formal definition.

Fix for now a family $f : \mathcal{X} \to S$ of compact Kähler manifolds, S a connected manifold. The first feature is a sheaf $R^n f_* \mathbb{C}$. Formally speaking, $R^n f_* \mathbb{C}$ is the sheaf associated to the presheaf $U \mapsto H^n(f^{-1}(U), \mathbb{C})$. However, $R^n f_* \mathbb{C}$ has more structure: it is in fact a local system.

What is a local system? Recall that a constant sheaf associated to a group G on a topological space X is the sheaf $\mathcal{G}$ with

$$\mathcal{G}(U) = \{f : U \to G | f \text{ is continuous}\},$$

where U is any open subset of X and G is given the discrete topology. Thus, in particular, if U is connected, then $\mathcal{G}(U) = G$. A *local system* $\mathcal{F}$ with coefficient group G is a sheaf on X for which each point $x \in X$ has an open neighbourhood U with $\mathcal{F}|_U$ isomorphic to the constant sheaf associated to G on U. Such a sheaf is also called a *locally constant sheaf*.

Applying Theorem 14.5, we note that for an arbitrary family of compact complex manifolds $f : \mathcal{X} \to S$, $R^n f_* \mathbb{C}$ is then a local system with coefficient group $H^n(X, \mathbb{C})$, where X is any fibre of f. Indeed, if $U \subseteq S$ is a connected contractible open subset, then $f^{-1}(U) \cong U \times X$, and then $H^n(f^{-1}(U), \mathbb{C}) \cong H^n(X, \mathbb{C})$. Thus $(R^n f_* \mathbb{C})|_U$ is just the constant sheaf with group $H^n(X, \mathbb{C})$.

It is important to keep in mind that $R^n f_* \mathbb{C}$ is not a vector bundle, even though it has fibres which are complex vector spaces. This is the difference between the constant sheaf $\mathbb{C}^r$, say, on a connected space S whose space of global sections is just $\mathbb{C}^r$, and the sheaf of holomorphic functions with values in $\mathbb{C}^r$ whose space of global sections is the space of all r-tuples of holomorphic functions on S.

In fact, there is a correspondence between local systems $\mathcal{E}$ with coefficients in $\mathbb{C}^r$ and pairs $(\mathcal{F}, \nabla)$, where $\mathcal{F}$ is a rank r holomorphic vector bundle and ∇ is a flat connection on $\mathcal{F}$ compatible with the holomorphic structure on $\mathcal{F}$. Given a local system $\mathcal{E}$ on S, set $\mathcal{F} = \mathcal{E} \otimes \mathcal{O}_S$. Locally, on a contractible open set U this has the effect of replacing the constant sheaf $\mathbb{C}^r$ on U with the trivial rank r holomorphic vector bundle on U. Define a connection ∇ on

$\mathcal{F}$ as follows. If locally $e_1, \ldots, e_n$ form a basis of sections for $\mathcal{E}$, any section σ of $\mathcal{F}$ can be written as $\sum f_i e_i$, where the f_i are holomorphic functions. Then we put $\nabla \sigma = \sum df_i \otimes e_i$. In particular, the e_i give a frame field of flat sections since $\nabla e_i = 0$.

Conversely, given $(\mathcal{F}, \nabla)$ with ∇ a flat holomorphic connection, one can recover $\mathcal{E}$ via

$$\mathcal{E}(U) = \{\sigma \in \mathcal{F}(U) | \nabla \sigma = 0\}.$$

In the context of our map $f : \mathcal{X} \to S$, we now obtain a holomorphic bundle $\mathcal{H}^n = (R^n f_* \mathbb{C}) \otimes \mathcal{O}_S$ and a flat holomorphic connection ∇ on $\mathcal{H}^n$. The connection ∇ is usually called the *Gauss–Manin connection*.

It is very important to keep in mind that $R^n f_* \mathbb{C}$ may not be a constant sheaf. Here is the typical picture to consider. Suppose $S = \Delta^*$, the punctured unit disk. Let $\gamma : [0, 1] \to \Delta^*$ be the loop based at $s = 1/2$ given by $\gamma(\theta) = \frac{1}{2} e^{2\pi i \theta}$. The stalk $(R^n f_* \mathbb{C})_s$ is naturally identified with $H^n(\mathcal{X}_s, \mathbb{C})$. By parallel transport about the loop γ using the flat connection ∇, one obtains a linear transformation $T : H^n(\mathcal{X}_s, \mathbb{C}) \to H^n(\mathcal{X}_s, \mathbb{C})$ called the *monodromy transformation*. (You would get the same transformation no matter what representative for the generator of $\pi_1(\Delta^*)$ you chose; this is because the Gauss–Manin connection ∇ is flat). This transformation is zero if and only if the local system is constant.

Another way to visualize the monodromy transformation is as follows. Pull back the family $\mathcal{X} \to \Delta^*$ via γ to obtain a family $\mathcal{X}' \to [0, 1]$. Since $[0, 1]$ is contractible, we can trivialise: i.e. there is an isomorphism $\psi : \mathcal{X}' \cong X \times [0, 1]$ compatible with the map $\mathcal{X}' \to [0, 1]$. Using the fact that $\mathcal{X}'_0 = \mathcal{X}'_1$, we obtain a diffeomorphism φ of $X \cong \mathcal{X}'_0$ being the composition

$$\mathcal{X}'_0 \xrightarrow{\psi} X \times \{0\} = X \times \{1\} \xrightarrow{\psi^{-1}} \mathcal{X}'_1.$$

If $S^1 \subseteq \Delta^*$ is the image of γ, then to recover the fibration $f^{-1}(S^1) \to S^1$ topologically, one glues $X \times \{0\}$ to $X \times \{1\}$ via φ. The diffeomorphism φ is often called a monodromy diffeomorphism. Since the trivialization is not unique, the monodromy diffeomorphism is only determined up to isotopy. However, φ does induce a unique map on cohomology, and this is the monodromy transformation.

Example 16.3 Consider the family of elliptic curves over Δ^* with coordinate t on Δ^* with period $\tau = \frac{1}{2\pi i} \log t$. In other words, we divide $\mathbb{C} \times \Delta^*$ by the action $(z, t) \mapsto (z + 1, t)$ and $(z, t) \mapsto (z + \frac{1}{2\pi i} \log t, t)$ to obtain a complex manifold $\mathcal{X}$. While $\frac{1}{2\pi i} \log t$ is multi-valued, the lattice spanned by 1 and $\frac{1}{2\pi i} \log t$ is well-defined. This gives us a family $f : \mathcal{X} \to \Delta^*$ of elliptic curves.

One can see the monodromy of the local system $R^1 f_* \mathbb{C}$ as arising from the multi-valuedness of $\tau(t)$. Indeed, if $E \cong \mathbb{C}/\Lambda$ is an elliptic curve, then there is a canonical isomorphism $H_1(E, \mathbb{Z}) \cong \Lambda$, and thus by Poincaré duality, $H^1(E, \mathbb{Z}) \cong \Lambda$. Thus we can think of the local system $R^1 f_* \mathbb{Z}$ as the family of lattices in $\mathbb{C} \times \Delta^*$ spanned by 1 and $\tau(t)$. If we fix a point $t_0 \in \Delta^*$ and go around a closed loop representing $\pi_1(\Delta^*)$, we can understand the monodromy. Fix a branch of the logarithm to obtain a basis $1, \tau(t_0)$ for $H^1(\mathcal{X}_{t_0}, \mathbb{Z})$. Going around the closed loop, follow our basis in the local system. The cohomology class 1 stays constant, but when we return to t_0, the branch of the logarithm has changed, and so $\tau(t_0)$ is replaced by $\tau(t_0) + 1$. Thus the matrix for the monodromy transformation T in this basis is

$$\begin{pmatrix} 1 & 1 \\ 0 & 1 \end{pmatrix}.$$

This is the monodromy of the local system $R^1 f_* \mathbb{Z}$, but it is the same for $R^1 f_* \mathbb{C} = R^1 f_* \mathbb{Z} \otimes_{\mathbb{Z}} \mathbb{C}$.

We can also work out the monodromy diffeomorphism explicitly. Let $\gamma(\theta) = \frac{1}{2} e^{2\pi i \theta}$, $0 \leqslant \theta \leqslant 1$. Then $\tau(\gamma(\theta)) = \frac{i \log 2}{2\pi} + \theta$. We can define a map $\alpha : \mathbb{R}^2 / \mathbb{Z}^2 \times [0,1] \to \mathcal{X}$ by

$$\alpha(x, y, \theta) = \left(x + y \left(\frac{i \log 2}{2\pi} + \theta \right), \gamma(\theta) \right).$$

Then the monodromy diffeomorphism $\varphi : \mathcal{X}_{\gamma(0)} \to \mathcal{X}_{\gamma(1)}$ is defined by

$$\mathcal{X}_{\gamma(0)} \xrightarrow{\alpha^{-1}} \mathbb{R}^2 / \mathbb{Z}^2 \times \{0\} = \mathbb{R}^2 / \mathbb{Z}^2 \times \{1\} \xrightarrow{\alpha} \mathcal{X}_{\gamma(1)}$$

or $\varphi(x, y) = (x + 2\pi y / \log 2, y)$, or after identifying $\mathcal{X}_{\gamma(0)}$ with $\mathbb{R}^2 / \mathbb{Z}^2$ via α^{-1}, $\varphi : \mathbb{R}^2 / \mathbb{Z}^2 \to \mathbb{R}^2 / \mathbb{Z}^2$ is given by $\varphi(x, y) = (x + y, y)$.

This diffeomorphism of a torus is known as a *Dehn twist*.

Returning to the general case of $f : \mathcal{X} \to S$, each fibre of the holomorphic vector bundle $\mathcal{H}^n = R^n f_* \mathbb{C} \otimes \mathcal{O}_S$ has a Hodge filtration, and this yields a filtration of $\mathcal{H}^n$ by subbundles

$$\mathcal{H}^n = \mathcal{F}^0 \supseteq \mathcal{F}^1 \supseteq \cdots \supseteq \mathcal{F}^n,$$

which can be proved to be holomorphic subbundles of $\mathcal{H}^n$. The reason that we use the Hodge filtration instead of the subspaces $H^{p,q}$ is that these latter spaces do not vary holomorphically inside $\mathcal{H}^n$. However, one can still define holomorphic bundles $\mathcal{H}^{p,n-p} = \mathcal{F}^p / \mathcal{F}^{p+1}$.

An important aspect of this structure is the relationship between the Gauss–Manin connection and the Hodge filtration. This is known as *Griffiths transversality*:

Theorem 16.4 $\nabla \mathcal{F}_p \subseteq \mathcal{F}_{p-1} \otimes \Omega^1_S$.

Proof. Let α be a form of type $(p, n - p)$ on X, so we can write

$$\alpha = \sum_{\substack{\#I = p \\ \#J = n-p}} f_{IJ} dz_I \wedge d\bar{z}_J,$$

in local coordinates. Now we deform the complex structure by moving in some direction on S, say in the direction indicated by a tangent vector $\partial/\partial s \in T_{S,0}$. In our coordinate patch, we should think of this as meaning that we are changing holomorphic coordinates as a function of s, say

$$z_i(s) = z_i + s f_i(z_1, \ldots, z_n, \bar{z}_1, \ldots, \bar{z}_n) + \bar{s} g_i(z_1, \ldots, z_n, \bar{z}_1, \ldots, \bar{z}_n) + O(s^2).$$

Suppose α deforms to a family of forms $\alpha(s)$ so that $\alpha(s)$ is of type $(p, n-p)$ on X_s. Then in our original coordinates,

$$\alpha(s) = \sum_{\substack{\#I=p \\ \#J=n-p}} f_{IJ}(s) dz_I(s) \wedge d\bar{z}_J(s).$$

Here I and J are multi-index sets. We can then compute $\frac{\partial \alpha(s)}{\partial s}\big|_{s=0}$, which, in expanding out $\alpha(s)$ as a Taylor series in s, is the coefficient of s. In this coefficient, we have at most one extra dz_i or $d\bar{z}_j$ in each term. Thus $\frac{\partial \alpha(s)}{\partial s}\big|_{s=0}$ is a form of type $(p-1, n-p+1) + (p, n-p) + (p+1, n-p-1)$, and the cohomology class represented by this form is then in $\mathcal{F}_{p-1}$. Note that what we have calculated is $\nabla_{\partial/\partial s}\alpha(s)$. This gives the result. $\square$

This allows us to give an abstract definition of *variation of Hodge structures*.

Definition 16.5 A *variation of Hodge structures* is a quadruple $(\mathcal{H}_\mathbb{Z}, \nabla, S, \mathcal{F})$, where S is a complex manifold, $\mathcal{H}_\mathbb{Z}$ a local system on S with coefficient group $\mathbb{Z}^h$ for some h, and ∇ a flat holomorphic connection on the holomorphic vector bundle $\mathcal{H} = \mathcal{H}_\mathbb{Z} \otimes \mathcal{O}_S$ such that sections of $\mathcal{H}_\mathbb{Z}$ are flat for ∇. Furthermore, $\mathcal{F}$ must be a Hodge filtration on $\mathcal{H}$ and $\nabla \mathcal{F}^p \subseteq \mathcal{F}^{p-1} \otimes \Omega_S^1$.

One can go on from here and study many interesting features of variations of Hodge structure in the abstract. Instead, we are going to simplify things by working only with Calabi–Yau manifolds (and usually only 3-folds), and by trying to keep everything as concrete as possible.

We now restrict attention to the 3-fold case.

Definition 16.6 Given a family of Calabi–Yau 3-folds $f : X \to S$ and a holomorphically varying family of holomorphic three-forms on X, $\Omega(s)$, we define a cubic form on the tangent bundle of S called the $(1,2)$-Yukawa coupling. Given a local trivialisation of f, $f^{-1}(U) = U \times X$, we can define it as follows. Let $\partial/\partial s_1, \partial/\partial s_2, \partial/\partial s_3 \in T_{S,s}$. Set

$$\langle \partial/\partial s_1, \partial/\partial s_2, \partial/\partial s_3 \rangle = \int_X \Omega(s) \wedge \frac{\partial}{\partial s_1} \frac{\partial}{\partial s_2} \frac{\partial}{\partial s_3} \Omega(s).$$

A more invariant definition is as follows. Thinking of $\Omega(s)$ as giving a section of $R^3 f_* \mathbb{C} \otimes \mathcal{O}_S$, we have

$$\langle \partial/\partial s_1, \partial/\partial s_2, \partial/\partial s_3 \rangle = \int_X \Omega(s) \wedge \nabla_{\frac{\partial}{\partial s_1}} \nabla_{\frac{\partial}{\partial s_2}} \nabla_{\frac{\partial}{\partial s_3}} \Omega(s).$$

This cubic form is clearly linear in each variable, and is symmetric because the connection is flat, or equivalently because mixed partials commute.

This is the cubic form which we will match up with the cubic form we defined on the Kähler moduli space in Definition 15.9. However, the story is not complete because in order to get any useful information one has to know a priori the isomorphism between complex and Kähler moduli space. To do this, we will have to study the structures we have defined in greater detail.

Remark 16.7 1. One reason taking three derivatives is natural is as follows. By Griffiths transversality, $\nabla_{\partial/\partial s_1} \Omega(s)$ is of type $(3,0) + (2,1)$, so $\Omega(s) \wedge \nabla_{\partial/\partial s_1} \Omega(s) = 0$. Similarly if one takes two derivatives the same still holds. It is only on taking the third derivative that one gets a bit of type $(0,3)$, which is non-zero when wedging with $\Omega(s)$, of type $(3,0)$.
   2. This coupling depends on the choice of $\Omega(s)$. Again, we could change it by multiplying by a holomorphic function $f(s)$ on the base S. If we write the Yukawa coupling coming from $\Omega^*(s) = f(s)\Omega(s)$ with a star, and take the above remark into account, we see that

$$\langle \partial/\partial s_1, \partial/\partial s_2, \partial/\partial s_3 \rangle^* = f^2(s)\langle \partial/\partial s_1, \partial/\partial s_2, \partial/\partial s_3 \rangle.$$

   This dependence on $\Omega(s)$ cannot be removed, and we will later have to specify a natural normalised choice for $\Omega(s)$.
   3. Taking the three covariant derivatives requires extending the tangent vectors $\partial/\partial s_1, \ldots, \partial/\partial s_3$ to vector fields in a neighbourhood. One checks easily that this doesn't depend on the particular choice of extension, so that we really obtain a trilinear form on the tangent space at a given point $s \in S$.

16.2 Period Maps

Definition 16.8 Let $f : \mathcal{X} \to S$ be a family of Calabi–Yau 3-folds. We define the *period map* as follows. Let $\tilde{S} \to S$ be the universal cover of S, and let $\tilde{f} : \tilde{\mathcal{X}} \to \tilde{S}$ be the pull-back of $\mathcal{X} \to S$. The local system $R^3 \tilde{f}_* \mathbb{C}$ on $\tilde{S}$ is constant, since $\tilde{S}$ is simply connected. Thus we can identify $H^3(\tilde{\mathcal{X}}_s, \mathbb{C})$ canonically with $H^3(X, \mathbb{C})$ for X some fixed fibre of $\tilde{f}$. Choosing any family of holomorphic 3-folds $\Omega(s)$ on $\tilde{\mathcal{X}}$, define the *period map* $\mathcal{P} : \tilde{S} \to \mathbb{P}(H^3(X, \mathbb{C}))$ by $\mathcal{P}(s) = [\Omega(s)]$, which denotes the class of a line generated by the cohomology class of a holomorphic form on $\tilde{\mathcal{X}}_s$.

One should think of the period map as a multi-valued map on S, and this multi-valuedness is crucial. The kinds of S we will be interested in will never be simply connected.

First let us consider the differential of the period map

$$\mathcal{P}_* : \mathcal{T}_{S,s} \to \mathcal{T}_{\mathbb{P}(H^3(X, \mathbb{C})), \mathcal{P}(s)}.$$

If V is a finite dimensional $\mathbb{C}$-vector space, the differential of the map

$$\pi : V - \{0\} \to \mathbb{P}(V) = (V - \{0\})/\,\mathbb{C}^{*}$$

identifies, for a point $v \in V - \{0\}$, $V/\,\mathbb{C}\,v$ with $\mathcal{T}_{\mathbb{P}(V),[v]}$. Thus we may identify $\mathcal{T}_{\mathbb{P}(H^3(X,\mathbb{C})),[\Omega(s)]}$ with $H^3(X,\mathbb{C})/\,\mathbb{C}\,\Omega(s) = H^3(X,\mathbb{C})/H^{3,0}(\mathcal{X}_s)$. By the Hodge decomposition, this is isomorphic to $H^{2,1}(\mathcal{X}_s) \oplus H^{1,2}(\mathcal{X}_s) \oplus H^{0,3}(\mathcal{X}_s)$. In particular,

$$\mathcal{P}_*(\partial/\partial s) = \partial\Omega(s)/\partial s \bmod H^{3,0}(\mathcal{X}_s)$$

so $\mathcal{P}_*$ factors as

$$\mathcal{T}_{S,s} \to H^{2,1}(\mathcal{X}_s) \to \mathcal{T}_{\mathbb{P}(H^3(X,\mathbb{C})),\mathcal{P}(s)},$$

where the first map is the Kodaira–Spencer map as defined in §14.1 and the second map is the inclusion. This gives us the *local Torelli theorem*:

Theorem 16.9 *If $f : \mathcal{X} \to S$ is a family of Calabi–Yau 3-folds and the Kodaira–Spencer map is an isomorphism for each $s \in S$, then the differential of the period mapping is injective.*

This tells us that for a sufficiently small neighbourhood U of any $s \in S$, the complex structure of $\mathcal{X}_{s'}$ for $s' \in U$ is determined by $\mathcal{P}(s')$. (Assuming as always that the Kodaira–Spencer map is an isomorphism).

In contrast, a global Torelli theorem would tell us essentially that the complex structure of a Calabi–Yau 3-fold is determined by its period. However, global Torelli theorems are known to fail for Calabi–Yau 3-folds. See [187], [188], [186] for examples and further discussion.

Remark 16.10 One nice observation, due to Bryant and Griffiths [32], is that from the knowledge of the period map it is possible to reconstruct the corresponding variation of Hodge structures. This is worth mentioning because what is usually termed the period map in the literature is a map from $\tilde{S}$ to a classifying space of Hodge structures of weight 3, rather than the projective space classifying one-dimensional subspaces of $H^3(X,\mathbb{C})$. (See for example [76, §1].) So it is nice to know that one can reconstruct the Hodge filtration from the more limited information provided by just knowledge of F^3, since F^3 is spanned by the holomorphic three-form.

To reconstruct the Hodge filtration locally, consider a family $\mathcal{X} \to S$ for which the Kodaira–Spencer map is always an isomorphism, and Ω a family of holomorphic three-forms on $\mathcal{X}$. If $s_1,\ldots,s_n$ are local coordinates on S, then $\Omega(s)$ spans F^3, while $\Omega(s)$ and its derivatives $\partial\Omega(s)/\partial s_1,\ldots,\partial\Omega(s)/\partial s_n$ will span F^2. Again by Griffiths transversality, the elements above and all second derivatives $\partial^2\Omega(s)/\partial s_i\partial s_j$ are contained in F^1, but in fact one can show they span F^1. Finally F^0 is all of $H^3(X,\mathbb{C})$. The point is that all these derivatives are recoverable from the period map, and hence so is the Hodge filtration. See [32] for details.

Remark 16.11 Another observation is about the image of the period map. Again take a family $\mathcal{X} \to S$ where the Kodaira–Spencer map is always an isomorphism, and pull back the image of the period map to $H^3(X, \mathbb{C})$ to obtain a cone $\mathcal{L}$. If $n = h^{2,1}(X) = \dim S$, then $\dim H^3(X, \mathbb{C}) = 2n + 2$, and $\dim \mathcal{L} = n + 1$. Now $H^3(X, \mathbb{C})$ has the skew-symmetric cup product given by $(\alpha, \beta) = \int_X \alpha \wedge \beta$. This yields a non-degenerate 2-form ω on the space $H^3(X, \mathbb{C})$, (not to be confused with a Kähler form).

Let $x \in \mathcal{L}$ correspond to some complex structure on X. The tangent space to $\mathcal{L}$ at x then is $H^{3,0}(X) \oplus H^{2,1}(X)$, where the Dolbeault cohomology groups are defined with respect to that particular complex structure on X. Now the wedge of two forms of type $(3, 0)$ or $(2, 1)$ is zero, so the tangent space to $\mathcal{L}$ is isotropic with respect to the skew-symmetric pairing. This means that $\mathcal{L}$ is *Lagrangian* with respect to the symplectic form ω, i.e. $\omega|_{\mathcal{L}} = 0$.

We can make this more explicit as follows. Again, $H_3(X, \mathbb{Z})$ comes along with a unimodular skew-symmetric pairing by Poincaré duality. Thus we can choose a symplectic basis of 3-cycles $\alpha_0, \ldots, \alpha_n, \beta_0, \ldots, \beta_n$ with $\alpha_i . \beta_j = \delta_{ij}$, and $\alpha_i . \alpha_j = \beta_i . \beta_j = 0$. This gives us coordinates on $H^3(X, \mathbb{C})$ by integration over the above cycles. In particular, the *period vector* of a Calabi–Yau 3-fold X with holomorphic three-form Ω is

$$\left(\int_{\alpha_0} \Omega, \ldots, \int_{\alpha_n} \Omega, \int_{\beta_0} \Omega, \ldots, \int_{\beta_n} \Omega \right).$$

Locally near X, think of $\mathcal{L}$ as a graph of a function over the coordinate plane spanned by the first $n + 1$ coordinates. One can do this as long as the projection $\mathcal{L} \to \mathbb{C}^{n+1}$ onto the first $n + 1$ coordinates induces an isomorphism on tangent spaces at the point $x \in \mathcal{L}$, in which case by the implicit function theorem we can locally consider $\mathcal{L}$ as the graph of a function. In other words, we can write the above period vector as

$$(t_0, \ldots, t_n, f_0(t_0, \ldots, t_n), \ldots, f_n(t_0, \ldots, t_n))$$

where $t_i = \int_{\alpha_i} \Omega$ and $f_0, \ldots, f_n$ are functions of $t_0, \ldots, t_n$. Now in coordinates

$$(t_0, \ldots, t_n, y_0, \ldots, y_n)$$

on $H^3(X, \mathbb{C})$, given by integration over $\alpha_0, \ldots, \alpha_n, \beta_0, \ldots, \beta_n$, we can write the symplectic form ω as $\sum_{i=0}^{n} dt_i \wedge dy_i$, since the pairing on H^3 is dual to the one on H_3. Thus we can translate the condition of being Lagrangian into something more precise. Restricting ω to $\mathcal{L}$ by setting $y_i = f_i(t_0, \ldots, t_n)$, one obtains the form

$$\omega|_{\mathcal{L}} = \sum_{i,j} \frac{\partial f_i}{\partial t_j} dt_i \wedge dt_j.$$

This is only zero if $\frac{\partial f_i}{\partial t_j} = \frac{\partial f_j}{\partial t_i}$ for all i and j. This means that, again locally, there exists a complex valued function $F(t_0, \ldots, t_n)$ such that $\partial F / \partial t_i = f_i$. This function F is referred in the physics literature as the *prepotential*.

16.3   Degenerations of Hodge Structures

The next aspect of variation of Hodge structures we would like to study is the following important situation. Suppose we are given a family of compact Kähler manifolds $\mathcal{X} \to S$ where $S = (\Delta^*)^h$, the product of h copies of the punctured disk. Set $\bar{S} = \Delta^h$, where Δ is the disk. Sometimes one can extend the family $\mathcal{X} \to S$ to a family $\bar{\mathcal{X}} \to \bar{S}$, but not always. Usually the most one can hope for is a family $\bar{\mathcal{X}} \to \bar{S}$ for which the fibres $\mathcal{X}_s$ for $s \in \bar{S} - S$ are singular. Thus there is no reason to expect in particular that we can extend the variation of Hodge structures on S to one on $\bar{S}$. However, there is a way to define some limiting structure at points of $\bar{S} - S$, called a *mixed Hodge structure*. This is a complicated notion that we will not define completely, though we will get most of the way there.

To simplify matters, we will first consider families $\mathcal{X} \to \Delta^*$. One should keep in mind some of the following examples.

Example 16.12 (1) Consider the family of elliptic curves defined by the equation $y^2 z = x^3 + x^2 z + t z^3$ contained in $\mathbb{P}^2 \times \Delta$, with coordinate t on Δ^*. For $t = 0$, we obtain a nodal cubic, while for $t \neq 0$, t small, we have a non-singular elliptic curve. This gives us a family $f : \mathcal{X} \to \Delta^*$ with coordinate t on Δ^*. There is no way to extend this family to a family of *non-singular* elliptic curves over Δ.

(2) A family of quintics in $\mathbb{P}^4$, which we will see again, parametrized by a variable t:

$$t(x_0^5 + x_1^5 + x_2^5 + x_3^5 + x_4^5) + x_0 x_1 x_2 x_3 x_4 = 0.$$

For $t \neq 0$, t small, this gives a non-singular quintic, but for $t = 0$, we get a union of coordinate hyperplanes.

(3) The family of elliptic curves given in Example 16.3. This is closely related to the first example.

Returning to a general family of compact Kähler manifolds $f : \mathcal{X} \to \Delta^*$, we have a local system $\mathcal{H}_{\mathbb{Z}} = R^n f_* \mathbb{Z}$ on Δ^*. Fixing a point $t_0 \in \Delta^*$, we have the stalk $(R^n f_* \mathbb{Z})_{s_0} \cong H^n(X, \mathbb{Z})$ with $X \cong \mathcal{X}_{t_0}$. Let $T : H^n(X, \mathbb{Z}) \to H^n(X, \mathbb{Z})$ be the monodromy transformation given by a loop generating $\pi_1(\Delta^*)$. We state some basic properties of T.

Theorem 16.13   *1. T is an isomorphism. If $\dim X = n$, then $H^n(X, \mathbb{Z})$ is equipped with an intersection pairing, and T respects this intersection pairing.*

2. All the eigenvalues of T are roots of unity. Thus there exists an N and k such that $(T^N - I)^k = 0$. Such a matrix is called quasi-unipotent.

3. In (2), k can be taken to be less than or equal to $n + 1$.

Proof. 1 is obvious, since T is induced by a diffeomorphism of X. For 2 and 3, see [76], page 41 and [180]. These are much more involved.   □

Definition 16.14 We say T is *unipotent* if $(T - I)^k = 0$ for some k.

We are interested in the unipotent case. The general case can always be reduced to this case by pulling back the family under the map $\Delta^* \to \Delta^*$ given by $t \mapsto t^N$. By pulling back the family f via this map, we get a new family $f_N : \mathcal{X}' \to \Delta^*$. Since a loop generating $\pi_1(\Delta^*)$ is taken via $t \mapsto t^N$ to N times the generator of π_1, the monodromy associated with a generator of $\pi_1(\Delta^*)$ for the family f_N is T^N, which is then unipotent. Thus without a significant loss of generality, we will only consider the unipotent case.

Example 16.15 In Example 16.12, (3) it is clear from Example 16.3 that the monodromy is unipotent. This is also true of the other two examples of Example 16.12, but it is less obvious.

If T is unipotent, we can define the logarithm of the monodromy using the Taylor series, which is now finite:

$$N = \log(T) = (T - I) - (T - I)^2/2 + \cdots + (-1)^{n+1}(T - I)^n/n.$$

We can then use N to define a structure on $H_{\mathbb{Q}} = H^n(X, \mathbb{Q})$ called the monodromy weight filtration. The definition can be very confusing at first, but one should keep in mind that it is just linear algebra arising from the Jordan decomposition of T. Whenever $N : V_{\mathbb{Q}} \to V_{\mathbb{Q}}$ is a nilpotent operator on a $\mathbb{Q}$-vector space $V_{\mathbb{Q}}$ with $N^{n+1} = 0$, one has a weight filtration which is a filtration

$$0 \subseteq W_0 \subseteq \cdots \subseteq W_{2n} = V_{\mathbb{Q}}$$

satisfying

$$N(W_i) \subseteq W_{i-2}$$

and

$$N^k : W_{n+k}/W_{n+k-1} \xrightarrow{\cong} W_{n-k}/W_{n-k-1}.$$

The weight filtration is completely determined by these two properties. To see this and to actually construct it, we see first of all that

$$N^n : W_{2n}/W_{2n-1} \cong W_0,$$

so

$$W_{2n-1} = \ker N^n$$

and

$$W_0 = \operatorname{Im} N^n.$$

We now proceed inductively by setting

$$V'_{\mathbb{Q}} = W_{2n-1}/W_0.$$

Then N induces an operator N' on $V_{\mathbb{Q}}'$ (since $N(W_0) = 0$). Also $(N')^n = 0$. So inductively we get a filtration

$$0 \subseteq \quad W_0' \quad \subseteq \cdots \subseteq \quad W_{2n-2}' \quad = V_{\mathbb{Q}}'$$
$$\qquad \downarrow \cong \qquad\qquad\qquad \downarrow \cong$$
$$W_1/W_0 \subseteq \cdots \subseteq W_{2n-1}/W_0.$$

We lift the primed filtration up to complete the monodromy weight filtration on $V_{\mathbb{Q}}$. Thus we have shown both existence and uniqueness of the weight filtration. In particular, for $V_{\mathbb{Q}} = H_{\mathbb{Q}}$, we obtain the monodromy weight filtration.

Example 16.16 For the family of elliptic curves of Example 16.3, $T = \begin{pmatrix} 1 & 1 \\ 0 & 1 \end{pmatrix}$, and $(T - I)^2 = 0$. This gives a weight filtration

$$0 \subseteq W_0 \subseteq W_1 \subseteq W_2 = H^1(E, \mathbb{Q}) = \mathbb{Q}^2.$$

Here W_0 is the image of $N = \begin{pmatrix} 0 & 1 \\ 0 & 0 \end{pmatrix}$, hence one-dimensional, while W_1 is the kernel of N, hence also one-dimensional. Thus $W_0 = W_1$ are both one-dimensional. Note that W_0 consists precisely of those cohomology classes left invariant by T.

Definition 16.17 A family $\mathcal{X} \to \Delta^*$ of Calabi–Yau n-folds is said to be *maximally unipotent* if the monodromy transformation $T : H^n(X, \mathbb{Z}) \to H^n(X, \mathbb{Z})$ is unipotent and $(T - I)^n \neq 0$.

Example 16.18 The family of elliptic curves of Example 16.3 is maximally unipotent. In fact all examples of Example 16.12 are maximally unipotent, but this is harder to check.

The monodromy weight filtration is one of the two key ingredients of the concept of mixed Hodge structure. The other is the notion of a limiting Hodge filtration. We will only define a part of that filtration here: see [76, §IV] for the full definition.

Suppose we have a family $\mathcal{X} \to \Delta^*$ of Calabi–Yau 3-folds with unipotent monodromy. Let $\mathcal{H}$ denote the upper half-plane, with coordinate z on $\mathcal{H}$. Then $\mathcal{H}$ is naturally the universal covering space of Δ^* via the map $z \mapsto t = e^{2\pi i z}$, t the coordinate on Δ^*. The period map is a map $\mathcal{P} : \mathcal{H} \to \mathbb{P}(H^3(X, \mathbb{C}))$. Because of the action of monodromy, this does not descend to a single-valued map on Δ^*. In particular, $\mathcal{P}(z + 1) = T\mathcal{P}(z)$.

Define a new map $\tilde{\phi} : \mathcal{H} \to \mathbb{P}(H^3(X, \mathbb{C}))$ by

$$\tilde{\phi}(z) = e^{-zN}\mathcal{P}(z).$$

Here N is a nilpotent matrix, so e^{-zN} can be defined using a finite piece of the Taylor series for the exponential. Now

$$\tilde{\phi}(z+1) = e^{-zN}e^{-N}\mathcal{P}(z+1)$$
$$= e^{-zN}T^{-1}T\mathcal{P}(z)$$
$$= \tilde{\phi}(z).$$

Thus $\tilde{\phi}$ descends to a map $\phi : \Delta^* \to \mathbb{P}(H^3(X,\mathbb{C}))$, which is single-valued.

Theorem 16.19 *The map ϕ extends across the origin to give a map $\phi : \Delta \to \mathbb{P}(H^3(X,\mathbb{C}))$.*

Proof. [76], page 65 for a sketch, or [78], page 104.   □

This gives us a point $\phi(0)$ in $\mathbb{P}(H^3(X,\mathbb{C}))$, and hence a one-dimensional vector space $F^3_\infty \subseteq H^3(X,\mathbb{C})$. It is possible to define similar spaces F^p_∞ for $p \leqslant 2$ and get a filtration

$$F^3_\infty \subseteq \cdots \subseteq F^0_\infty = H^3(X,\mathbb{C}),$$

which is called the *limiting Hodge filtration*. The limiting Hodge filtration and monodromy weight filtration together make up what is known as a *mixed Hodge structure*. For completeness, we give the definition of an abstract mixed Hodge structure, but keep in mind we have only defined one piece of the Hodge filtration as this is all we need.

Definition 16.20 A *mixed Hodge structure* of weight n on a rational vector space $H_\mathbb{Q}$ consists of an ascending weight filtration

$$0 \subseteq W_0 \subset \cdots \subseteq W_{2n} = H_\mathbb{C}$$

defined over $\mathbb{Q}$, and a descending filtration

$$H_\mathbb{C} = F^0 \supseteq \cdots \supseteq F^n \supseteq 0,$$

such that if $Gr_m = W_m/W_{m-1}$ then the filtration

$$F^p(Gr_m) = (F^p \cap W_m)/(F^p \cap W_{m-1})$$

is a Hodge filtration on Gr_m of weight m in the usual sense (with $F^p = 0$ for $p > n$).

The general existence result is then

Theorem 16.21 *If $f : \mathcal{X} \to \Delta^*$ is a family of compact Kähler manifolds with unipotent monodromy, then there is a limiting Hodge filtration*

$$F_\infty^n \subseteq \cdots \subseteq F_\infty^0 = H^n(X, \mathbb{C})$$

which along with the monodromy weight filtration

$$W_0 \subseteq \cdots \subseteq W_{2n} = H^n(X, \mathbb{C})$$

forms a mixed Hodge structure. In case f is a family of Calabi–Yau 3-folds, $n = 3$, F_∞^3 coincides with the vector space defined above.

Proof. See [76, §IV]. □

Remark 16.22 We can organize the numerical data implicit in a mixed Hodge structure as follows. The filtration $F^\bullet$ induces Hodge structures on Gr_m of weight m, $0 \leqslant m \leqslant 2n$, hence numbers

$$h^{p,q} = \dim_{\mathbb{C}} F^p(Gr_{p+q})/F^{p+1}(Gr_{p+q})$$

satisfying $h^{p,q} = h^{q,p}$. Note also that for $p > n$, $h^{p,q} = 0$, so the numbers $h^{p,q}$ are non-zero only for $0 \leqslant p, q \leqslant n$, and these numbers $h^{p,q}$ can be represented in a Hodge diamond. In particular,

$$\sum_{p+q=m} h^{p,q} = \dim_{\mathbb{C}} W_m/W_{m-1}.$$

One of the important reasons for defining the limiting Hodge structure is the nilpotent orbit theorem. This tells us that, given the limiting Hodge structure, one can approximate the period map to high order. More specifically, continuing with the above notation, define a map

$$\mathcal{O} : \mathcal{H} \to \mathbb{P}(H^3(X, \mathbb{C}))$$

by

$$\mathcal{O}(z) = e^{zN}\phi(0).$$

Note that

$$\mathcal{O}(z + 1) = Te^{zN}\phi(0) = T\mathcal{O}(z).$$

$\mathcal{O}$ is called the nilpotent orbit of the family $\mathcal{X} \to \Delta^*$. We then have the *nilpotent orbit theorem:*

Theorem 16.23 *The nilpotent orbit osculates the period map to very high order. More precisely, with a reasonable metric on $\mathbb{P}(H^3(X, \mathbb{C}))$, there are constants A and B such that for $\operatorname{Im} z \geqslant A > 0$, the distance between $\mathcal{O}(z)$ and $\mathcal{P}(z)$ is bounded by $(\operatorname{Im} z)^B e^{-2\pi \operatorname{Im} z}$.*

Proof. See [180].   □

Going on to the multi-dimensional case, we pass to families $\mathcal{X} \to (\Delta^*)^s$, and go through the same constructions. This is much the same, but the additional complexity made it worth doing the one parameter case first.

The universal covering space of $(\Delta^*)^s$ is $\mathcal{H}^s$, with coordinates $(z_1, \ldots, z_s)$ and $t_j = e^{2\pi i z_j}$ coordinates on Δ^*. We have a period map $\mathcal{P} : \mathcal{H}^s \to \mathbb{P}(H^3(X, \mathbb{C}))$. Denote by $\gamma_1, \ldots, \gamma_s$ the counterclockwise loops around the origin of each copy of Δ^*; $\gamma_1, \ldots, \gamma_s$ generate $\pi_1((\Delta^*)^s) \cong \mathbb{Z}^s$. Let $T_1, \ldots, T_s$ denote the monodromy about $\gamma_1, \ldots, \gamma_s$. Assume that each monodromy transformation T_i is unipotent, and set $N_i = \log T_i$. Because the fundamental group of $(\Delta^*)^s$ is abelian, $[T_i, T_j] = 0$, so $[N_i, N_j] = 0$. If we wish to define a weight filtration, the first question is which N_i should we use. The correct answer is provided by a theorem of Cattani and Kaplan:

Theorem 16.24 *Let*

$$\Sigma = \left\{ \sum \lambda_i N_i \mid \lambda_i \in \mathbb{R}, \lambda_i > 0 \right\}.$$

Σ *is called the monodromy cone. Then each $N \in \Sigma$ defines the same monodromy weight filtration on $H^3(X, \mathbb{Q})$.*

Proof. See [76, §V.2].   □

Thus we can use any $N \in \Sigma$ to define the weight filtration. This should remind you of the cone Σ in the Kähler side introduced in §15.1. Take note that if N is in the boundary of the closure $\overline{\Sigma}$, N may not define the same weight filtration.

Next we can define a map $\tilde{\phi} : \mathcal{H}^s \to \mathbb{P}(H^3(X, \mathbb{Q}))$ by

$$\tilde{\phi}(z_1, \ldots, z_s) = e^{-\sum z_i N_i} \mathcal{P}(z_1, \ldots, z_s).$$

As before,

$$\tilde{\phi}(z_1, \ldots, z_i + 1, \ldots z_s) = e^{-\sum z_i N_i} T_i^{-1} T_i \mathcal{P}(z_1, \ldots, z_s) = \tilde{\phi}(z_1, \ldots, z_s),$$

and we can thus descend to a map $\phi : (\Delta^*)^s \to \mathbb{P}(H^3(X, \mathbb{Q}))$. Theorem 16.19 generalises, and we can extend to $\phi : \Delta^s \to \mathbb{P}(H^3(X, \mathbb{Q}))$ to get a limiting $F^3_\infty = \phi(0)$. Finally, the nilpotent orbit theorem also generalises and tells us that

$$\mathcal{O}(z_1, \ldots, z_s) = e^{\sum z_i N_i} \phi(0)$$

is a good approximation to the period map $\mathcal{P}$.

We now come to the most important definition of the section:

Definition 16.25 Let $\mathcal{X} \to (\Delta^*)^s \subseteq \Delta^s$, $s = h^{1,2}(X)$ be a family of Calabi–Yau 3-folds with the Kodaira–Spencer map an isomorphism at each point in $(\Delta^*)^s$. We say $0 \in \Delta^s$ is a *large complex structure limit point*, if

1. The monodromy transformations T_j around γ_j are all unipotent.
2. Let $N_j = \log T_j$, $N = \sum a_j N_j$ for some $a_j > 0$,

$$0 \subseteq W_0 \subseteq W_1 \subseteq W_2 \subseteq \cdots \subseteq W_6 = H^3(X, \mathbb{Q})$$

the induced weight filtration. Then $\dim W_0 = \dim W_1 = 1$ and $\dim W_2 = \dim W_3 = 1 + s$.
3. Let $\alpha_0^*, \ldots, \alpha_s^*$ be a basis of W_2 such that α_0^* spans W_0, and define m_{jk} by $N_j \alpha_k^* = m_{jk} \alpha_0^*$ for $1 \leqslant j, k \leqslant s$. Then the matrix $M := (m_{jk})$ is invertible over $\mathbb{Q}$.

This notion was made precise by D. Morrison in [159].

The main point is that $(\Delta^*)^s$ should also be the Kähler moduli space of a mirror to X. The $(1, 2)$-Yukawa coupling on $(\Delta^*)^s$ induced by the family $\mathcal{X} \to (\Delta^*)^s$ should be related to the $(1, 1)$-Yukawa coupling on the Kähler moduli space of the mirror. To make this clearer, we will end the section by showing how to approximate the $(1, 2)$-Yukawa coupling using the nilpotent orbit theorem.

Let $\mathcal{X} \to (\Delta^*)^s$ be a family with $0 \in \Delta^s$ a large complex structure limit. Then by item 2 of Definition 16.25, the weight filtration takes the form

$$0 \subset W_0 = W_1 \subset W_2 = W_3 \subset W_4 = W_5 \subset W_6$$

with $\dim W_0 = \dim W_6/W_4 = 1$ and $\dim W_2/W_0 = \dim W_4/W_2 = s$. We can forget about W_i for i odd.

The space $H_3(X, \mathbb{Q})$ is dual to $H^3(X, \mathbb{Q})$. We denote the transpose action of T_i and N_i on $H_3(X, \mathbb{Q})$ by the same symbols. In addition, we have the skew-symmetric cup product $(\cdot, \cdot)$ on $H^3(X, \mathbb{Q})$ (and also on $H_3(X, \mathbb{Q})$), and this allows us to identify the spaces $H_3(X, \mathbb{Q})$ and $H^3(X, \mathbb{Q})$ by Poincaré duality. Using this identification, the weight filtration induces a filtration on $H_3(X, \mathbb{Q})$. Let

$$0 \subseteq S_0 \subseteq \cdots \subseteq S_6 = H_3(X, \mathbb{Q})$$

be this filtration.

Proposition 16.26 *Under the identification of $H^3(X, \mathbb{Q})$ as the dual space of $H_3(X, \mathbb{Q})$, S_{2i} is the annihilator of W_{4-2i}.*

Proof. It is sufficient to show that $W_{2i}^{\perp} = W_{4-2i}$ under the pairing $(,)$. Of course T_i preserves this pairing, and hence $N_i = \log T_i$ is in the corresponding Lie algebra, i.e.

$$(x, N_i y) + (N_i x, y) = 0. \tag{30}$$

Thus N satisfies the same property. Now $W_0 = N^3(W_6)$, so if $x \in W_4$, $N^3 y \in W_0$, and $(x, N^3 y) = -(N^3 x, y) = 0$ since $N^3(W_4) = 0$. Thus $W_4 = W_0^\perp$. Similarly, $N(W_4) = W_2$, so if $Nx, y \in W_2$, $(Nx, y) = -(x, Ny) = 0$ since $x \in W_4$, and $Ny \in W_0$. Thus $W_2 = W_2^\perp$. $\square$

This can be used to produce a nice basis for $H_3(X, \mathbb{Q})$.

Proposition 16.27 *There exists an integral basis*

$$\alpha_0, \ldots, \alpha_s, \beta_0, \ldots, \beta_s$$

of $H_3(X, \mathbb{Q})$ with $\beta_0 \in S_0$, $\beta_1, \ldots, \beta_s \in S_2$, $\alpha_1, \ldots, \alpha_s \in S_4$ and $\alpha_0 \in S_6 = H_3(X, \mathbb{Q})$, and also

$$(\alpha_i, \alpha_j) = 0, (\beta_i, \beta_j) = 0 \text{ and } (\alpha_i, \beta_j) = \delta_{ij}.$$

Proof. Choose an integral generator β_0 of S_0 (unique up to sign) and extend to an integral basis $\beta_0, \ldots, \beta_s$ of S_2. Then $(\beta_i, \beta_j) = 0$, as S_2 is a Lagrangian subspace of $H_3(X, \mathbb{Q})$ with respect to the cup product by Proposition 16.26. Thus in particular, since the pairing is unimodular, $\beta_0, \ldots, \beta_s$ can be extended to an integral symplectic basis $\beta_0, \ldots, \beta_s, \alpha_0, \ldots, \alpha_s$ of $H_3(X, \mathbb{Q})$, i.e. with $(\alpha_i, \alpha_j) = 0$ and $(\alpha_i, \beta_j) = 0$ if $i \neq j$. Since the pairing is unimodular, we must have $\alpha_i \cdot \beta_i = \pm 1$. After a change of sign we can assume $(\alpha_i, \beta_i) = 1$. Since $(\alpha_i, \beta_0) = 0$ for $1 \leqslant i \leqslant s$, $\alpha_1, \ldots, \alpha_s \in S_4$ by Proposition 16.26 as desired. $\square$

Continuing on in the same situation, consider the space $F_\infty^3 \subseteq H^3(X, \mathbb{C})$. Note that in the notation of Remark 16.22, $h^{3,3} = \dim W_6 / W_5 = 1$, so $F_\infty^3 \not\subseteq W_5$. Thus we can choose a class $\Omega_{\lim} \in F_\infty^3$ such that $\int_{\beta_0} \Omega_{\lim} = 1$. We can then approximate the period map using the nilpotent orbit, putting

$$\Omega_{\text{nil}}(z_1, \ldots, z_s) = e^{\sum z_i N_i} \Omega_{\lim}$$

$$= \left(1 + \left(\sum z_i N_i \right) + \frac{1}{2} \left(\sum z_i N_i \right)^2 + \frac{1}{6} \left(\sum z_i N_i \right)^3 \right) \Omega_{\lim}.$$

Note that

$$\int_{\beta_0} \Omega_{\text{nil}}(z_1, \ldots, z_s) = \int_{\beta_0} \Omega_{\lim} = 1$$

since $N_i \Omega_{\lim} \subseteq W_4 = \beta_0^\perp$.

We can now compare this nilpotent orbit with the period map. Also, if we choose a family of holomorphic three-forms $\Omega(t_1, \ldots, t_s)$ on the family $\mathcal{X} \to (\Delta^*)^s$, then as the cohomology class of $\Omega(z_1, \ldots, z_s)$ is close to $\Omega_{\text{nil}}(z_1, \ldots, z_s)$ in $\mathbb{P}(H^3(X, \mathbb{C}))$, it is clear there is an open neighbourhood U of 0 in Δ^s such that for $(t_1, \ldots, t_s) \in U \cap (\Delta^*)^s$, $\int_{\beta_0} \Omega(t_1, \ldots, t_s) \neq 0$. Thus we can normalize the family Ω by insisting that

$$\int_{\beta_0} \Omega(t_1, \ldots, t_s) = 1.$$

This makes sense since β_0 is monodromy invariant, i.e. $T_i \beta_0 = \beta_0$. This choice of family of holomorphic three-forms gives us a lifting of the period map $\mathcal{P}$: $\mathcal{H}^s \to \mathbb{P}(H^3(X, \mathbb{C}))$ to $\tilde{\mathcal{P}} : \mathcal{H}^s \to H^3(X, \mathbb{C})$ by $\tilde{\mathcal{P}}(z_1, \ldots, z_s) = \Omega(z_1, \ldots, z_s)$. Furthermore, the nilpotent orbit theorem tells us that

$$\Omega(z_1, \ldots, z_s) \text{ and } \Omega_{\text{nil}}(z_1, \ldots, z_s)$$

are very close to each other as $\operatorname{Im} z_i \to \infty$. Thus, we can use $\Omega_{\text{nil}}(z_1, \ldots, z_s)$ to approximate the Yukawa coupling by defining

$$\left\langle \frac{\partial}{\partial z_i}, \frac{\partial}{\partial z_j}, \frac{\partial}{\partial z_k} \right\rangle_{\text{nil}} = \int_X \Omega_{\text{nil}} \wedge \frac{\partial^3}{\partial z_i \partial z_j \partial z_k} \Omega_{\text{nil}}(z).$$

This should be a good approximation to the $(1, 2)$-Yukawa coupling defined using the normalised family of 3-forms $\Omega(z)$. We can calculate this approximation by noting that

$$\frac{\partial}{\partial z_i} \frac{\partial}{\partial z_j} \frac{\partial}{\partial z_k} \Omega_{\text{nil}} = N_i N_j N_k \Omega_{\text{lim}}.$$

If $\alpha_0^*, \ldots, \beta_s^*$ is the dual basis to $\alpha_0, \ldots, \beta_s$, then $\Omega_{\text{lim}} = \beta_0^* \bmod W_4$, so that $N_i N_j N_k \Omega_{\text{lim}} = N_i N_j N_k \beta_0^*$, and then

$$\left\langle \frac{\partial}{\partial z_i}, \frac{\partial}{\partial z_j}, \frac{\partial}{\partial z_k} \right\rangle_{\text{nil}} = (\beta_0^*, N_i N_j N_k \beta_0^*). \tag{31}$$

Of course, this should not agree precisely with the $(1, 2)$-Yukawa coupling, but by the nilpotent orbit theorem we should expect the correction to be small, $O(e^{-2\pi \operatorname{Im} z})$ as $z \to \infty$, so this fits with the possibility that the correction could be, for example, a power series in $t_j = e^{2\pi i z_j}$.

If this is the case, then what we have determined is the constant term in the Taylor expansion of the $(1, 2)$-Yukawa coupling. In fact these constant terms are integers, and this should be (and will be shortly) compared to the constant term of the $(1, 1)$-Yukawa coupling.

16.4   Canonical Coordinates

There is one more structure we require before we can clearly state what we mean by mirror symmetry. These are canonical coordinates.

We continue as in the previous section with a family of Calabi–Yau 3-folds $\mathcal{X} \to (\Delta^*)^s$, with $0 \in \Delta^s$ a large complex structure limit point, and $\alpha_0, \ldots, \alpha_n, \beta_0, \ldots, \beta_n$ an integral symplectic basis of $H_3(X, \mathbb{Q})$ given by Proposition 16.27. Let $\mathcal{H}^s \to (\Delta^*)^s$ be the universal cover, with coordinates $z_1, \ldots, z_s$ on $\mathcal{H}^s$. Let $\Omega(z_1, \ldots, z_n)$ be a family of holomorphic three-forms normalised so that

$$\int_{\beta_0} \Omega(z_1, \ldots, z_s) = 1.$$

We then let

$$w_i(z_1, \ldots, z_s) = \int_{\beta_i} \Omega(z_1, \ldots, z_s).$$

Note that w_i satisfies

$$\begin{aligned} w_i(z_1, \ldots, z_j + 1, \ldots, z_s) &= \int_{\beta_i} \Omega(z_1, \ldots, z_j + 1, \ldots, z_s) \\ &= \int_{\beta_i} T_j \Omega(z_1, \ldots, z_s) \\ &= \int_{T_j \beta_i} \Omega \\ &= \int_{\beta_i - m'_{ji}\beta_0} \Omega \\ &= w_i(z_1, \ldots, z_s) - m'_{ji}, \end{aligned}$$

where $m'_{ij} \in \mathbb{Z}$ since we are using an integral basis. Note that the m'_{ij} coincide with the m_{ij} in Definition 16.25. Indeed,

$$\begin{aligned} -m'_{ji} = \int_{N_j \beta_i} \beta_0^* = \int_{\beta_i} N_j \beta_0^* &= (\alpha_i^*, N_j \beta_0^*) \\ &= -(N_j \alpha_i^*, \beta_0^*) \\ &= -(m_{ji} \alpha_0^*, \beta_0^*) \\ &= -m_{ji} \end{aligned}$$

using equation (30). Thus in particular, the matrix (m'_{ij}) is invertible.

We can now set

$$q_j = e^{2\pi i w_j};$$

these are well-defined functions on $(\Delta^*)^s$. These are *canonical coordinates* on $(\Delta^*)^s$.

Again, the nilpotent orbit theorem allows us to approximate the canonical coordinates. Put

$$w_j^{\text{nil}}(z_1, \ldots, z_s) = \int_{\beta_i} \Omega_{\text{nil}}(z_1, \ldots, z_s).$$

If we write $\Omega_{\text{lim}} = \beta_0^* + \sum_{j=1}^{s} b_j \beta_j^* \bmod W_2$, then

$$
\begin{aligned}
w_i^{\text{nil}}(z_1, \ldots, z_s) &= \int_{\beta_i} \sum_{j=1}^{s} z_j N_j \beta_0^* + \sum_{j=1}^{s} b_j \beta_j^* \\
&= b_i + \sum_{j=1}^{s} z_j \int_{\beta_i} N_j \beta_0^* \\
&= b_i + \sum_{j=1}^{s} z_j \int_{N_j \beta_i} \beta_0^* \\
&= b_i - \sum_{j=1}^{s} m_{ji} z_j.
\end{aligned}
$$

Thus the w_j^{nil} are just an affine linear combination of the coordinates z_j, and in particular give a coordinate system on $\mathcal{H}^s$. Thus the functions

$$
q_j^{\text{nil}} = e^{2\pi i b_j} \prod_{k=1}^{s} t_k^{-m_{kj}}
$$

give local coordinates on $(\Delta^*)^s$. Now if the m_{kj}'s are all non-positive so that only non-negative powers of the coordinates t_i occur, then by the nilpotent orbit theorem the genuine canonical coordinates q_j are close to q_j^{nil} near $0 \in \Delta^s$. Thus in this case, the canonical coordinates are genuine coordinates in a neighbourhood of 0.

These canonical coordinates of course depend on the choice of $\beta_0, \ldots, \beta_s$, so when we say we choose canonical coordinates we mean we have chosen these cycles.

17 A Mirror Conjecture

We can now make a mirror symmetry conjecture, which, while not totally precise, is the closest we will come to a precise statement. This will be enough for working out the best-known example of the quintic.

Conjecture 17.1 Let $f : \mathcal{X} \to (\Delta^*)^s$ be a family of Calabi–Yau 3-folds with $0 \in \Delta^s$ a large complex structure limit point. Then there exists a Calabi–Yau 3-fold $\check{X}$, called the *mirror* of the family f, a choice of framing $\Sigma \subseteq \mathcal{K}_{\check{X}}$ generated by a basis $e_1, \ldots, e_s \in \overline{\mathcal{K}}_{\check{X}}$, and a choice of canonical coordinates $q_1, \ldots, q_s$ on $(\Delta^*)^s$ with the following property: The basis $e_1, \ldots, e_s$ determines coordinates $\check{q}_1, \ldots, \check{q}_s$ on $\mathcal{M}_{Kah,\Sigma}(\check{X})$ and hence a map $m : (\Delta^*)^s \to \mathcal{M}_{Kah,\Sigma}(\check{X})$ taking a point with coordinates $(q_1, \ldots, q_s) \in (\Delta^*)^s$ to the point in $\mathcal{M}_{Kah,\Sigma}(\check{X})$ with coordinates $(\check{q}_1, \ldots, \check{q}_s) = (q_1, \ldots, q_s)$. Then

$$\left\langle \frac{\partial}{\partial q_i}, \frac{\partial}{\partial q_j}, \frac{\partial}{\partial q_k} \right\rangle_p = \left\langle \frac{\partial}{\partial \check{q}_i}, \frac{\partial}{\partial \check{q}_j}, \frac{\partial}{\partial \check{q}_k} \right\rangle_{m(p)}.$$

Here the left hand side refers to the $(1,2)$-Yukawa coupling for X at $p \in (\Delta^*)^s$, while the right hand side refers to the $(1,1)$-Yukawa coupling on Kähler moduli space of $\check{X}$ at $m(p)$. The map m is called the *mirror map*.

Keep in mind that the choice of canonical coordinates means a choice of basis for W_2/W_0, which amounts to choosing an integral isomorphism between W_2/W_0 and $H^2(\check{X}, \mathbb{Q})$. However, the conjecture as stated doesn't specify how one chooses this isomorphism. In the toric situation there is a canonical isomorphism, but when we deal with the quintic, it won't be much of a problem since these spaces are one-dimensional.

The reader should not regard this version of mirror symmetry as being the final word defining mirror symmetry; it is in fact an early definition, first phrased by Morrison in [159], and can be refined. Ultimately, however, this conjecture is only really reflecting one symptom of mirror symmetry, and an eventual mathematical definition of mirror symmetry may prove to be quite different from the above conjecture, with the equality of Yukawa couplings being deduced in some deep way from the hypothetical eventual definition.

18   Mirror Symmetry in Practice

18.1   The Basic Approach

Suppose we have a Calabi–Yau 3-fold X, and we wish to use the mirror symmetry conjecture to count the number of rational curves on X. Here is the basic procedure we need to follow in order to compute the Gromov–Witten invariants of X.

1. Identify, through some magic, a family of mirror Calabi–Yau manifolds $f : \check{X} \to S$, with $\dim S = s = h^{1,1}(X)$. We will use the original Greene-Plesser orbifold construction for the mirror quintic [74], but more generally one would use the Batyrev construction or generalisations thereof mentioned in Example 14.23.
2. Study a compactification $\overline{S}$ of S, and identify some point P of $\overline{S} \setminus S$ as a large complex structure limit point. In other words, there is a neighbourhood of P in $\overline{S}$ isomorphic to Δ^s such that $\Delta^s \cap S = (\Delta^*)^s$, and $f^{-1}((\Delta^*)^s) \to (\Delta^*)^s$ is a family of Calabi–Yau manifolds with $P = 0 \in \Delta^s$ being a large complex structure limit point. Since we are working the mirror symmetry conjecture of §17 backwards, there is no reason to expect there is a unique large complex structure limit point. If there is not a unique one, one should not believe that every one would yield predictions

for X. In fact, in general there might be many different large complex structure limit points, each one with its own mirror. Fortunately, this will not be the case in the quintic example.

3. Choosing $\beta_0, \ldots, \beta_s$ as in Proposition 16.27, compute the period integrals $\int_{\beta_i} \Omega$ in a neighbourhood of P in order to obtain canonical coordinates on this neighbourhood.

4. Compute the $(1,2)$-coupling, and try to match this up with the $(1,1)$-Yukawa coupling on Kähler moduli space. Read off the resulting predictions for the number of rational curves.

This is easier said than done, and we need to go through the example of the quintic in order to understand the difficulties involved and see how they are resolved. However, since the case of the quintic is a one-parameter example, even this case does not exhibit all of the difficulties of the general case.

18.2   The Mirror to the Quintic

The first step is to find the mirror family to the quintic. This was one of the very first examples studied by the physicists. The initial construction was inspired by conformal field theory, and we will not explain here how this construction was found but see [74]. We can however describe the construction.

We will focus on the following one-parameter family of quintic 3-folds:

$$X_\psi = \{(x_0, \ldots, x_4) \in \mathbb{P}^4 | f_\psi = 0\}$$

where

$$f_\psi = x_0^5 + x_1^5 + x_2^5 + x_3^5 + x_4^5 - 5\psi x_0 x_1 x_2 x_3 x_5.$$

This can be viewed as a family $\mathcal{X} \to \mathbb{A}^1$ with $\mathcal{X} \subseteq \mathbb{P}^4 \times \mathbb{A}^1$ and ψ a coordinate on $\mathbb{A}^1$. We can also projectivize the $\mathbb{A}^1$ and consider a family $\mathcal{X} \to \mathbb{P}^1$ with

$$\mathcal{X}_\infty = X_\infty = \{(x_0, \ldots, x_4) | \prod x_i = 0\}.$$

Consider the group $(\mathbb{Z}/5\mathbb{Z})^5$ acting on $\mathbb{P}^4$ diagonally by multiplication by fifth roots of unity, i.e. $(a_0, \ldots, a_4) \in (\mathbb{Z}/5\mathbb{Z})^5$ acts on $\mathbb{P}^4$ by

$$(x_0, \ldots, x_4) \mapsto (\xi^{a_0} x_0, \ldots, \xi^{a_4} x_4)$$

where ξ is a fixed fifth root of unity. The subgroup $\mathbb{Z}/5\mathbb{Z} = \{(a, a, a, a, a) | a \in \mathbb{Z}\}$ acts as the identity on $\mathbb{P}^4$, so we really have an action of $(\mathbb{Z}/5\mathbb{Z})^5/(\mathbb{Z}/5\mathbb{Z})$ on $\mathbb{P}^4$. Note that the subgroup G of this group given by $\{(a_0, \ldots, a_4) | \sum a_i = 0\}$ in fact acts on X_ψ for each ψ, so it makes sense to consider

$$Y_\psi = X_\psi/G.$$

Y_ψ is quite singular, and we would like to understand the singularities to some extent. Before doing so, however, let us study some properties of the family X_ψ.

The Jacobian of X_ψ is $(5x_0^4 - 5\psi x_1 x_2 x_3 x_4, \ldots, 5x_4^4 - 5\psi x_0 x_1 x_2 x_3)$. Suppose that the Jacobian vanishes at a point $(x_0, \ldots, x_4)$. Then we see that $x_i^5 = \psi x_0 x_1 x_2 x_3 x_4$, and hence

$$x_0^5 = x_1^5 = \ldots = x_4^5 = \psi x_0 x_1 x_2 x_3 x_4,$$

so

$$\prod_i x_i^5 = \psi^5 \prod_i x_i^5.$$

Thus either $\psi^5 = 1$ or at least one of the x_i is zero. But if some x_i is zero, they all are, and thus $(x_0, \ldots, x_4)$ does not represent a point in $\mathbb{P}^4$. Thus we see that if $\psi^5 \neq 1$, then X_ψ is non-singular. If $\psi^5 = 1$, then X_ψ is singular in the points

$$(\xi^{a_0}, \ldots, \xi^{a_4})$$

with $\sum a_i = 0$. This consists of 125 distinct singular points.

In addition, if $\psi = \infty$, the above family degenerates to $x_0 x_1 x_2 x_3 x_4 = 0$, which is of course very singular. For $\psi^5 \neq 1$, we argued in Example 14.20 and Exercise 14.1 that $h^{1,1}(X_\psi) = 1$ and $h^{1,2}(X_\psi) = 101$.

The quotient X_ψ/G is quite singular. In fact, it is singular at each point $x \in X_\psi$ where the stabilizer of x in G is non-trivial. Notice that a point in $\mathbb{P}^4$ has a non-trivial stabilizer in G if at least two of the coordinates are zero. Thus the points of the curves

$$C_{ij} = \{x_i = x_j = 0\} \cap X_\psi$$

have stabilizer of order 5, while the points of the set

$$P_{ijk} = \{x_i = x_j = x_k = 0\} \cap X_\psi$$

have stabilizers of order 25. For example,

$$C_{12} = \{x_1 = x_2 = x_3^5 + x_4^5 + x_5^5 = 0\}$$

and

$$P_{123} = \{x_1 = x_2 = x_3 = x_4^5 + x_5^4 = 0\}.$$

From this one can see that $C_{12}/G = \mathbb{P}^1$, and P_{123}/G consists of one point. So after dividing out by G, and abusing notation by referring to C_{ij}/G as C_{ij} and P_{ijk}/G as P_{ijk}, we see that the singular locus of Y_ψ consists of 10 curves C_{ij}, each isomorphic to $\mathbb{P}^1$, with C_{ij}, C_{ik}, C_{jk} meeting at the point P_{ijk}.

We will not say a great deal about the way one resolves the singularities Y_ψ, since this requires experience with toric geometry. For those who know, we will just say the following. It is easy to see that away from the points

P_{ijk}, a neighbourhood of C_{ijk} on Y_ψ looks $\mathbb{C}^3/(\mathbb{Z}/5\mathbb{Z})$, where $\mathbb{Z}/5\mathbb{Z}$ acts by $(x_1, x_2, x_3) \mapsto (\xi^a x_1, \xi^{-a} x_2, x_3)$, and the quotient then is $(\mathbb{C}^2/(\mathbb{Z}/5\mathbb{Z})) \times \mathbb{C}$. Resolving this involves blowing up along the singular curve twice, and producing four exceptional divisors over each curve C_{ij}. To resolve the singularities at P_{ijk}, which locally look like $\mathbb{C}^3/(\mathbb{Z}/5\mathbb{Z})^2$, one needs to use toric methods. The singularity is toric, given by the cone spanned by $(1,0,0)$, $(0,1,0)$, and $(0,0,1)$ in the lattice $\mathbb{Z}^3 + \frac{1}{5}(1,-1,0) + \frac{1}{5}(1,0,-1)$. One then subdivides this cone to find six exceptional divisors over each point P_{ijk}.

We now obtain a resolution $\check{X}_\psi \to Y_\psi$ of singularities. In total, one obtains $10 \times 4 + 10 \times 6$ exceptional divisors. This, along with the hyperplane section, provides 101 divisors. This is not a proof, but is suggestive that $h^{1,1}(\check{X}_\psi) = 101$. Similarly, the fact that we have produced a one-dimensional family suggests that $h^{1,2}(\check{X}_\psi) = 1$. This was originally proved by S.S. Roan [174], and now follows easily from toric methods introduced by Batyrev [5]. However, to prove this is beyond the scope of this book. One must also prove that $\check{X}_\psi$ is a Calabi–Yau 3-fold. One first shows that Y_ψ is a singular Calabi–Yau 3-fold by observing that a holomorphic three-form on X_ψ is left invariant under the action, hence descends to a nowhere zero holomorphic three-form on Y_ψ. Then one makes sure the resolution of Y_ψ constructed is crepant.

The details of this resolution are not particularly important for us, because on $\check{X}_\psi$ the computations we will be doing will involve integration over 3-cycles on $\check{X}_\psi$. If these 3-cycles are disjoint from the exceptional locus of the resolution, we can perform the integration on Y_ψ. In fact these integrals can just as well be performed by pulling back to X_ψ, if we then remember to divide by a suitable number to take into account the action of G.

Summarizing, we state

Theorem 18.1 *For $\psi^5 \neq 1, \infty$, there exists a resolution of singularities $\check{X}_\psi \to Y_\psi$ such that $\check{X}_\psi$ is a Calabi–Yau 3-fold, and*

$$h^{1,1}(\check{X}_\psi) = 101, \quad h^{1,2}(\check{X}_\psi) = 1.$$

The family of 3-folds $\check{X}_\psi$ is the desired family of mirror quintics.

18.3 Complex Moduli Space of $\check{X}$

We now have a family of Calabi–Yau 3-folds $\check{X}_\psi$ parametrised by ψ. This is not the real complex moduli space. Indeed, for ξ a fifth root of unity, $X_\psi \cong X_{\xi\psi}$ via the map $(x_0, \ldots, x_4) \mapsto (\xi x_0, \ldots, x_4)$, and hence $\check{X}_\psi \cong \check{X}_{\xi\psi}$. We will use the coordinate $z = (5\psi)^{-5}$ on what should be the true moduli space. There are several important points in the z-plane:

- $z = 0$: This corresponds to the quotient of the very singular $x_0 \cdots x_4 = 0$.
- $z = 5^{-5}$: This Calabi–Yau is singular, with one singular point, as X_1 has 125 singular points forming one G-orbit. This singular point is an ordinary double point, i.e. locally of the form $x^2 + y^2 + z^2 + w^2 = 0$. (This

is because while the first derivatives of the defining equation vanish at the point, the matrix of second derivatives is non-singular.) Physicists often refer to this as a "conifold," and the point $z = 1$ is the "conifold point."

- $z = \infty$: This is known as an orbifold point. While it corresponds to the non-singular Calabi–Yau 3-fold X_0, if we try to form a family over the z-plane by dividing out the family $\mathcal{X} \to \mathbb{C}$ parametrised by ψ by the additional action of $\mathbb{Z}/5\mathbb{Z}$, we would get a quotient of X_0 as the fibre over $z^{-1} = 0$. What has happened is that X_0 has an unexpectedly large automorphism group, and this causes problems in complex moduli space. The result is that one expects monodromy about $z = \infty$, but the monodromy should be of order 5. Indeed, if one lifts a loop about $z = \infty$ up to the ψ-plane, the family $\mathcal{X}$ is locally trivial at $\psi = 0$, and hence has no monodromy.

The general principle is that large complex structure limit points correspond to the most degenerate Calabi–Yaus. The orbifold point is certainly not maximally unipotent, and the monodromy about ordinary double point degenerations is well-understood and is known not to be maximally unipotent. (The Picard–Lefschetz formula tells us what this monodromy is.) So $z = 0$ is the only remaining candidate for a large complex structure limit point.

18.4   The Periods of $\check{X}_\psi$

There are a number of ways to approach the problem of computing the period map. We will give the most explicit method, historically the first, as carried out in the fundamental paper of Candelas, de la Ossa, Greene and Parkes [36], and elaborated in [158]. For hypersurfaces in toric varieties, life is simpler these days as the Picard–Fuchs equations we introduce below can be understood from combinatorics. However, if one wants to compute periods of other families of Calabi–Yau manifolds, one may need to resort to the earlier methods given here.

We will select a 3-cycle β_0 in X_ψ. Denote by f_ψ the polynomial defining X_ψ. Let β_0 be the set of points $(x_0, \ldots, x_4)$ in $\mathbb{P}^4$ with

$$x_4 = 1, |x_0| = |x_1| = |x_2| = \delta,$$

and x_3 given by the solution to $f_\psi = 0$ that tends to zero as $\psi \to \infty$. To make sense of this, fix x_0, x_1 and x_2 and define y by

$$x_3 = (\psi x_0 x_1 x_2)^{1/4} y.$$

The equation $f_\psi = 0$ then becomes, with $x_4 = 1$,

$$x_0^5 + x_1^5 + x_2^5 + (\psi x_0 x_1 x_2)^{5/4} y^5 + 1 - 5(\psi x_0 x_1 x_2)^{5/4} y = 0,$$

or

$$y = \frac{y^5}{5} + \frac{1 + x_0^5 + x_1^5 + x_2^5}{5(\psi x_0 x_1 x_2)^{5/4}}.$$

We see that as $\psi \to \infty$, there is one solution to this equation going to zero like $\psi^{-5/4}$, and four solutions which approach fourth roots of 5. Therefore the corresponding values of x_3 approach zero like ψ^{-1}, and the others go to ∞ like $\psi^{1/4}$. Thus there is always, for large ψ, a well-defined branch of x_3 to choose from. In particular, β_0 is a three-torus. In addition, G acts on this 3-torus, and descends to a three-torus on Y_ψ, which is disjoint from the singular locus. Thus we can think of β_0 as a torus on $\check{X}_\psi$.

Note also that this torus β_0 is a deformation of

$$\{(x_0, \ldots, x_4) \in \mathbb{P}^4 | x_4 = 1, |x_0| = |x_1| = |x_2| = \delta, x_3 = 0\}/G \subseteq Y_\infty.$$

Next, we need the family of holomorphic three-forms. Again we first give the holomorphic three-form on X_ψ. We can represent holomorphic three-forms by restricting three-forms on projective space to X_ψ. Take the open affine subset of X_ψ defined by $x_4 = 1$, and take on this open subset

$$\Omega(\psi) = 5\psi \frac{dx_0 \wedge dx_1 \wedge dx_2}{\partial f_\psi / \partial x_3}\bigg|_{X_\psi}$$

$$= 5\psi \frac{dx_0 \wedge dx_1 \wedge dx_2}{5x_3^4 - 5\psi x_0 x_1 x_2}\bigg|_{X_\psi}.$$

This is a priori a meromorphic three-form on X_ψ. Note that when $\partial f_\psi / \partial x_3 \neq 0$, the projection of the open affine subset of X_ψ given by $x_4 = 1$ onto the first three coordinates (x_0, x_1, x_2) is locally an isomorphism, so the form has no zeroes or poles away from $\partial f_\psi / \partial x_3 = 0$ and $x_4 = 0$. To see that $\Omega(\psi)$ does not have a pole on X_ψ along the locus $\partial f_4 / \partial x_3 = 0$, let's use the fact that we can write, say, locally, x_2 as a function of x_0, x_1 and x_3. Write this as $x_2 = g(x_0, x_1, x_3)$. Note that $\frac{\partial g}{\partial x_3}$ can be computed implicitly by differentiating $0 = f_\psi(x_0, x_1, g, x_3)$ with respect to x_3 and obtaining

$$\frac{\partial g}{\partial x_3} = -\frac{\partial f_\psi / \partial x_3}{\partial f_\psi / \partial x_2},$$

and thus we can also write

$$\Omega(\psi) = -5\psi \frac{dx_0 \wedge dx_1 \wedge dx_3}{\partial f_\psi / \partial x_2}.$$

We could have done this with any of the variables x_0, x_1 or x_2, thus seeing that $\Omega(\psi)$ has a pole only if all the $\partial f_\psi / \partial x_i$ vanish, which doesn't happen if X_ψ is non-singular. Thus $\Omega(\psi)$ defines a holomorphic three-form on an affine open subset of X_ψ whose complement is a divisor given by $\{x_4 = 0\} \cap X_\psi$. Since $\omega_{X_\psi} = \mathcal{O}_{X_\psi}$, the divisor of zeroes and poles of $\Omega(\psi)$ is zero, and hence

$\Omega(\psi)$ extends to a nowhere vanishing holomorphic three-form on X_ψ, rather than a meromorphic three-form.

It is easy to see that $\Omega(\psi)$ is invariant under G, and hence descends to a nowhere vanishing holomorphic three-form on the non-singular part of Y_ψ. This three-form will necessarily extend to a holomorphic three-form on $\check{X}_\psi$, the resolution of Y_ψ. Thus we have our family of three-forms on the mirror family. Now we can calculate the integral of this holomorphic three-form over β_0, or, up to a factor of 5^3, perform the integral over $\beta_0 \subseteq X_\psi$.

Thus we compute in X_ψ

$$\int_{\beta_0} \frac{dx_0 dx_1 dx_2}{-x_0 x_1 x_2 + \psi^{-1} x_3^4}.$$

For this computation we will use residues liberally. First, we claim that this integral is the same as

$$\frac{1}{2\pi i} \int_{T^4} 5\psi \frac{dx_0 dx_1 dx_2 dx_3}{f_\psi}$$

with $T^4 = \{(x_0, x_1, x_2, x_3, 1) | \|x_i\| = \delta\}$. To see this, first perform the latter integral around the x_3 loop. Fix x_0, x_1 and x_2. The integrand then has poles when $f_\psi(x_0, \ldots, x_3, 1) = 0$. However, we showed that for ψ large there was only one such value of x_3 near 0. Thus the integrand has one pole, and its residue is $1/(\partial f_\psi / \partial x_3)$, showing equality between the two integrals. So we need to compute

$$\int_{T^4} \frac{dx_0 dx_1 dx_2 dx_3}{(5\psi)^{-1}(1 + x_0^5 + \cdots + x_3^5) - x_0 x_1 x_2 x_3}$$
$$= \int_{T^4} \frac{dx_0 dx_1 dx_2 dx_3}{x_0 x_1 x_2 x_3} \frac{1}{(5\psi)^{-1}(1 + x_0^5 + \cdots + x_3^5)/(x_0 x_1 x_2 x_3) - 1}.$$

Expanding in a Taylor series we get

$$-\sum_{n=0}^{\infty} \int_{T^4} \frac{dx_0 dx_1 dx_2 dx_3}{x_0 x_1 x_2 x_3} \frac{(1 + x_0^5 + \cdots + x_3^5)^n}{(5\psi)^n (x_0 x_1 x_2 x_3)^n}.$$

We now use residues to evaluate this integral. The only terms which will contribute will be those terms in the second factor not involving any x_i's; therefore the only relevant terms of the numerator are those proportional to $x_0^n x_1^n x_2^n x_3^n$. But in the numerator all exponents are divisible by 5, so one only gets contributions when n is divisible by 5. We now have

$$-\sum_{n=0}^{\infty} \int_{T^4} \frac{dx_0 dx_1 dx_2 dx_3}{x_0 x_1 x_2 x_3} \frac{(1 + x_0^5 + \cdots + x_3^5)^{5n}}{(5\psi)^{5n} (x_0 x_1 x_2 x_3)^{5n}}.$$

The number of terms of the form $x_0^{5n} x_1^{5n} x_2^{5n} x_3^{5n}$ in the numerator is then $(5n)!/(n!)^5$. Thus the above integral is

$$\int_{\beta_0} \frac{dx_0 dx_1 dx_2}{-x_0 x_1 x_2 + \psi^{-1} x_3^4} = (2\pi i)^3 \sum_{n=0}^{\infty} \frac{(5n)!}{(n!)^5 (5\psi)^{5n}}.$$

It is now convenient to use the parameter $z = (5\psi)^{-5}$ and write

$$\phi_0(z) = \sum_{n=0}^{\infty} \frac{(5n)!}{(n!)^5} z^n.$$

This is proportional to the period we computed above.

To continue, we need to compute more periods. However, we don't want to compute any other integrals of the above sort. We will follow here [158], which simplified the period calculations of [36]. Let's begin with the following elementary observation. If $a_n = \frac{(5n)!}{(n!)^5}$, then the a_n obey a recurrence relation

$$(n+1)^5 a_{n+1} = (5n+5)(5n+1)(5n+2)(5n+3)(5n+4)a_n$$

or

$$(n+1)^4 a_{n+1} = 5(5n+1)(5n+2)(5n+3)(5n+4)a_n.$$

This translates into a differential equation for $\phi_0(z)$. Let Θ denote the differential operator $z\frac{\partial}{\partial z}$. Then we can rewrite the above recurrence relation as saying that if

$$\mathcal{D} = \Theta^4 - 5z(5\Theta + 1)(5\Theta + 2)(5\Theta + 3)(5\Theta + 4),$$

then $\phi_0(z)$ is a solution to the differential equation

$$\mathcal{D}\phi(z) = 0. \tag{32}$$

Candelas et al in [36] proved that all period integrals of our holomorphic three-form satisfy this same differential equation. The proof given here will use ideas introduced by Morrison [158]; but it is now known that the above differential equation can be constructed from very general principles without having to go through the process we did. This differential equation is a special case of a GKZ (Gelfand-Kapranov-Zelevinsky) hypergeometric system of partial differential equations. The approach of GKZ systems is followed, e.g. in [44], and we follow the historical approach to give a more hands-on experience.

18.5   Picard–Fuchs Equations

The periods of a family of Calabi–Yau manifolds always satisfy a system of differential equations called the Picard–Fuchs equations. In the current example, we have a family of three-forms in cohomology, $\Omega(z)$. Since $H^3(\check{X}, \mathbb{C})$ is four-dimensional, the cohomology classes

$$\Omega(z), \frac{\partial \Omega}{\partial z}, \frac{\partial^2 \Omega}{\partial z^2}, \frac{\partial^3 \Omega}{\partial z^3}, \frac{\partial^4 \Omega}{\partial z^4}$$

are linearly dependent, giving a fourth order differential equation. This equation also must be satisfied by any period integral $\int_\alpha \Omega(z)$ for any 3-cycle α.

The fourth order differential equation we wrote down above is a good candidate for the Picard–Fuchs equation, since it is satisfied by at least one period and is fourth order. Still, we need to prove this. To do so, we must be a bit more precise about representing three-forms on $\check{X}_\psi$, so that we can differentiate our family and find the linear dependence relation. To do so, we sketch a technique due to Griffiths [75] for representing the middle cohomology of a hypersurface.

Let $f = 0$ define a quintic 3-fold X in $\mathbb{P}^4$. Denote by Ω the four-form

$$\Omega := \sum_{j=0}^4 (-1)^j x_j \mathrm{d}x_0 \wedge \cdots \wedge \widehat{\mathrm{d}x_j} \wedge \cdots \wedge \mathrm{d}x_4.$$

This is not actually a genuine four-form on $\mathbb{P}^4$ because it is not homogeneous of degree 0. However, any rational four-form on $\mathbb{P}^4$ can be written as $P\Omega/Q$, where P and Q are homogeneous polynomials with $\deg P + 5 = \deg Q$. Thus, in particular, the forms $g\Omega/f^l$ where $\deg g = 5(l-1)$ define rational four-forms with poles along $f = 0$. The *residue* of such a form is a cohomology class in $H^3(X, \mathbb{C})$ as follows. We will define this cohomology class by indicating what is its integral over each three-cycle in X. If γ is a three-cycle in X, denote by Γ the *tube over* γ. This is an $\mathcal{S}^1$ bundle over γ inside the normal bundle of X in $\mathbb{P}^4$, and is a four-cycle in $\mathbb{P}^4$. We then define the cohomology class $\mathrm{Res}_X(g\Omega/f^l) \in H^3(X, \mathbb{C})$ by demanding that

$$\int_\gamma \mathrm{Res}_X(g\Omega/f^l) = \frac{1}{2\pi i} \int_\Gamma \frac{g\Omega}{f^l}.$$

Note we have already used this concept in the previous section in passing between an integral over a 3-cycle on X and an integral over a four-cycle on $\mathbb{P}^4$.

Slightly more abstractly, this can be viewed as follows: we have the long exact sequence of relative cohomology

$$H^4(\mathbb{P}^4, \mathbb{C}) \to H^4(\mathbb{P}^4 \setminus X, \mathbb{C}) \to H^5(\mathbb{P}^4, \mathbb{P}^4 \setminus X, \mathbb{C}) \to H^5(\mathbb{P}^4, \mathbb{C}) = 0.$$

Now it follows from Lefschetz duality that $H^5(\mathbb{P}^4, \mathbb{P}^4 \setminus X, \mathbb{C}) \cong H^3(X, \mathbb{C})$, and in fact under this identification the boundary map

$$H^4(\mathbb{P}^4 \setminus X, \mathbb{C}) \to H^5(\mathbb{P}^4, \mathbb{P}^4 \setminus X, \mathbb{C})$$

is precisely the residue map, with any four-form on $\mathbb{P}^4$ with poles only along X defining an element of $H^3(X, \mathbb{C})$.

Of course, two different rational four-forms on $\mathbb{P}^4$ might give rise to the same cohomology class on X. To see how this works, let $g_0, \ldots, g_4$ be homogeneous polynomials of degree $5l - 4$, and consider the rational three-form on $\mathbb{P}^4$ given by

$$\varphi = \frac{1}{f^l} \sum_{i<j} (-1)^{i+j} (x_i g_j - x_j g_i) dx_0 \wedge \cdots \wedge \hat{dx_i} \wedge \cdots \wedge \hat{dx_j} \wedge \cdots \wedge dx_4.$$

We calculate

$$d\varphi = \frac{(l \sum g_j \frac{\partial f}{\partial x_j} - f \sum \frac{\partial g_j}{\partial x_j}) \Omega}{f^{l+1}}.$$

Thus if a form is of the form

$$\sum l g_j \frac{\partial f}{\partial x_j} \Omega / f^{l+1},$$

then modulo exact forms, it can be rewritten as a form of lower degree. This is Griffiths' method of reduction of pole order.

Rephrasing this, let $J(f)$ denote the *Jacobian ideal* of f, i.e. the ideal in $\mathbb{C}[x_0, \ldots, x_4]$ generated by the $\partial f / \partial x_i$'s. Then if the numerator of a rational form $\frac{g\Omega}{f^l}$ is in $J(f)$, the form is equivalent modulo exact forms to one whose numerator is of smaller degree.

Thus our strategy for finding the Picard–Fuchs equations is as follows. We start with the family of holomorphic three-forms on X_ψ, which we have seen before, and express this in terms of residues, namely

$$\Omega(\psi) = \text{Res}_{X_\psi} \frac{5\psi\Omega}{f_\psi}.$$

We then proceed to differentiate with respect to z. After taking four derivatives, we have obtained something of the form $g\Omega/f_\psi^5$, which we then reduce modulo $J(f_\psi)$ to obtain a form which is a linear combination of $\Omega(\psi)$ and its first three derivatives.

It is slightly more convenient to proceed as follows. Let

$$\omega_l = \frac{5(-1)^{l-1}(l-1)!\psi^l (\prod x_i^{l-1}) \Omega}{f_\psi^l}.$$

Note that these forms are all invariant under G, and hence descend to forms on Y_ψ. Using the relationship $z = (5\psi)^{-5}$, we note that

$$\psi \frac{d}{d\psi} = \psi \frac{dz}{d\psi} \frac{d}{dz} = \psi(-5)(5)(5\psi)^{-6} \frac{d}{dz} = -5z \frac{d}{dz},$$

so $-\frac{\psi}{5} \frac{d}{d\psi} = z \frac{d}{dz} = \Theta$. We calculate

$$-\frac{\psi}{5} \frac{d}{d\psi} \omega_l = \frac{-\psi}{5} \left(\frac{5(-1)^{l-1}(l-1)! l \psi^{l-1} (\prod x_i^{l-1}) f_\psi^l \Omega}{f_\psi^{2l}} \right)$$

$$+ \frac{\psi}{5} \left(\frac{5(-1)^{l-1}(l-1)! \psi^l (\prod x_i^{l-1}) (-5 \prod x_i) l f_\psi^{l-1} \Omega}{f_\psi^{2l}} \right)$$

$$= -\frac{l}{5} \omega_l + \omega_{l+1}.$$

Thus in particular, $(5\Theta + l)\omega_l = -l\omega_l + 5\omega_{l+1} + l\omega_l = 5\omega_{l+1}$.

To verify that $\Omega(z) = \omega_1(z)$ satisfies the differential equation (32) we have to show that in $H^3(\check{X}_\psi, \mathbb{C})$, the forms ω_l satisfy the relation

$$\Theta^4\omega_1 - 5^5 z\omega_5 = 0.$$

Now

$$\Theta^4\omega_1 = \Theta^3\left(-\frac{1}{5}\omega_1 + \omega_2\right)$$

$$= \Theta^2\left(\frac{1}{25}\omega_1 - \frac{3}{5}\omega_2 + \omega_3\right)$$

$$= \Theta\left(-\frac{1}{125}\omega_1 + \frac{7}{25}\omega_2 - \frac{6}{5}\omega_3 + \omega_4\right)$$

$$= \frac{1}{625}\omega_1 - \frac{15}{125}\omega_2 + \frac{25}{25}\omega_3 - \frac{10}{5}\omega_4 + \omega_5.$$

Thus we need to show that the relation

$$\frac{1}{625}\omega_1 - \frac{3}{25}\omega_2 + \omega_3 - 2\omega_4 + (1 - 5^5 z)\omega_5 = 0$$

holds in cohomology, or equivalently,

$$\frac{1}{625}\frac{1}{1 - 5^5 z}\omega_1 - \frac{3}{25}\frac{1}{1 - 5^5 z}\omega_2 + \frac{1}{1 - 5^5 z}\omega_3 - 2\frac{1}{1 - 5^5 z}\omega_4 + \omega_5 = 0. \quad (33)$$

The strategy we will take is as follows. First, we use the fact that the numerator of ω_5 is contained in $J(f_\psi)$. Thus we can use the reduction of pole order method to rewrite ω_5 as a rational four-form with a pole of lower order. This gives a new form η_4, which is equivalent to ω_5. Second, we find an $\epsilon_4(\psi)$ such that $\eta_4 - \epsilon_4\omega_4$ has its numerator in $J(f_\psi)$. Here one must use Gröbner basis algorithms; for example one can use Macaulay 2 [72] to do this. One thus obtains again a rational form η_3 such that ω_5 is equivalent to $\eta_3 + \epsilon_4\omega_4$. We then find an ϵ_3 such that $\eta_3 - \epsilon_3\omega_3$ has its numerator in $J(f_\psi)$, and so on. Continuing in this fashion, one gets an expression

$$\omega_5 - \sum_i \epsilon_i\omega_i = d\varphi$$

for some φ. The ϵ_i turn out to be the coefficients we needed. We give some details on how to carry this out in Exercise 18.1. Suffice it to say, this gives a computational proof that the Picard–Fuchs equation (32) holds for the periods of $\Omega(\psi)$, i.e. for any cycle ω, the function of ψ

$$\int_\gamma \Omega(\psi)$$

satisfies (32).

18.6   Equations with Regular Singular Points

Definition 18.2 The differential equation

$$\frac{d^s f}{dz^s} + \sum_{j=0}^{s-1} C_j(z)\frac{d^j f}{dz^j} = 0$$

with $C_j(z)$ meromorphic functions is said to have a *regular singular point* at $z = 0$ if, after multiplying by z^s and with $\Theta = z\frac{d}{dz}$ rewriting the equation as

$$\Theta^s f + \sum_{j=0}^{s-1} B_j(z)\Theta^j f = 0,$$

the $B_j(z)$ are holomorphic at $z = 0$.

For example, the Picard–Fuchs equation (32) has a regular singular point at $z = 0$. This is in fact a general property of Picard–Fuchs equations of degenerations: see for example [76], page 46. We can rewrite an equation with regular singular points as follows. Set

$$A(z) = \begin{pmatrix} 0 & 1 & 0 & \cdots & 0 \\ 0 & 0 & 1 & \cdots & 0 \\ & & & \vdots & 1 \\ -B_0(z) & -B_1(z) & -B_2(z) & \cdots & -B_{s-1}(z) \end{pmatrix}$$

and put

$$w(z) = \begin{pmatrix} f(z) \\ \Theta f(z) \\ \vdots \\ \Theta^{s-1} f(z) \end{pmatrix}.$$

Then we can write our equation as the system of equations

$$\Theta w(z) = A(z)w(z).$$

The key result in the theory of equations with regular singular points is

Theorem 18.3 *There exists a constant $s \times s$ matrix R and an $s \times s$ matrix $S(z)$ of holomorphic functions such that*

$$\Phi(z) = S(z)z^R$$

is a fundamental system of solutions for $\Theta w(z) = A(z)w(z)$, where

$$z^R = I + (\log z)R + \frac{(\log z)^2}{2!}R^2 + \cdots$$

is a multi-valued matrix function of z. Note that $z \mapsto e^{2\pi i}z$ transforms a solution $\Phi(z)$ to a solution $\Phi(z)e^{2\pi iR}$, so $e^{2\pi iR}$ represents the monodromy of the system. In addition, if the distinct eigenvalues of $A(0)$ do not differ by integers then we can take $R = A(0)$, and so the monodromy is $T = e^{2\pi iA(0)}$, and $N = \log T = 2\pi i A(0)$.

Proof. See for example [43]. $\qquad\qquad\qquad\qquad\qquad\qquad\qquad\qquad\qquad$ $\square$

For the operator $D = \Theta^4 - 5z(5\Theta+1)\cdots(5\Theta+4)$, by dividing the equation $D\phi = 0$ by $1 - 5^5z$, we obtain

$$\Theta^4\phi - \frac{5z}{1-5^5z}(\text{a polynomial of degree 3 in } \Theta \text{ not involving } z)\phi = 0$$

from which it follows in this case that

$$A(0) = \begin{pmatrix} 0 & 1 & 0 & 0 \\ 0 & 0 & 1 & 0 \\ 0 & 0 & 0 & 1 \\ 0 & 0 & 0 & 0 \end{pmatrix}.$$

Since all eigenvalues of $A(0)$ are zero, the hypotheses necessary to conclude that we can take $R = A(0)$ are satisfied. Thus we learn that the monodromy of the fundamental system of solutions to our equation is

$$T = e^{2\pi iA(0)} = \begin{pmatrix} 1 & 2\pi i & (2\pi i)^2/2 & (2\pi i)^3/6 \\ 0 & 1 & 2\pi i & (2\pi i)^2/2 \\ 0 & 0 & 1 & 2\pi i \\ 0 & 0 & 0 & 1 \end{pmatrix}.$$

The moral of this is the following: we know that any period $\omega(z) = \int_\beta \Omega$ is a solution to our Picard–Fuchs equation $D\omega(z) = 0$. Thus it must be a linear combination of the fundamental solutions, which are found in the first row of the matrix $\Phi(z)$. Therefore there exists a basis $\alpha_1, \ldots, \alpha_4$ of $H_3(X, \mathbb{C})$ such that $\int_{\alpha_i} \Omega = \Phi(z)_{1i}$. In particular, the monodromy transformation about $z = 0$ in this basis is then represented by the matrix T. It follows easily from this that $z = 0$ is a large complex structure limit point. We have found the correct point in complex moduli of the mirror quintic about which we should compute canonical coordinates and the $(1,2)$-Yukawa coupling.

18.7 Canonical Coordinates

As the next step, we wish to find another solution to the Picard–Fuchs equation (32). We already have the solution

$$\phi_0(z) = \sum_{n=0}^{\infty} \frac{(5n)!}{(n!)^5}z^n,$$

which is single-valued. By the results of the previous section, we know that up to a multiplicative factor, this is the unique solution of $\mathcal{D}\phi = 0$ that is single-valued in a neighbourhood of $z = 0$. We wish to find a solution ϕ_1 of (32) that is multi-valued, with $\phi_1(e^{2\pi i}z) = \phi_1(z) + 2\pi i \phi_0(z)$, which should exist again by results of the previous section. Furthermore, it is clear ϕ_1 is unique up to adding multiples of ϕ_0.

To find such a ϕ_1, one looks for a solution of the type

$$\phi_1(z) = \phi_0(z)\log z + \psi(z).$$

To find $\psi(z)$, observe first that using the product rule inductively,

$$\Theta^i(f(z)\log z) = (\Theta^i f(z))\log z + i\Theta^{i-1}f(z),$$

from which it follows that if we write $\mathcal{D} = F(\Theta)$ for F the polynomial

$$F(x) = x^4 - 5z(5x + 1)(5x + 2)(5x + 3)(5x + 4),$$

we see that

$$\mathcal{D}\phi_1(z) = (\mathcal{D}\phi_0(z))\log z + F'(\Theta)\phi_0(z) + \mathcal{D}\psi(z) = F'(\Theta)\phi_0(z) + \mathcal{D}\psi(z).$$

Setting this to be zero and expanding ϕ_0 and ψ in Taylor series gives a recurrence relation on the coefficients of the Taylor series of $\psi(z)$. Note that the constant term of $\psi(z)$ can be set to be anything we want, since we can always add a multiple of $\phi_0(z)$ to $\psi(z)$. So we normalize by setting $\psi(0) = 0$. This will completely determine ψ. It is then easy to see that we have

$$\psi(z) = 5\sum_{n=1}^{\infty} \frac{(5n)!}{(n!)^5}\left(\sum_{j=n+1}^{5n}\frac{1}{j}\right)z^n.$$

Now let us examine canonical coordinates. Consider z as a coordinate on Δ^* over which we have a family $\check{X}_z$ of mirror quintics, inducing a monodromy transformation $T : H^3(\check{X}_{z_0}) \to H^3(\check{X}_{z_0})$ around the origin for some point z_0 near $z = 0$. We can then choose an integral symplectic basis $\beta_0, \beta_1, \alpha_0, \alpha_1$ of $H_3(\check{X}_{z_0}, \mathbb{Q})$ by Proposition 16.27, with $T(\beta_0) = \beta_0$ and $T(\beta_1) = \beta_1 + n\beta_0$ for some n. Thus $\int_{\beta_0}\Omega$ is a single-valued solution to (32), so

$$\int_{\beta_0}\Omega = C\phi_0$$

for some constant C, while

$$\int_{\beta_1}\Omega = D_0\phi_0 + D_1\phi_1$$

for some constants D_0, D_1. Now

$$\int_{T(\beta_1)} \Omega = n \int_{\beta_0} \Omega + \int_{\beta_1} \Omega$$

$$= (nC + D_0)\phi_0 + D_1\phi_1,$$

and

$$D_0\phi_0(e^{2\pi i}z) + D_1\phi_1(e^{2\pi i}z) = (D_0 + 2\pi i D_1)\phi_0(z) + D_1\phi_1(z),$$

so we must have $nC = 2\pi i D_1$.

Now we will assume that we can take $n = 1$. This can be proved, but we will not show this here. Then the canonical coordinate is

$$w = \frac{\int_{\beta_1} \Omega}{\int_{\beta_0} \Omega}$$

$$= \frac{D_0}{C} + \frac{1}{2\pi i} \frac{\phi_1(z)}{\phi_0(z)}$$

$$= \frac{1}{2\pi i} \log z + \frac{1}{2\pi i} \log c_2 + \frac{1}{2\pi i} \frac{\psi(z)}{\phi_0(z)},$$

where c_2 is a constant depending on D_0 and C. We can then write

$$q = e^{2\pi i w} = c_2 z e^{\psi/\phi_0}.$$

We don't know c_2 because we don't know the precise cycles we are integrating over. We will be able to determine c_2 later via the mirror symmetry conjecture. In any event, as predicted in §16.4, the canonical coordinate is a pertubation of $c_2 z$. If desired, we can expand q as a power series in z, or, vice versa, write z as a power series in q.

18.8   The Yukawa Coupling

Having succesfully computed the canonical coordinate q, we next have to compute the Yukawa coupling, which is not particularly difficult. For this we follow [36].

Let

$$W_k = \int_{\check{X}_z} \Omega(z) \wedge \frac{d^k}{dz^k} \Omega(z),$$

where we still use the same family $\Omega(z)$ of holomorphic three-forms as always. If one rewrites the Picard–Fuchs equation $\mathcal{D}\Omega(z) = 0$ as

$$\left(\frac{d^4}{dz^4} + \sum_{k=0}^{3} C_k(z) \frac{d^k}{dz^k} \right) \Omega(z) = 0,$$

we get the relation

$$W_4 + \sum_{k=0}^{3} C_k W_k = 0.$$

However, we know that $W_0 = W_1 = W_2 = 0$ by Griffiths transversality. In addition,

$$0 = \frac{d^2}{dz^2} W_2 = \int \frac{d^2}{dz^2}\Omega \wedge \frac{d^2}{dz^2}\Omega + 2\int \frac{d}{dz}\Omega \wedge \frac{d^3}{dz^3}\Omega + \int \Omega \wedge \frac{d^4}{dz^4}\Omega$$
$$= 0 + 2(W_3' - W_4) + W_4$$

so $W_4 = 2W_3'$, and we obtain a differential equation

$$W_3' + \frac{1}{2}C_3 W_3 = 0.$$

Now looking at the coefficient of d^3/dz^3 in the Picard–Fuchs equation we find, with some calculation, that

$$C_3(z) = \frac{6}{z} - \frac{2 \cdot 5^5}{1 - 5^5 z},$$

and that

$$W_3 = \frac{c_1}{(2\pi i)^3 z^3 (5^5 z - 1)}$$

is the general solution to the differential equation $W_3' + \frac{1}{2}C_3(z)W_3 = 0$. Here $c_1/(2\pi i)^3$ is the constant of integration, suitably normalised. This gives us, up to a constant, the Yukawa coupling

$$\left\langle \frac{\partial}{\partial z}, \frac{\partial}{\partial z}, \frac{\partial}{\partial z} \right\rangle.$$

We have only to normalise this suitably and express this in the canonical coordinate q. We follow Morrison's treatment [158]. To normalise this we need to do two things:

- We need to compute the Yukawa coupling with respect not to the family of three-forms $\Omega(z)$ but rather to the family of normalised three-forms

$$\frac{\Omega(z)}{\int_{\beta_0} \Omega(z)}.$$

- We actually need to compute

$$\left\langle \frac{\partial}{\partial w}, \frac{\partial}{\partial w}, \frac{\partial}{\partial w} \right\rangle$$

To make these changes, first recall that if we replace $\Omega(z)$ by $f(z)\Omega(z)$, the Yukawa coupling changes by a factor of $f^2(z)$. Now since $\int_{\beta_0} \Omega(z)$ is proportional to ϕ_0, the normalised Yukawa coupling is

$$\left\langle \frac{\partial}{\partial z}, \frac{\partial}{\partial z}, \frac{\partial}{\partial z} \right\rangle = \frac{c_1}{(2\pi i)^3 z^3 (5^5 z - 1)\phi_0^2(z)}$$

with the constant of proportionality absorbed into the constant c_1.

We get

$$\left\langle \frac{\partial}{\partial w}, \frac{\partial}{\partial w}, \frac{\partial}{\partial w} \right\rangle = \left(\frac{dz}{dw}\right)^3 \left\langle \frac{\partial}{\partial z}, \frac{\partial}{\partial z}, \frac{\partial}{\partial z} \right\rangle$$

$$= \left(\frac{dw}{dz}\right)^{-3} \left\langle \frac{\partial}{\partial z}, \frac{\partial}{\partial z}, \frac{\partial}{\partial z} \right\rangle$$

$$= \frac{c_1}{(5^5 z - 1)\phi_0^2(z)\delta^3(z)},$$

where

$$\delta(z) = 1 + z\frac{d}{dz}\left(\frac{\psi(z)}{\phi_0(z)}\right).$$

The last step is to expand this in a power series in q. We have

$$dq/dz = c_2 e^{\psi/\phi_0} + c_2 z\frac{d(\psi/\phi_0)}{dz}e^{\psi/\phi_0} = c_2 \delta(z)e^{\psi/\phi_0}.$$

Thus

$$\frac{d}{dq} = \frac{dz}{dq}\frac{d}{dz} = \frac{1}{c_2 \delta(z)e^{\psi/\phi_0}}\frac{d}{dz}.$$

So if we set

$$h_0(z) = \frac{1}{(5^5 z - 1)\phi_0^2(z)\delta^3(z)},$$

and

$$h_j(z) = \frac{1}{\delta(z)e^{\psi/\phi_0}}\frac{dh_{j-1}(z)}{dz},$$

then

$$h_j(z) = \frac{(c_2)^j}{c_1}\frac{d^j}{dq^j}\left\langle \frac{\partial}{\partial w}, \frac{\partial}{\partial w}, \frac{\partial}{\partial w} \right\rangle$$

and hence the Yukawa coupling is expressed as a power series in q as

$$\left\langle \frac{\partial}{\partial w}, \frac{\partial}{\partial w}, \frac{\partial}{\partial w} \right\rangle = \sum_{j=0}^{\infty} \frac{c_1}{c_2^j}\frac{h_j(0)}{j!}q^j.$$

These coefficients $h_j(0)$ then can be computed using the power series expansions for ϕ_0 and ψ, somewhat laboriously, to any order. Thus we get

$$h_0(0) = -1, \qquad\qquad h_1(0) = -575, h_2(0) = -1950750,$$
$$h_3(0) = -10277490000, \qquad\qquad h_4(0) = -74486048625000.$$

Therefore the power series expansion in q for the Yukawa coupling is

$$-c_1 - 575\frac{c_1}{c_2}q - \frac{1950750}{2}\frac{c_1}{c_2^2}q^2 - \frac{10277490000}{6}\frac{c_1}{c_2^3}q^3 - \frac{-74486048625000}{24}\frac{c_1}{c_2^4}q^4 - \cdots.$$

Now this is predicted to coincide with

$$5 + \sum_{d=1}^{\infty} d^3 N_d \frac{q^d}{1-q^d} = 5 + N_1 q + (8N_2 + N_1)q^2 + (27N_3 + N_1)q^3$$
$$+ (64N_4 + 8N_2 + N_1)q^4 + \cdots$$

where N_d is interpreted as the number of rational curves of degree d. Thus we must have $c_1 = -5$ and $N_1 = (-575)(-5)/c_2 = 2875/c_2$. Since we know classically that N_1 should be 2875, we should have $c_2 = 1$. Obtaining predictions for higher degree, we get

$$N_2 = 609250, N_3 = 317206375, N_4 = 242467530000, \ldots$$

At the time of [36], the number N_2 was already known, having been calculated by Sheldon Katz in 1986 [128]. The number N_3 was computed in 1990 by Ellingsrud and Strømme [53] around the same time as [36] was written. That even these first three numbers are correct seemed like a miracle. However, it took more than six years for the complete set of predictions for the quintic to be proved valid by work of Givental [64], Lian, Liu and Yau [144] and Bertram [19].

18.9 Exercises

18.1 Verify the relation of equation (33) using Macaulay 2 [72] as follows. Work over the field $k = \mathbb{Q}(\psi)$, where ψ is a transcendental, and then carry out computations in the ring $R = k[x_0, \ldots, x_4]$. Then f_ψ can be viewed as a polynomial in R, and let $J = J(f_\psi)$ denote the Jacobian ideal of f_ψ, generated by the partials of f_ψ with respect to $x_0, \ldots, x_4$. Represent ω_l by the polynomial $g_l = 5(-1)^{l-1}(l-1)!\psi^l(\prod x_i^{l-1})$. Now we apply Griffiths' reduction of pole order method by trying to find $h_0, \ldots, h_4 \in R$ such that

$$\sum_{i=0}^{4} \frac{\partial f_\psi}{\partial x_i} h_i = g_5.$$

Find the appropriate Macaulay 2 command to find such h_i. This allows us to write ω_5 as being equivalent to some $\eta_4 \Omega / f_\psi^5$ for some η_4. Now again use Macaulay 2 to find an $\epsilon_4 \in k$ such that $\eta_4 - \epsilon_4 g_4 \in J$. Repeat the process to get the form corresponding to this function equivalent to some $\eta_3 \Omega / f_\psi^3$. Continue in this fashion, obtaining ϵ_3, ϵ_2 and ϵ_1 such that

$$\omega_5 + \sum \epsilon_i \omega_i$$

is equivalent to 0. This will give equation (33). To do this is a good exercise in learning the basic uses of the immensely useful tool Macaulay 2.

18.2 Using your favorite symbolic computation package, check the numbers given above for N_d.

18.3 Study the analytic continuation of the single-valued period ϕ_0 to the z-plane. It does not have a single-valued extension. Explain the monodromy of this analytic extension in terms of the singularities of the family Y_ψ.

19   The Strominger–Yau–Zaslow Approach to Mirror Symmetry

In this last section, we will discuss some more recent ideas about mirror symmetry surrounding the Strominger–Yau–Zaslow conjecture. The original conjecture involves special Lagrangian submanifolds, but currently, as hinted at in §12.4, the conjecture probably needs to be viewed in a limiting sense if we are to continue to consider the special Lagrangian condition. Instead, we will greatly simplify the discussion here by only considering "half" the conjecture. We will view mirror symmetry, as we have done throughout, as a process which identifies complex moduli of a Calabi–Yau manifold X and Kähler moduli of the mirror $\check{X}$. So we will ignore metric issues by only considering X as a complex manifold and $\check{X}$ as a symplectic manifold. We will give an introduction to the Strominger–Yau–Zaslow point of view which we hope will clarify some basic mysteries of mirror symmetry as studied so far. These include

- What is the topological relationship between X and $\check{X}$? In particular, in the three-dimensional case, why does $h^{1,3-i}(X) = h^{1,i}(\check{X})$?
- Why does complex and Kähler (symplectic) structure get interchanged?
- What is the B-field?
- What is the role of monodromy and the weight filtration?
- What does mirror symmetry for the quintic really "look like"?

19.1   Affine Manifolds and Torus Bundles

We will start by giving some simple notions concerning affine manifolds. For basic information on affine manifolds, see for example [65].

We fix $M = \mathbb{Z}^n$, $N = \mathrm{Hom}_{\mathbb{Z}}(M, \mathbb{Z})$, $M_{\mathbb{R}} = M \otimes_{\mathbb{Z}} \mathbb{R}$, $N_{\mathbb{R}} = N \otimes_{\mathbb{Z}} \mathbb{R}$. We set

$$\mathrm{Aff}(M_{\mathbb{R}}) = M_{\mathbb{R}} \rtimes \mathrm{GL}_n(\mathbb{R})$$

to be the group of affine transformations of $M_{\mathbb{R}}$, i.e. maps $T : M_{\mathbb{R}} \to M_{\mathbb{R}}$ of the form $T(x) = Ax + b$ for $A : M_{\mathbb{R}} \to M_{\mathbb{R}}$ invertible and linear and $b \in M_{\mathbb{R}}$. This has a subgroup

$$\mathrm{Aff}(M) = M \rtimes \mathrm{GL}_n(\mathbb{Z}).$$

Definition 19.1 Let B be an n-dimensional manifold. An *affine structure* on B is given by an open cover $\{U_i\}$ along with coordinate charts $\psi_i : U_i \to M_{\mathbb{R}}$, whose transition functions $\psi_i \circ \psi_j^{-1}$ lie in $\mathrm{Aff}(M_{\mathbb{R}})$. The affine structure is *integral* if the transition functions lie in $\mathrm{Aff}(M)$. If B and B' are affine manifolds of the same dimension, and $f : B \to B'$ is an immersion, then we say f is an *(integral) affine map* if f is given by (integral) affine maps on each (integral) affine coordinate chart.

Proposition 19.2 *Let* $\pi : \tilde{B} \to B$ *be the universal covering of an (integral) affine manifold* B, *inducing an (integral) affine structure on* $\tilde{B}$. *Then there is an (integral) affine map* $d : \tilde{B} \to M_{\mathbb{R}}$ *called the developing map, and any two such maps differ only by an (integral) affine transformation.*

Proof. This is standard, see [65, p. 641] for a proof. One simply patches together affine coordinate charts. $\qquad\square$

Note that there is no need for the developing map to be either injective or a covering space; it is only an immersion in general.

Definition 19.3 The fundamental group $\pi_1(B)$ acts on $\tilde{B}$ by deck transformations; for $\gamma \in \pi_1(B)$, let $T_{\gamma} : \tilde{B} \to \tilde{B}$ be the corresponding deck transformation. Then by the uniqueness of the developing map there exists a $\rho(\gamma) \in \mathrm{Aff}(M_{\mathbb{R}})$ such that $\rho(\gamma) \circ d \circ T_{\gamma} = d$. The map $\rho : \pi_1(B) \to \mathrm{Aff}(M_{\mathbb{R}})$ is called the *holonomy representation*. If the affine structure is integral, then $\mathrm{Im}\, \rho \subseteq \mathrm{Aff}(M)$.

Remark 19.4 Note conversely we can define an affine structure on B by giving an immersion $d : \tilde{B} \to M_{\mathbb{R}}$ and a representation $\rho : \pi_1(B) \to \mathrm{Aff}(M_{\mathbb{R}})$ such that $\rho(\gamma) \circ d \circ T_{\gamma} = d$. This can be a useful construction method.

Example 19.5 A torus $T^n = M_{\mathbb{R}}/\Lambda$ for a lattice $\Lambda \subseteq M_{\mathbb{R}}$ has a natural affine structure induced by $d = I : M_{\mathbb{R}} \to M_{\mathbb{R}}$. Here for $\lambda \in \pi_1(T_n) = \Lambda$, $\rho(\lambda)$ is translation by $-\lambda$. Thus the affine structure is integral if and only if $\Lambda \subseteq M$.

Definition 19.6 Let $\mathrm{Lin} : \mathrm{Aff}(M_{\mathbb{R}}) \to \mathrm{GL}_n(\mathbb{R})$ and $\mathrm{Trans} : \mathrm{Aff}(M_{\mathbb{R}}) \to M_{\mathbb{R}}$ be the projections. Only Lin is a group homomorphism. If $\rho : \pi_1(B) \to \mathrm{Aff}(M_{\mathbb{R}})$ is a representation, we denote by $\bar{\rho}$ the composition $\mathrm{Lin} \circ \rho$; this is the *linear part* of the representation. Note that $\bar{\rho}$ makes $M_{\mathbb{R}}$ into a left $\pi_1(B)$-module, which we denote by $M_{\mathbb{R}}^{\bar{\rho}}$. If in addition $\mathrm{Im}(\bar{\rho}) \subseteq \mathrm{GL}_n(\mathbb{Z})$, then M is also a $\pi_1(B)$-module as well as $M_{\mathbb{R}}/M$, written as $M^{\bar{\rho}}$ and $(M_{\mathbb{R}}/M)^{\bar{\rho}}$ respectively.

Definition 19.7 If B is an affine manifold, there is a flat connection ∇ on $\mathcal{T}_B$, where if $y_1, \ldots, y_n$ are local affine coordinates, $\partial/\partial y_1, \ldots, \partial/\partial y_n$ are a frame of flat sections of $\mathcal{T}_B$. Denote by $\Lambda_{\mathbb{R}} \subseteq \mathcal{T}_B$ the local system of flat sections, and $\check{\Lambda}_{\mathbb{R}} \subseteq \mathcal{T}_B^*$ the dual local system of flat sections of the dual connection (locally of the form $dy_1, \ldots, dy_n$). Furthermore if the holonomy of B takes values in $M_{\mathbb{R}} \rtimes \mathrm{GL}_n(\mathbb{Z})$, then there exists integral subsystems $\Lambda \subseteq \Lambda_{\mathbb{R}}$ and $\check{\Lambda} \subseteq \check{\Lambda}_{\mathbb{R}}$ coming from the inclusions $M \subseteq M_{\mathbb{R}}$ and $N \subseteq N_{\mathbb{R}}$.

We now move on to torus bundles. Given an affine manifold with holonomy contained in $M_{\mathbb{R}} \rtimes \mathrm{GL}_n(\mathbb{Z})$, the local system $\Lambda \subseteq \mathcal{T}_B$ is defined, and $\mathcal{T}_B/\Lambda$ is a torus bundle over B. Similarly, $\mathcal{T}_B^*/\check{\Lambda}$ is a different torus bundle over B. These torus bundles are dual in an obvious sense, and it is this duality which the Strominger–Yau–Zaslow approach views as mirror symmetry. Let us be clearer about the structures on these bundles.

First recall that if G is a group and M is a G-module (i.e. there is an action of G on M from the left), then $H^1(G, M)$ is the group

$$\frac{\{\varphi : G \to M \,|\, \varphi(g_1 g_2) = g_1 \varphi(g_2) + \varphi(g_1)\}}{\{\varphi : G \to M \,|\, \text{there exists } m \in M \text{ such that } \varphi(g) = gm - m \text{ for all } g \in G\}}.$$

A map $\varphi : G \to M$ satisfying $\varphi(g_1 g_2) = g_1 \varphi(g_2) + \varphi(g_1)$ is usually called a crossed homomorphism.

Definition 19.8 Let B be an affine manifold with holonomy contained in $M_{\mathbb{R}} \rtimes \mathrm{GL}_n(\mathbb{Z})$. Let $\mathbf{B} : \pi_1(B) \to M_{\mathbb{R}}^{\check\rho}$ be a crossed homomorphism. Identifying $\mathcal{T}_{M_{\mathbb{R}}} = M_{\mathbb{R}} \times M_{\mathbb{R}}$ with $M_{\mathbb{R}} \oplus i M_{\mathbb{R}} = M_{\mathbb{R}} \otimes_{\mathbb{R}} \mathbb{C}$, this tangent space can be viewed as a complex manifold. The complex structure on $\mathcal{T}_{M_{\mathbb{R}}}$ pulls back to give a complex structure on $\mathcal{T}_{\tilde{B}} = \tilde{B} \times M_{\mathbb{R}}$ via the differential of the developing map, which descends to a complex structure on $\tilde{B} \times (M_{\mathbb{R}}/M)$. Let $\gamma \in \pi_1(B)$ act on $\tilde{B} \times (M_{\mathbb{R}}/M)$ by

$$\gamma(b, m) = (T_{\gamma^{-1}}(b), \tilde{\rho}(\gamma)(m) + \mathbf{B}(\gamma)).$$

Then this action is holomorphic, and the quotient is a torus bundle $f : X(B, \mathbf{B}) \to B$ with a complex structure. We will write $X(B) := X(B, 0)$ for the case when $\mathbf{B} = 0$.

Definition 19.9 Let B be an affine manifold with holonomy contained in $M_{\mathbb{R}} \rtimes \mathrm{GL}_n(\mathbb{Z})$. Let $\check{X}(B) = \mathcal{T}_B^*/\check{\Lambda}$. Then the canonical symplectic form on $\mathcal{T}_B^*$ descends to a symplectic form $\check{\omega}$ on $\check{X}(B)$, making $\check{X}(B) \to B$ into a Lagrangian torus bundle.

Remark 19.10 (1) In local affine coordinates $y_1, \ldots, y_n$ on B, we obtain coordinates $(x_1, \ldots, x_n, y_1, \ldots, y_n)$ on $\mathcal{T}_B$ where the point with coordinates $(x_1, \ldots, x_n, y_1, \ldots, y_n)$ corresponds to the tangent vector $\sum_{i=1}^{n} x_i \partial/\partial y_i$ at the point $(y_1, \ldots, y_n)$. Then

$$z_j = e^{2\pi i(x_j + i y_j)}$$

defines holomorphic coordinates on $\mathcal{T}_B/\Lambda$. One can easily check that if $\mathbf{B} = 0$ then $X(B,\mathbf{B}) = \mathcal{T}_B/\Lambda$, with the complex structure given by these coordinates. It is also easy to check that $X(B,\mathbf{B}) \cong X(B,\mathbf{B}')$ as complex manifolds if $\mathbf{B}$ and $\mathbf{B}'$ represent the same element of $H^1(\pi_1(B), (M_\mathbb{R}/M)^{\tilde{\rho}})$.

(2) Similarly, given local affine coordinates $y_1, \ldots, y_n$ on B, we obtain canonical coordinates $(q_1, \ldots, q_n, y_1, \ldots, y_n)$ on $\mathcal{T}_B^*$, where the point with coordinates $(q_1, \ldots, q_n, y_1, \ldots, y_n)$ corresponds to the one-form $\sum_{i=1}^n q_i dy_i$ at $(y_1, \ldots, y_n)$. The canonical symplectic form is $\sum_j dq_j \wedge dy_j$.

We view $X(B, \mathbf{B})$ and $\check{X}(B)$ as a mirror pair. The manifold $X(B, \mathbf{B})$ has a complex structure, $\check{X}(B)$ a symplectic structure. Here we see the origins of the B-field: we obtain one symplectic manifold, corresponding perhaps to some point in the Kähler cone, but a family of complex manifolds parametrised by $H^1(\pi_1(B), (M_\mathbb{R}/M)^{\tilde{\rho}})$. We think of $\mathbf{B}$ as a B-field.

Example 19.11 Let $B = T^n = M_\mathbb{R}/L$. Then $X(B, \mathbf{B})$ is a complex torus and $\check{X}(B)$ is a symplectic torus.

Let's look at the elliptic curve case in more detail. In this case, we take $B = \mathcal{S}^1 = \mathbb{R}/\tau_2\mathbb{Z}$, where τ_2 is a positive real number. Now the B-field lives in $H^1(\pi_1(B), (M_\mathbb{R}/M)^{\tilde{\rho}})$. But $\pi_1(B) = \mathbb{Z}$, $M_\mathbb{R}/M = \mathbb{R}/\mathbb{Z}$, and $\tilde{\rho}$ is the trivial representation. Thus $H^1(\mathbb{Z}, \mathbb{R}/\mathbb{Z})$ is just the set of group homomorphisms $\{\mathbf{B} : \mathbb{Z} \to \mathbb{R}/\mathbb{Z}\}$, and such a homomorphism is determined by $\tau_1 = \mathbf{B}(1)$.

What is $X(B, \mathbf{B})$? This is a quotient of $\tilde{B} \times (M_\mathbb{R}/M) = \mathbb{R} \times \mathcal{S}^1$, with the standard complex structure with coordinate $z = e^{2\pi i(x+iy)}$, where x is a periodic coordinate on $\mathcal{S}^1$ and y is the coordinate on B. This in fact identifies $\tilde{B} \times M_\mathbb{R}/M$ with $\mathbb{C}^*$. We then divide $\tilde{B} \times M_\mathbb{R}/M$ by the $\mathbb{Z}$-action, for $n \in \mathbb{Z}$,

$$(y, x) \mapsto (y + \tau_2 n, x + \tau_1 n).$$

In particular,

$$z \mapsto \omega^n z,$$

where $\omega = e^{2\pi i \tau}$, $\tau = \tau_1 + i\tau_2$. Thus

$$X(B, \mathbf{B}) = \mathbb{C}^*/(z \mapsto \omega^n z).$$

This can also be identified as the elliptic curve $\mathbb{C}/\langle 1, \tau \rangle$ via the exponential map. In particular, we get all elliptic curves this way, with τ living in $\mathcal{H}/\mathbb{Z}$.

What about large complex structure limits? In some cases, if we have a cone of affine structures on B, we can use the above construction of torus bundles to obtain an entire family of complex manifolds.

Let B be a manifold, $\pi : \tilde{B} \to B$ the universal cover, and suppose we are given a subspace $V_\mathbb{R}$ of the vector space of continuous maps $d : \tilde{B} \to M_\mathbb{R}$. Suppose there exists a representation $\tilde{\rho} : \pi_1(B) \to \mathrm{GL}_n(\mathbb{Z})$ such that for each $d \in V_\mathbb{R}$, there exists a representation $\rho_d : \pi_1(B) \to \mathrm{Aff}(M_\mathbb{R})$ such that

$\tilde{\rho}_d = \text{Lin} \circ \rho_d = \tilde{\rho}$, and $\rho_d(\gamma) \circ d \circ T_\gamma = d$ for all $d \in V_\mathbf{R}$. For a given γ, we also then obtain a map

$$\text{Trans}_\gamma : V_\mathbf{R} \to M_\mathbf{R}$$

given by $\text{Trans}_\gamma(d) = \text{Trans} \circ \rho_d(\gamma)$, i.e. the translational part of $\rho_d(\gamma)$. Note that Trans_γ is actually a linear transformation. To see this, observe that if $d_1, d_2 \in V_\mathbf{R}$, $a_1, a_2 \in \mathbf{R}$, then using the fact that the inverse of the affine transformation $x \mapsto Ax + b$ is $x \mapsto A^{-1}x - A^{-1}b$, we get

$$
\begin{aligned}
(a_1 d_1 + a_2 d_2) \circ T_\gamma &= a_1 \rho_{d_1}(\gamma)^{-1} d_1 + a_2 \rho_{d_2}(\gamma)^{-1} d_2 \\
&= a_1 (\tilde{\rho}(\gamma)^{-1} d_1 - \tilde{\rho}(\gamma)^{-1} \text{Trans}_\gamma(d_1)) \\
&\quad + a_2 (\tilde{\rho}(\gamma)^{-1} d_2 - \tilde{\rho}(\gamma)^{-1} \text{Trans}_\gamma(d_2)) \\
&= \tilde{\rho}(\gamma)^{-1} (a_1 d_1 + a_2 d_2) \\
&\quad - \tilde{\rho}(\gamma)^{-1} (a_1 \text{Trans}_\gamma(d_1) + a_2 \text{Trans}_\gamma(d_2)).
\end{aligned}
$$

Thus

$$\rho_{a_1 d_1 + a_2 d_2}(\gamma)(x) = \tilde{\rho}(\gamma)(x) + a_1 \text{Trans}_\gamma(d_1) + a_2 \text{Trans}_\gamma(d_2),$$

so

$$\text{Trans}_\gamma(a_1 d_1 + a_2 d_2) = a_1 \text{Trans}_\gamma(d_1) + a_2 \text{Trans}_\gamma(d_2).$$

Now suppose $V_\mathbf{R}$ contains an (open) cone Σ for which $d \in \Sigma$ implies d is an immersion. Then $d \in \Sigma$ induces an affine structure on B. Thus on the symplectic side, each $d \in \Sigma$ produces a symplectic manifold $\check{X}(B)$. Of course these are all topologically equivalent, but the symplectic form varies, and we can view Σ as analagous to the Kähler cone or a subcone coming from a framing.

On the complex structure side, we can use Σ to produce a family of complex structures analagous to a large complex structure limit. Explicitly, let $B' = B \times \Sigma$. We define an affine structure on B' by the developing map $d' : \tilde{B} \times \Sigma \to M_\mathbf{R} \times V_\mathbf{R}$ given by

$$d'(b, d) = (d(b), d).$$

It is easy to check this is an immersion, and so we just need to calculate how this transforms under T_γ:

$$
\begin{aligned}
d' \circ T_\gamma^{-1}(b, d) &= d'(T_\gamma^{-1}(b), d) \\
&= (d(T_\gamma^{-1}(b)), d) \\
&= (\rho_d(\gamma)(d(b)), d) \\
&= (\tilde{\rho}(\gamma)(d(b)) + \text{Trans}_\gamma(d), d) \\
&= \begin{pmatrix} \tilde{\rho}(\gamma) & \text{Trans}_\gamma \\ 0 & 1 \end{pmatrix} \begin{pmatrix} d(b) \\ d \end{pmatrix}.
\end{aligned}
$$

Thus

$$\rho'(\gamma) \circ d' \circ T_\gamma = d'$$

where $\rho'(\gamma) : M_{\mathbb{R}} \oplus V_{\mathbb{R}} \to M_{\mathbb{R}} \oplus V_{\mathbb{R}}$ is the linear map given by the matrix

$$\begin{pmatrix} \tilde{\rho}(\gamma) & \mathrm{Trans}_\gamma \\ 0 & 1 \end{pmatrix}.$$

In particular, we obtain an affine structure on B'.

When can we talk about $X(B')$? We need $\rho'(\gamma)$ to be integral, and in particular we need an integral structure on $V_{\mathbb{R}}$. Let $V \subseteq V_{\mathbb{R}}$ be a subgroup of

$$\{d \in V_{\mathbb{R}} | \mathrm{Trans}_\gamma(d) \in M \text{ for all } \gamma \in \pi_1(B)\}.$$

In general this group could be trivial, but if the rank of V and the dimension of $V_{\mathbb{R}}$ coincide, then V defines an integral structure on $V_{\mathbb{R}}$. This happens, for example, if $V_{\mathbb{R}}$ has a basis of integral affine structures. Assuming V is of full rank, then using the integral structure $M \oplus V \subseteq M_{\mathbb{R}} \oplus V_{\mathbb{R}}$, it is clear d' is in fact an integral affine transformation.

Thus we obtain $\mathcal{X} := X(B')$. In addition, the projection $B' \to \Sigma$ induces a map on the total spaces $T_{B'} \to T_\Sigma$, and then a map $f : X(B') \to X(\Sigma)$. Note that

$$X(\Sigma) = (V_{\mathbb{R}} + i\Sigma)/V;$$

this looks like the Kähler moduli space of §15.1.

Proposition 19.12 *Let $d_1 + id_2$ represent a point in $X(\Sigma)$, $d_1 \in V_{\mathbb{R}}/V$, $d_2 \in \Sigma$. Then $f^{-1}(d_1 + id_2) \cong X(B, \mathbf{B})$, where B has the affine structure induced by d_2, and $\mathbf{B}$ is the crossed homomorphism given by*

$$\gamma \in \pi_1(B) \mapsto \mathrm{Trans}_\gamma(d_1).$$

Proof. First we check $\mathbf{B}$ is a crossed homomorphism:

$$\begin{aligned} \mathbf{B}(\gamma_1\gamma_2) &= \mathrm{Trans}_{\gamma_1\gamma_2}(d_1) \\ &= \mathrm{Trans}(\rho_{d_1}(\gamma_1\gamma_2)) \\ &= \mathrm{Trans}(\rho_{d_1}(\gamma_1) \circ \rho_{d_1}(\gamma_2)) \\ &= \tilde{\rho}(\gamma_1)\mathrm{Trans}_{\gamma_2}(d_1) + \mathrm{Trans}_{\gamma_1}(d_1), \end{aligned}$$

the last equality following from the behaviour of the translational part of affine transformations under composition. Thus $\mathbf{B}$ is a crossed homomorphism. Now $X(B')$ is a quotient of $(\tilde{B} \times M_{\mathbb{R}}/M) \times (V_{\mathbb{R}}/V \times \Sigma)$ by $\pi_1(B)$, the action being (using the calculated form for ρ')

$$(b, m, d_1, d_2) \mapsto (T_{\gamma^{-1}}(b), \tilde{\rho}(\gamma)(m) + \mathrm{Trans}_\gamma(d_1), d_1, d_2).$$

Thus it is clear that the fibre $f^{-1}(d_1 + id_2)$ is $\tilde{B} \times M_{\mathbb{R}}/M$ modulo the action

$$(b, m) \mapsto (T_{\gamma^{-1}}(b), \tilde{\rho}(\gamma)(m) + \mathbf{B}(\gamma))$$

as desired.   □

We can also study the monodromy of $f : X(B') \to X(\Sigma)$. Note that

$$\pi_1(X(\Sigma)) = \pi_1(V_{\mathbb{R}}/V) \cong V,$$

so, given any $d_1 \in V$, $d_2 \in \Sigma$, we obtain a circle $(\mathbb{R}d_1 / \mathbb{Z}\, d_1) + id_2$ in $X(\Sigma)$. Fix as a basepoint of this circle the point $0 + id_2$. We will determine the monodromy diffeomorphism of $f^{-1}(id_2)$ around this circle. To do this, we need to define a continuous family of diffeomorphisms for $0 \leqslant r \leqslant 1$

$$\phi_r : f^{-1}(id_2) \to f^{-1}(rd_1 + id_2)$$

with ϕ_0 the identity. Then ϕ_1 will be the desired monodromy diffeomorphism. We define ϕ_r by defining

$$\tilde{\phi}_r : B \times M_{\mathbb{R}}/M \to B \times M_{\mathbb{R}}/M$$

by

$$\tilde{\phi}_r(b, m) = (b, m + rd_1(b)).$$

This commutes with the action of γ, i.e.

$$\begin{aligned}
\tilde{\phi}_r(T_{\gamma^{-1}}(b), \tilde{\rho}(\gamma)(m)) &= (T_{\gamma^{-1}}(b), \tilde{\rho}(\gamma)(m) + rd_1(T_{\gamma^{-1}}(b))) \\
&= (T_{\gamma^{-1}}(b), \tilde{\rho}(\gamma)(m) + r\rho_{d_1}(\gamma)(d_1(b))) \\
&= (T_{\gamma^{-1}}(b), \tilde{\rho}(\gamma)(m + rd_1(b)) + \mathrm{Trans}_{\gamma}(rd_1));
\end{aligned}$$

hence $\tilde{\phi}_r$ induces the desired diffeomorphism ϕ_r on the quotient by the actions of $\pi_1(B)$ on the two different copies of $\pi_1(B) \times M_{\mathbb{R}}/M$. Thus the monodromy diffeomorphism which we write as

$$T_{d_1} = \phi_1 : f^{-1}(id_2) \to f^{-1}(id_2)$$

can be interpreted as follows: The graph $\Gamma_{d_1} \subseteq \tilde{B} \times M_{\mathbb{R}}$ of the map $d_1 : \tilde{B} \to M_{\mathbb{R}}$ is easily checked to descend to a section of the torus bundle $f^{-1}(id_2) \to B$, which we denote by $\sigma_{d_1} : B \to f^{-1}(id_2)$. Then T_{d_1} is given by fibrewise translation by this section σ_{d_1}: i.e. if $b \in B$, $v \in T_{B,b}/\Lambda_b$, then

$$T_{d_1}(v) = v + \sigma_{d_1}(b).$$

Example 19.13 Again we look at the elliptic curve. Let $B = S^1 = \mathbb{R}/\mathbb{Z}$, inducing an integral affine structure given by $d : \tilde{B} \to \mathbb{R}$ the identity. This map d generates a one-dimensional space $V_{\mathbb{R}}$ of maps, and we take $V = n\mathbb{Z}\, d$ to define an integral structure on $V_{\mathbb{R}}$. The map d generates a ray Σ in $V_{\mathbb{R}}$. We can take as a basepoint in $X(\Sigma)$ the point id, so $f^{-1}(id) = \mathbb{C}/\langle 1, i\rangle$. Then thinking of the S^1 fibration $f^{-1}(id) \to S^1$ as given by $\mathbb{C}/\langle 1, i\rangle \xrightarrow{\mathrm{Re}} \mathbb{R}/\mathbb{Z}$, nd defines a section defined by the line of slope n through the origin in $\mathbb{C}$. Fibrewise translation then corresponds to the Dehn twist repeated n times. (See Example 16.3.)

19.2   Compactification

So far we have seen how, given a cone $\Sigma \subseteq V_{\mathbb{R}}$ of affine structures on a manifold B and a suitable integral structure $V \subseteq V_{\mathbb{R}}$, we obtain both a family of symplectic manifolds $\check{X}(B)$ parametrized by Σ and a family of complex manifolds $f : X(B') \to X(\Sigma)$. Thus the data B and Σ can be viewed as the underlying data governing both the Kähler side and the complex structure side of mirror symmetry. Unfortunately this doesn't give any really interesting examples. It explains mirror symmetry successfully for complex tori, but does little else. The basic problem is torus bundles by themselves are not particularly complicated topologically. In particular, if X is a simply connected six-manifold, it cannot be realised as a torus bundle $f : X \to B$, since then B would also be simply connected and f must be the trivial bundle. This would contradict X being simply connected.

So in general, we have to allow singular fibres. We approach this as follows:

Definition 19.14 An affine manifold with singularities is a C^0 (topological) manifold B along with a set $\Delta \subseteq B$, which is a finite union of locally closed submanifolds of codimension at least 2, and an affine structure on $B_0 = B \backslash \Delta$.

Given B an affine manifold with singularities with holonomy contained in $\mathbb{R}^n \rtimes \mathrm{GL}_n(\mathbb{Z})$, we obtain fibrations $f_0 : X(B_0) \to B_0$ and $\check{f}_0 : \check{X}(B_0) \to B_0$. We hope to have compactifications

$$
\begin{array}{ccc}
X(B_0) & \subseteq & X(B) \\
\downarrow f_0 & & \downarrow f \\
B_0 & \subseteq & B
\end{array}
$$

and

$$
\begin{array}{ccc}
\check{X}(B_0) & \subseteq & \check{X}(B) \\
\downarrow \check{f}_0 & & \downarrow \check{f} \\
B_0 & \subseteq & B
\end{array}
$$

The fibres of f and $\check{f}$ will not always be submanifolds of $X(B)$ or $\check{X}(B)$, but some singular objects. It is precisely the introduction of these singular fibres which will provide a clear distinction between $X(B)$ and $\check{X}(B)$; in particular, we expect we might have to compactify using different singular fibres for f and $\check{f}$. We will see this explicitly in the next section.

For the moment, let us make some assumptions, and draw some consequences, about these compactifications. We are not yet claiming such compactifications exist, and will only do so in certain cases as *topological* compactifications; we do not discuss here the question of symplectic or complex compactifications.

Let $f : X \to B$ be any continuous map with an open set $B_0 \subseteq B$ such that $f_0 : f^{-1}(B_0) \to B_0$ is a torus bundle. We call such a map a *torus fibration*. Let $i : B_0 \hookrightarrow B$ be the inclusion.

Just as in §16.1, we can consider the sheaves $R^p f_{0*}\mathbb{Q}$ and $R^p f_*\mathbb{Q}$. As in §16.1, $R^p f_{0*}\mathbb{Q}$ is a local system, and a fibre (or stalk) of this local system at a point $b \in B_0$ can be identified with $H^p(f^{-1}(b), \mathbb{Q})$. However, $R^p f_*\mathbb{Q}$ fails to be locally constant at points $b \in B \setminus B_0$, but it is still true that the stalk

$$(R^p f_*\mathbb{Q})_b \cong H^p(f^{-1}(b), \mathbb{Q}).$$

The point is, if $f^{-1}(b)$ is singular, we might expect this cohomology group to be different from that of a non-singular torus.

Definition 19.15 If f is a torus fibration, we say $f : X \to B$ is *simple* if

$$i_* R^p f_{0*}\mathbb{Q} = R^p f_*\mathbb{Q}$$

for all p.

Note by definition, if $U \subseteq B$ is open, then

$$\Gamma(U, i_* R^p f_{0*}\mathbb{Q}) = \Gamma(U \cap B_0, R^p f_{0*}\mathbb{Q}).$$

Thus simplicity essentially requires that if $b \in B \setminus B_0$, then

$$H^p(f^{-1}(b), \mathbb{Q}) \cong \Gamma(U \cap B_0, R^p f_{0*}\mathbb{Q})$$

for U a sufficiently small open neighbourhood of b.

This is quite a powerful assumption. Let's see how this helps us. First consider the cohomology of a single torus V/Λ, where V is an n-dimensional vector space and Λ a rank n lattice. Then

$$H_1(V/\Lambda, \mathbb{Z}) \cong \Lambda$$
$$H^1(V/\Lambda, \mathbb{Z}) \cong \Lambda^\vee$$
$$H^p(V/\Lambda, \mathbb{Z}) \cong \textstyle\bigwedge^p \Lambda^\vee.$$

On the other hand, viewing $V^\vee/\Lambda^\vee$ as the dual torus, where

$$\Lambda^\vee = \{v \in V^\vee | v(\Lambda) \subseteq \mathbb{Z}\},$$

we see $H^p(V^\vee/\Lambda^\vee, \mathbb{Z}) \cong \bigwedge^p \Lambda$. If we choose a generator of $\bigwedge^n \Lambda^\vee \cong \mathbb{Z}$, then the pairing

$$\textstyle\bigwedge^p \Lambda^\vee \times \bigwedge^{n-p} \Lambda^\vee \to \bigwedge^n \Lambda^\vee \xrightarrow{\cong} \mathbb{Z}$$

induces an isomorphism

$$\textstyle\bigwedge^p \Lambda^\vee \cong \bigwedge^{n-p} \Lambda.$$

Thus, up to a choice of generator of $\bigwedge^n \Lambda^\vee$, there is a canonical isomorphism

$$H^p(V/\Lambda, \mathbb{Z}) \cong H^{n-p}(V^\vee/\Lambda^\vee, \mathbb{Z}).$$

This can be done relatively, i.e. let B be an affine manifold with singularities, giving $f_0 : X(B_0) \to B_0$, $\check{f}_0 : \check{X}(B_0) \to B_0$. Here $X(B_0) = T_B/\Lambda$, $\check{X}(B_0) =$

$T_B^*/\Lambda^\vee$, and fibrewise we are in the above situation. Then after a choice of section of $R^n f_{0*}\mathbb{Z}$ (which exists if there is a uniform choice of orientation of the fibres), we obtain isomorphisms

$$R^p f_{0*}\mathbb{Z} \cong R^{n-p} \check{f}_{0*}\mathbb{Z},$$

which also holds over $\mathbb{Q}$,

$$R^p f_{0*}\mathbb{Q} \cong R^{n-p} \check{f}_{0*}\mathbb{Q}.$$

If f_0 and $\check{f}_0$ have simple compactifications $f : X(B) \to B$ and $\check{f} : \check{X}(B) \to B$, then the induced isomorphisms

$$i_* R^p f_{0*}\mathbb{Q} \cong i_* R^{n-p} \check{f}_{0*}\mathbb{Q}$$

become

$$R^p f_*\mathbb{Q} \cong R^{n-p} \check{f}_*\mathbb{Q}.$$

This is immensely useful for determining the relationship between the cohomology of $X(B)$ and $\check{X}(B)$ using the Leray spectral sequence.

For $f : X \to B$ any continuous map, the Leray spectral sequence is

$$E_2^{p,q} = H^p(B, R^q f_*\mathbb{Q}) \Rightarrow E_\infty^n = H^n(X, \mathbb{Q}).$$

We will now explain in detail what this means. We can put the groups $H^p(B, R^q f_*\mathbb{Q})$ in an array:

$$\vdots$$
$$H^0(B, R^1 f_*\mathbb{Q}) \ H^1(B, R^1 f_*\mathbb{Q}) \ \cdots$$
$$H^0(B, R^0 f_*\mathbb{Q}) \ H^1(B, R^0 f_*\mathbb{Q}) \ \cdots$$

The key point of a spectral sequence is that there exist maps

$$d_2 : H^p(B, R^q f_*\mathbb{Q}) \to H^{p+2}(B, R^{q-1} f_*\mathbb{Q}).$$

This map is zero if p or $q-1$ is negative. But there is more: the composition of two of these maps is always zero, and if we set

$$E_3^{p,q} = \frac{\ker(d_2 : H^p(B, R^q f_*\mathbb{Q}) \to H^{p+2}(B, R^{q-1} f_*\mathbb{Q}))}{\mathrm{Im}(d_2 : H^{p-2}(B, R^{q+1} f_*\mathbb{Q}) \to H^p(B, R^q f_*\mathbb{Q}))},$$

then we also obtain maps

$$d_3 : E_3^{p,q} \to E_3^{p+3,q-2}.$$

(We do not explain where any of these maps come from; this is part of the magic of spectral sequences.) We can carry on like this, and at the nth step we have groups $E_n^{p,q}$ along with maps

$$d_n : E_n^{p,q} \to E_n^{p+n,q-n+1}$$

and

$$E_{n+1}^{p,q} = \frac{\ker(d_n : E_n^{p,q} \to E_n^{p+n,q-n+1})}{\mathrm{Im}(d_n : E_n^{p-n,q+n-1} \to E_n^{p,q})}.$$

Since $E_n^{p,q} = 0$ for $p < 0$ or $q < 0$, it follows that for a given p, q there is a sufficiently large n such that $E_n^{p,q} = E_{n+1}^{p,q} = \cdots$. If we set $E_\infty^{p,q}$ to be this fixed group, independent of n for large n, then the magic of spectral sequences says the following: for each n, there is a filtration

$$H^n(X, \mathbb{Q}) = F^0 \supseteq \cdots \supseteq F^n \supseteq 0$$

such that

$$F^p / F^{p+1} \cong E_\infty^{p,n-p}.$$

In other words, the Leray spectral sequence should be viewed as a machine that takes as input the cohomology groups of certain sheaves on B, and after turning a crank, produces the graded pieces of a filtration of the cohomology of X.

The general difficulty is that we do not know what these maps d_n are, and even the groups we begin with may be hard to compute.

Let's see what this gives us for the maps $f : X(B) \to B$, $\check{f} : \check{X}(B) \to B$ as above. First, $R^0 f_{0*} \mathbb{Q} = f_{0*} \mathbb{Q} = \mathbb{Q}$, and we also assume that $R^3 f_{0*} \mathbb{Q} \cong \mathbb{Q}$ also, so that we can choose a global section of $R^3 f_{0*} \mathbb{Q}$. Then by simplicity,

$$R^p f_* \mathbb{Q} = i_* R^p f_{0*} \mathbb{Q} = i_* \mathbb{Q} = \mathbb{Q}$$

for $p = 0, 3$. (Note that $i_* \mathbb{Q} = \mathbb{Q}$ follows from the fact that $B \setminus B_0$ is codimension two!) Second, if we further assume B is simply connected (say $B = S^3$), then we know $H^p(B, R^q f_* \mathbb{Q})$ for $q = 0, 3$. This gives us the E_2 terms in the Leray spectral sequence for f and $\check{f}$:

$$
\begin{array}{cccc}
\mathbb{Q} & 0 & 0 & \mathbb{Q} \\
H^0(B, R^2 f_* \mathbb{Q}) & H^1(B, R^2 f_* \mathbb{Q}) & H^2(B, R^2 f_* \mathbb{Q}) & H^3(B, R^2 f_* \mathbb{Q}) \\
H^0(B, R^1 f_* \mathbb{Q}) & H^1(B, R^1 f_* \mathbb{Q}) & H^2(B, R^1 f_* \mathbb{Q}) & H^3(B, R^1 f_* \mathbb{Q}) \\
\mathbb{Q} & 0 & 0 & \mathbb{Q}
\end{array}
$$

and

$$
\begin{array}{cccc}
\mathbb{Q} & 0 & 0 & \mathbb{Q} \\
H^0(B, R^2 \check{f}_* \mathbb{Q}) & H^1(B, R^2 \check{f}_* \mathbb{Q}) & H^2(B, R^2 \check{f}_* \mathbb{Q}) & H^3(B, R^2 \check{f}_* \mathbb{Q}) \\
H^0(B, R^1 \check{f}_* \mathbb{Q}) & H^1(B, R^1 \check{f}_* \mathbb{Q}) & H^2(B, R^1 \check{f}_* \mathbb{Q}) & H^3(B, R^1 \check{f}_* \mathbb{Q}) \\
\mathbb{Q} & 0 & 0 & \mathbb{Q}
\end{array}
$$

These two arrays have a symmetry, i.e. natural isomorphisms

$$H^p(B, R^q f_* \mathbb{Q}) \cong H^p(B, R^{3-q} \check{f}_* \mathbb{Q})$$

which reverses the order of the rows. This is analagous to the symmetry of the Hodge diamond for mirror pairs of Calabi–Yau 3-folds.

Now consider in the first diagram the $E_2^{0,1}$ term, $H^0(B, R^1 f_* \mathbb{Q})$. The two maps with domain or range $E_2^{0,1}$ are both zero, so $E_2^{0,1} = E_\infty^{0,1}$. Also, $E_\infty^{1,0} = 0$. Thus there exists a filtration of $H^1(X(B), \mathbb{Q}) = F^0 \supseteq F^1$ with $F^0/F^1 = E_2^{0,1}$ and $F^1 = E_2^{1,0} = 0$.

Suppose $H^1(X(B), \mathbb{Q}) = 0$. Then we conclude $H^0(B, R^1 f_* \mathbb{Q}) = 0$. A similar argument shows that, since $H^5(X(B), \mathbb{Q}) = 0$ by Poincaré duality, we also have $H^3(B, R^2 f_* \mathbb{Q}) = 0$.

If furthermore $H^1(\check{X}(B), \mathbb{Q}) = 0$, then similarly

$$H^0(B, R^1 \check{f}_* \mathbb{Q}) = H^3(B, R^2 \check{f}_* \mathbb{Q}) = 0.$$

Combining this with the symmetry between the groups for f and $\check{f}$, we see that E_2 terms for f and $\check{f}$ look like

$$
\begin{array}{cccc}
\mathbb{Q} & 0 & 0 & \mathbb{Q} \\
0 \;\; H^1(B, R^2 f_* \mathbb{Q}) & H^2(B, R^2 f_* \mathbb{Q}) & 0 \\
0 \;\; H^1(B, R^1 f_* \mathbb{Q}) & H^2(B, R^1 f_* \mathbb{Q}) & 0 \\
\mathbb{Q} & 0 & 0 & \mathbb{Q}
\end{array}
$$

and

$$
\begin{array}{cccc}
\mathbb{Q} & 0 & 0 & \mathbb{Q} \\
0 \;\; H^1(B, R^2 \check{f}_* \mathbb{Q}) & H^2(B, R^2 \check{f}_* \mathbb{Q}) & 0 \\
0 \;\; H^1(B, R^1 \check{f}_* \mathbb{Q}) & H^2(B, R^1 \check{f}_* \mathbb{Q}) & 0 \\
\mathbb{Q} & 0 & 0 & \mathbb{Q}
\end{array}
$$

The picture is almost complete. In the first spectral sequence there are at most two non-zero maps

$$d_2 : H^0(B, R^3 f_* \mathbb{Q}) \to H^2(B, R^2 f_* \mathbb{Q})$$

and

$$d_2' : H^1(B, R^1 f_* \mathbb{Q}) \to H^3(B, f_* \mathbb{Q}).$$

To show these are zero, we introduce one more assumption: suppose f has a section $\sigma : B \to X(B)$. Certainly $f_0 : X(B_0) \to B_0$ has a section, so this is a property of the compactification $X(B)$. Note that $E_\infty^{3,0}$ is a quotient of $E_2^{3,0} = H^3(B, \mathbb{Q})$ and $E_\infty^{3,0} \subseteq H^3(X, \mathbb{Q})$. A general fact about the Leray spectral sequence is that the composition

$$E_2^{3,0} \to E_\infty^{3,0} \hookrightarrow H^3(X, \mathbb{Q})$$

is the pull-back

$$f^* : H^3(B, \mathbb{Q}) \to H^3(X, \mathbb{Q}).$$

But the composition

$$H^3(B, \mathbb{Q}) \xrightarrow{f^*} H^3(X, \mathbb{Q}) \xrightarrow{\sigma^*} H^3(B, \mathbb{Q})$$

is the identity since $f \circ \sigma$ is the identity, so f^* is injective. Thus $d_2' = 0$.

Similarly, $E_\infty^{0,3}$ is a subgroup of $E_2^{0,3}$, and the composed map

$$H^3(X,\mathbb{Q}) \to E_\infty^{0,3} \to E_2^{0,3} = H^0(X, R^3 f_* \mathbb{Q})$$

is the natural map as follows: $R^3 f_* \mathbb{Q}$ is the sheaf associated to the presheaf

$$U \mapsto H^3(f^{-1}(U), \mathbb{Q}),$$

so in particular there is a natural map $H^3(f^{-1}(B), \mathbb{Q}) \to \Gamma(B, R^3 f_* \mathbb{Q})$. Clearly the Poincaré dual class of $\sigma(B)$ in $H^3(X, \mathbb{Q})$ restricts to a non-zero element of $H^0(B, R^3 f_{0*}\mathbb{Q})$. Thus $d_2 = 0$ also. So all the maps in E_2 are zero, and there is no room for maps in E_n for $n > 2$. Thus $E_2^{p,q} = E_\infty^{p,q}$. In such a case we say the spectral sequence *degenerates* at the E_2 term.

Putting this all together, we see that with the various assumptions on f and $\check{f}$ we have isomorphisms

$$H^0(B,\mathbb{Q}) \cong H^0(X(B),\mathbb{Q})$$
$$H^1(B, R^1 f_* \mathbb{Q}) \cong H^2(X(B),\mathbb{Q})$$
$$H^2(B, R^2 f_* \mathbb{Q}) \cong H^4(X(B),\mathbb{Q})$$
$$H^3(B, R^3 f_* \mathbb{Q}) \cong H^6(X(B),\mathbb{Q}).$$

In addition, we have a filtration

$$H^3(X(B),\mathbb{Q}) = F^0 \supseteq F^1 \supseteq F^2 \supseteq F^3 \supseteq 0$$

with

$$F^0/F^1 \cong H^0(B, R^3 f_* \mathbb{Q})$$
$$F^1/F^2 \cong H^1(B, R^2 f_* \mathbb{Q})$$
$$F^2/F^3 \cong H^2(B, R^1 f_* \mathbb{Q})$$
$$F^3 \cong H^3(B, f_* \mathbb{Q}).$$

Now $\dim H^2(\check{X}(B),\mathbb{Q}) = \dim H^4(\check{X}(B),\mathbb{Q})$ by Poincaré duality, so

$$\dim F^1/F^2 = \dim F^2/F^3$$

also. Putting this all together, we get

Theorem 19.16 *Let B be an affine manifold with singularities, and suppose there exist compactifications $f : X(B) \to B$ and $\check{f} : \check{X}(B) \to B$ of $f_0 : X(B_0) \to B_0$ and $\check{f}_0 : \check{X}(B_0) \to B_0$ that are simple. Suppose furthermore that $R^3 f_{0*}\mathbb{Q}$ has a non-trivial section, f and $\check{f}$ both have sections, and $H^1(X(B),\mathbb{Q}) = H^1(\check{X}(B),\mathbb{Q}) = 0$. Then*

$$\dim H^{2*}(X(B),\mathbb{Q}) = \dim H^3(\check{X}(B),\mathbb{Q})$$

and

$$\dim H^{2*}(\check{X}(B), \mathbb{Q}) = \dim H^3(X(B), \mathbb{Q}).$$

Here H^{2*} denotes even cohomology. In particular, if $X(B)$ and $\check{X}(B)$ have complex structures making them Calabi–Yau 3-folds, then

$$h^{1,1}(X(B)) = h^{1,2}(\check{X}(B))$$
$$h^{1,2}(X(B)) = h^{1,1}(\check{X}(B)).$$

This analysis of the cohomology of $X(B)$ and $\check{X}(B)$ raises an interesting question: what is the filtration on H^3 provided by the Leray spectral sequence? Looking at the monodromy weight filtration in families of compactifications provides an answer.

Suppose, as in §19.1, we have a cone $\Sigma \subseteq V_{\mathbb{R}}$ of affine structures on $B_0 \subseteq B$, with an integral structure $V \subseteq V_{\mathbb{R}}$. We then obtain a family of manifolds $g_0 : X(B_0') \to X(\Sigma)$. Assume that we can compactify this family

$$
\begin{array}{ccc}
X(B_0') & \subseteq & X(B') \\
\downarrow{\scriptstyle g_0} & & \downarrow{\scriptstyle g} \\
X(\Sigma) & = & X(\Sigma)
\end{array}
$$

so that a fibre $g^{-1}(d_1 + id_2)$ is a simple compactification of the torus bundle $X(B_0, \mathbf{B}) \to B_0$ for appropriate $\mathbf{B}$ determined by d_1. Now we saw that, if $d_1 \in V$, then the monodromy diffeomorphism around the circle $\mathbb{R}d_1 / \mathbb{Z} d_1 \times \{id_2\}$ is

$$T_{d_1} : g_0^{-1}(id_2) \to g_0^{-1}(id_2)$$

given by

$$T_{d_1}(v) = v + \sigma_{d_1}(f_0(v)),$$

where $f_0 : g_0^{-1}(id_2) \to B_0$ is the torus bundle structure on $g_0^{-1}(id_2)$, and $\sigma_{d_1} : B_0 \to g_0^{-1}(id_2)$ is the section of f_0 determined by d_1.

Let us assume that our compactification $X(B')$ of $X(B_0')$ is chosen so that

$$T_{d_1} : g_0^{-1}(id_2) \to g_0^{-1}(id_2)$$

extends continuously to a homeomorphism

$$T_{d_1} : g^{-1}(id_2) \to g^{-1}(id_2),$$

and that this is the monodromy diffeomorphism of the family $g : X(B') \to X(\Sigma)$.

This is in fact not as complicated an assumption as it may appear. As torus bundles over B_0, all the fibres of g_0 are diffeomorphic. Choose some $d_2 \in \Sigma$ and set $X_0 = X(B_0)$, where B_0 takes the affine structure given by d_2. If we have a simple compactification $X_0 \subseteq X$ such that $T_d : X_0 \to X_0$

extends to $T_d : X \to X$ for any $d \in V$, then we can construct $X(B')$ as the quotient of $X \times V_{\mathbb{R}} \times \Sigma$ by the action of V given by $d \in V$ acting by

$$(x, d_1, d_2) \mapsto (T_d(x), d_1 + d, d_2).$$

Thus, if we would like to understand the monodromy weight filtration of the family $g : X(B') \to X(\Sigma)$, we just need to understand the action of T_d on cohomology.

For simplicity we will restrict to the 3-fold case, but see [82] for somewhat more general results.

Theorem 19.17 *Let B, $f : X(B) \to B$, $\check{f} : \check{X}(B) \to B$ satisfy the hypotheses of Theorem 19.16, and suppose $\sigma : B \to X(B)$ is a section inducing a homeomorphism*

$$T_\sigma : X(B) \to X(B)$$

given by

$$T_\sigma(v) = v + \sigma(f_0(v))$$

on $X(B_0)$. Then

$$T_\sigma^* : H^3(X(B), \mathbb{Q}) \to H^3(X(B), \mathbb{Q})$$

satisfies

$$(T_\sigma^* - I)(F^i) \subseteq F^{i+1},$$

inducing maps

$$T_\sigma^* - I : F^i/F^{i+1} = H^i(B, R^{3-i} f_* \mathbb{Q}) \to F^{i+1}/F^{i+2} = H^{i+1}(B, R^{2-i} f_* \mathbb{Q}).$$

Furthermore, there exists an integral class D_σ in $H^2(\check{X}(B), \mathbb{Q})$ such that the diagram

$$
\begin{array}{ccc}
H^i(B, R^{3-i} f_* \mathbb{Q}) & \xrightarrow{T_\sigma^* - I} & H^{i+1}(B, R^{2-i} f_* \mathbb{Q}) \\
\downarrow{\scriptstyle \cong} & & \downarrow{\scriptstyle \cong} \\
H^{2i}(\check{X}(B), \mathbb{Q}) & \xrightarrow{\cdot D_\sigma} & H^{2i+2}(\check{X}(B), \mathbb{Q})
\end{array}
$$

is commutative up to sign.

Proof. See [82], Theorem 4.1.   □

This tells us how to interpret the Leray filtration on $H^3(X(B), \mathbb{Q})$. If T_σ is as in Theorem 19.17, let $N_\sigma = \log T_\sigma^*$. Then N_σ acts as $T_\sigma^* - I$ does on the graded pieces of the filtration $F^\bullet$. Let $D_\sigma \in H^2(\check{X}(B), \mathbb{Q})$ be the corresponding class given by Theorem 19.17. Suppose $\cdot D_\sigma : H^2(\check{X}(B), \mathbb{Q}) \to H^4(\check{X}(B), \mathbb{Q})$ is an isomorphism (which would be the case if D_σ were the class of an ample divisor), and $D_\sigma^3 \neq 0$. Then we see that using the construction of the weight filtration of §16.3,

$$W_0 = \operatorname{Im} N_\sigma^3 = F^3 = H^3(B, \mathbb{Q})$$
$$W_5 = \ker N_\sigma^3 = F^1.$$

We then get the induced map

$$N'_\sigma : W_5/W_0 = F^1/F^3 \to F^1/F^3,$$

and then

$$W_1/W_0 = \mathrm{Im}(N'_\sigma)^2 = W_0/W_0,$$
$$W_4/W_0 = \ker(N'_\sigma)^2 = W_5/W_0,$$

so $W_1 = W_0$ and $W_4 = W_5$. Finally, $W_2/W_1 = \mathrm{Im}(N'_\sigma) = F^2/F^3$ by the assumption that $\cdot D_\sigma$ is an isomorphism, and $W_3/W_1 = \ker(N'_\sigma) = F^2/F^3$, so $W_2 = W_3$. Thus, in this case, we see

$$W_{2i}/W_{2i-2} \cong F^{3-i}/F^{4-i}.$$

This gives the relationship between the Leray filtration and the weight filtration.

If we are in the situation of having a cone $\Sigma \subseteq V_\mathbb{R}$ of affine structures, and every $d \in V$ yields a section σ_d inducing a diffeomorphism $T_{\sigma_d} : X(B) \to X(B)$, then we get a corresponding element $D_d \in H^2(\check{X}(B), \mathbb{Q})$, and hence a map

$$V \to H^2(\check{X}(B), \mathbb{Q})$$

given by $d \mapsto D_d$. One can show this map is in fact linear. If this map is an isomorphism after tensoring with $\mathbb{Q}$, we should view the induced family $X(B') \to X(\Sigma)$ as a topological form of a large complex structure limit. Note that if $\beta_0^* \in H^3(X(B), \mathbb{Q})$ is Poincaré dual to a section of $X(B) \to B$ and α_0^* is Poincaré dual to a class of a fibre, then $(\beta_0^*, \alpha_0^*) = \pm 1$ (depending on orientation), β_0^* generates F^0/F^1, and α_0^* generates F^3. If $d_1, d_2, d_3 \in V$, then by Theorem 19.17

$$(\beta_0^*, N_{\sigma_{d_1}} N_{\sigma_{d_2}} N_{\sigma_{d_3}} \beta_0^*) = \pm D_{d_1} \cdot D_{d_2} \cdot D_{d_3}.$$

Comparing this with equation (31), we see that the Strominger–Yau–Zaslow conjecture explains why the constant term of the $(1,1)$ and $(1,2)$ Yukawa couplings agree in mirror pairs.

19.3   A Topological Construction of the Quintic and its Mirror

Finally, we end with an actual example of the phenomena discussed in §19.2. We construct B_0 and B as follows: Let $\Xi \subseteq \mathbb{R}^4$ be the convex hull of the points

$$P_0 = (-1, -1, -1, -1),$$
$$P_1 = (4, -1, -1, -1),$$
$$P_2 = (-1, 4, -1, -1),$$
$$P_3 = (-1, -1, 4, -1),$$
$$P_4 = (-1, -1, -1, 4).$$

So $\varXi$ is a simplex, and we set $B = \partial \varXi$, homeomorphic to $\mathcal{S}^3$.

For each two-face σ_{ijk} of $\varXi$ spanned by P_i, P_j, P_k, choose a triangulation of this two-face using only integral points as vertices. For convenience, we will choose the regular triangulation in Figure 3. Let $\varDelta_{ijk} \subseteq \sigma_{ijk}$ be the 1-skeleton

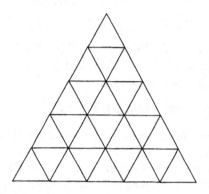

Figure 3. Triangulation of a 2-face of $\varXi$

of the dual cell complex of this triangulation, i.e. the graph shown superimposed on top of the triangulation in Figure 4. Finally, let $\varDelta = \bigcup_{\{i,j,k\}} \varDelta_{ijk}$,

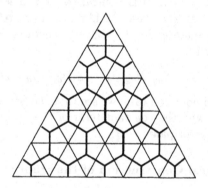

Figure 4. Triangulation and its dual complex

and put $B_0 = B \setminus \varDelta$.

Note that $\varDelta$ is a trivalent graph, and there are two sorts of vertices: 25 vertices in the interior of each face, which we will refer to as face vertices, and 5 vertices contained in each edge, which we will refer to as edge vertices.

Now we will define an affine structure on B_0 by giving an explicit set of charts. For each integral point x in a two-face σ_{ijk}, choose a small open neighbourhood U_x of x with the property that if

$$T = \bigcup_{\{i,j,k\}} \sigma_{ijk},$$

then $U_x \cap T$ is the connected component of $T \setminus \Delta$ containing x, and

$$U_x \subseteq \left(\bigcup_{\substack{\text{3-faces } \sigma \\ \text{containing } x}} Int(\sigma) \right) \cup (U_x \cap T).$$

Here $Int(\sigma)$ means the interior of a 3-face σ. Then the collection of open sets

$$\mathcal{U} = \{Int(\sigma) | \sigma \text{ is a 3-face of } \Xi\} \cup \{U_x | x \in T \cap \mathbb{Z}^4\}$$

form an open covering of B_0. Define charts

$$\psi_x : U_x \to \mathbb{R}^4 / \mathbb{R}\, x$$

via projection. It is easy to check ψ_x is a homeomorphism onto its image and hence ψ_x gives a coordinate chart on U_x.
We also have charts

$$\psi_\sigma : Int(\sigma) \to \mathbb{R}^3 = \text{affine span of } \sigma \text{ in } \mathbb{R}^4$$

given by inclusion. One can check these charts define an integral affine structure on B_0. (See Exercise 19.3 for details).

Theorem 19.18 *There exists compactifications $f : X(B) \to B$ and $\check{f} : \check{X}(B) \to B$ of $f_0 : X(B_0) \to B_0$ and $\check{f}_0 : \check{X}(B_0) \to B_0$ that satisfy the assumptions of Theorem 19.16. Furthermore, $\check{X}(B)$ is homeomorphic to a quintic 3-fold in $\mathbb{P}^4$, and $X(B)$ is homeomorphic to some crepant resolution of the 3-fold Y_ψ of §18.2.*

Proof. See [86].   □

We will not give many details of this compactification, but merely describe the singular fibres of f and $\check{f}$.
Over an edge of the graph Δ, the singular fibres of both f and $\check{f}$ take a generic form, homeomorphic to $S^1 \times F$, where F is obtained from a 2-torus $S^1 \times S^1$ by contracting $\{pt\} \times S^1$ to a point. Note that the topological Euler characteristic χ of such a fibre is 0.
There are two sorts of fibres appearing over vertices of Δ.
The first, with $\chi = 1$, which we call positive fibres, is obtained by contracting a $\{pt\} \times T^2$ on $S^1 \times T^2$ to a point.

The second, negative fibres, with $\chi = -1$, can be described as follows. Write $T^3 = \mathbb{R}^3 / \mathbb{Z}^3$, and divide T^3 by the equivalence relation $(x_1, x_2, x_3) \equiv (x_1, x_2, x_3')$ if and only if $x_3 = x_3'$ or $x_1 \in \mathbb{Z}$ or $x_2 \in \mathbb{Z}$. In this case the singular locus of the fibre is a figure 8.

The important point about the compactification is that $f : X(B) \to B$ has positive fibres over face vertices and negative fibres over the edge vertices, whereas for $\check{f} : \check{X}(B) \to B$ this is reversed. The dualizing process interchanges positive and negative fibres, and this is perhaps at the heart of the topological change between a Calabi–Yau 3-fold and its mirror.

Note that $\check{X}(B_0)$ is a symplectic manifold. I expect it should be possible to extend this symplectic structure to $\check{X}(B)$, obtaining a Lagrangian fibration $\check{f} : \check{X}(B) \to B$. This may coincide with the construction of Lagrangian fibrations proposed in [176]. However, one should not expect an extension of the complex structure on $X(B_0)$ to $X(B)$. Rather, only some "small" deformation of the complex structure on $X(B_0)$ should extend. It is precisely in this small deformation that the explanation, the secret, of the Yukawa couplings should lie. See [136], [62], and [87] for further ideas in this direction.

19.4   Exercises

19.1 Let B be an affine manifold with holonomy representation ρ such that $\tilde{\rho} = \text{Lin} \circ \rho$ maps $\pi_1(B)$ into $SL_n(\mathbb{Z})$. Show that $X(B, \mathbf{B})$ has a nowhere vanishing holomorphic n-form, locally given by $\frac{dz_1 \wedge \cdots \wedge dz_n}{z_1 \cdots z_n}$ in the coordinates of Remark 19.10.

19.2 Check the following details from the text.

(a) In Definition 19.8, show the action of γ on $\tilde{B} \times (M_\mathbb{R}/M)$ is indeed holomorphic.

(b) Check that $X(B, \mathbf{B}) \cong X(B, \mathbf{B}')$ if the compositions

$$\mathbf{B} : \pi_1(B) \to M_\mathbb{R}^{\tilde{\rho}} \to (M_\mathbb{R}/M)^{\tilde{\rho}}$$

and

$$\mathbf{B}' : \pi_1(B) \to M_\mathbb{R}^{\tilde{\rho}} \to (M_\mathbb{R}/M)^{\tilde{\rho}}$$

define the same element of $H^1(\pi_1(B), (M_\mathbb{R}/M)^{\tilde{\rho}})$.

19.3 Check the definition of the affine structure on B_0 in §19.3 gives a genuine integral structure by checking that

$$\psi_v \circ \psi_\sigma^{-1} : Span(Int(\sigma)) \to \mathbb{R}^4/v\mathbb{R}$$

is integral whenever v is an integral point in σ. What property of the polytope Ξ did this depend on?

19.4(a) Determine (part of) the linear part of the holonomy representation of the affine manifold B_0 defined in §19.3 as follows: Let σ, σ' be two 3-faces of Ξ, $v, v' \in \sigma \cap \sigma' \cap \mathbb{Z}^4$. Let γ be a loop that starts in $Int(\sigma)$, passes through v into σ', then through v' back into σ to the starting point, so that $\gamma \cap \sigma \cap \sigma' = \{v, v'\}$. Compute the linear part of the holonomy around γ.

(b) Using the answer of part (a), compare the holonomy about loops around three edges coming out of a face vertex and the holonomy about loops around three edges coming out of an edge vertex. What is the difference?

Part III

Compact Hyperkähler Manifolds

Daniel Huybrechts

Ricci-flat manifolds have been studied for many years and the interest in various aspects of their theory seems still to be growing. Their rich geometry has been explored with techniques from different branches of mathematics and the interplay of analysis, arithmetic, and geometry makes their theory highly attractive. At the same time, these manifolds play a prominent rôle in string theory and, in particular, in mirror symmetry. Physicists have come up with interesting and difficult mathematical questions and have suggested completely new directions that should be pursued.

An interesting class of Ricci-flat manifolds is provided by compact Kähler manifolds with vanishing first Chern class. Due to the decomposition theorem one knows that any such manifold (up to a finite cover) can be written as the product of complex tori, Calabi–Yau, and compact hyperkähler manifolds. In these notes we concentrate on the latter. From a differential geometric point of view these are compact Riemannian manifolds with holonomy $\mathrm{Sp}(n)$. From an algebraic point of view one studies simply-connected compact Kähler manifolds that admit a unique (up to scale) everywhere non-degenerate holomorphic two-form. Thus, the two names, compact hyperkähler and irreducible holomorphic symplectic, just emphasize different aspects of the same type of manifolds. A combination of the corresponding techniques allows one to develop a general theory which parallels to a large extent the classical theory of complex tori, abelian varieties, and K3 surfaces.

The main theme of these notes is the study of compact hyperkähler manifolds by means of their periods. The period of a compact hyperkähler manifold is given by the second cohomology endowed with its Hodge structure and the Beauville–Bogomolov quadratic form. It will also become apparent, I hope, that in many aspects compact hyperkähler manifolds behave very much like complex tori. Calabi–Yau manifolds are very different and often much harder to study. One striking difference between their theories is the number of examples that are available. This might be due to the lack of techniques to construct hyperkähler manifolds, but it might also just reflect the fact that compact hyperkähler manifolds are very rigid, like complex tori.

Although nothing really works without the hyperkähler metric, the metric itself will almost never appear explicitly. In fact, no explicit example on a compact manifold has ever been described. The reader familiar with e.g. hyperkähler quotients will soon realize that for the time being the theories of compact and non-compact hyperkähler manifolds have very little in common.

These notes grew out of eight lectures given at the Nordfjordeid summer school. I have tried to cover large parts of the existing theory. However, a number of important subjects have not be included. Almost nothing is said about moduli spaces of bundles (e.g. O'Grady's examples), Rozansky-Witten invariants, symplectic singularities, local aspects of the theory as developed by Kaledin, and moduli spaces of hyperkähler manifolds as they appear in CFT. I must also warn the reader that the structure of these notes is somewhat non-linear. Frequently, we make use of results that are only proved in later sections or sometimes not at all. Results for which a proof or at least a sketch of a proof is given are stated as 'Propositions'. Those results that are beyond these notes are called 'Theorems'.

The theory has many loose ends. E.g. the relation between compact hyperkähler manifolds and abelian varieties via integrable systems has not been explored, their topology is still a great mystery, a classification seems out of reach, and an appropriate version of the Global Torelli theorem is completely missing. I hope that this text might stimulate further research in these directions.

I would like to thank D. Kaledin, E. Markman, M. Nieper-Wißkirchen, D. Ploog, and especially J. Sawon for reading through preliminary versions of these notes.

21   Holomorphic Symplectic Manifolds

Definition 21.1 An *irreducible holomorphic symplectic manifold* is a simply-connected compact Kähler manifold X, such that $H^0(X, \Omega_X^2)$ is generated by an everywhere non-degenerate holomorphic two-form σ.

Since an irreducible holomorphic symplectic manifold is in particular a compact Kähler manifold, Hodge decomposition holds, i.e.

$$H^k(X, \mathbb{C}) = \bigoplus_{p+q=k} H^{p,q}(X)$$

with $H^{p,q}(X) = H^q(X, \Omega_X^p)$. Since X is simply-connected, this yields in particular $H^0(X, \Omega_X) = H^1(X, \mathcal{O}_X) = 0$ and in degree two

$$H^2(X, \mathbb{C}) = H^{2,0}(X) \oplus H^{1,1}(X) \oplus H^{0,2}(X)$$
$$= \mathbb{C}\sigma \oplus H^{1,1}(X) \oplus \mathbb{C}\bar{\sigma}.$$

The existence of an everywhere non-degenerate two-form $\sigma \in H^0(X, \Omega_X^2)$ implies that the manifold has even complex dimension $\dim_{\mathbb{C}}(X) = 2n$. Moreover, σ induces an alternating homomorphism $\sigma : \mathcal{T}_X \to \Omega_X$. Since the two-form is everywhere non-degenerate, this homomorphism is bijective. Thus, the tangent bundle and cotangent bundle of an irreducible holomorphic symplectic manifold are isomorphic. Moreover, the canonical bundle $K_X = \Omega_X^{2n}$ is trivialized by the $(2n, 0)$-form σ^n. Thus, an irreducible holomorphic symplectic manifold has trivial canonical bundle and, therefore, vanishing first Chern class $c_1(X)$.

A general irreducible holomorphic symplectic manifold is not projective. However, those that are projective are dense in the moduli space (cf. Proposition 26.6). From this point of view, irreducible holomorphic symplectic manifold are very different from Calabi–Yau manifolds and behave more like complex tori. Indeed, as a Calabi–Yau manifold is Kähler and satisfies $H^{2,0} = 0$, it is automatically projective (cf. Exercise 26.1).

21.1 K3 Surfaces

Two-dimensional examples of irreducible holomorphic symplectic manifolds are called K3 surfaces. The official definition of a K3 surface goes as follows.

Definition 21.2 A *K3 surface* is a compact complex surface with trivial canonical bundle and $H^1(X, \mathcal{O}_X) = 0$.

It is a deep result due to Siu [183] that any K3 surface is actually Kähler. In higher dimensions the Kähler condition cannot be dropped. Counterexamples have been found by Guan [88] (cf. [22]) and Yoshioka [208] (cf. Example 21.9). As the canonical bundle $K_X = \Omega_X^2$ of a K3 surface X is trivial, there exists a unique (up to scale) everywhere non-degenerate holomorphic symplectic structure. Note that the definition does not explicitly require that a K3 surface is simply-connected. This can be shown be deforming any given K3 surface to a quartic hypersurface in $\mathbb{P}^3$ or to a Kummer surface.

Example 21.3 The adjunction formula shows that any smooth hypersurface of degree four in $\mathbb{P}^3$ has trivial canonical bundle. By the Lefschetz hyperplane theorem one knows that any such hypersurface has vanishing $H^1(X, \mathcal{O}_X)$ or, more precisely, that it is simply-connected. Thus, a smooth quartic hypersurface in $\mathbb{P}^3$ is an example of a (projective) K3 surface. The most prominent example of such a hypersurface is the so called *Fermat quartic*

$$\{(x_0 : x_1 : x_2 : x_3) \mid x_0^4 + x_1^4 + x_2^4 + x_3^4 = 0\} \subset \mathbb{P}^3.$$

It can be shown that any K3 surface X such that $\mathrm{Pic}(X) = H^2(X, \mathbb{Z}) \cap H^{1,1}(X)$ is generated by a class α with $\alpha^2 = 4$ is isomorphic to a quartic hypersurface in $\mathbb{P}^3$ (see [16]).

Example 21.4 A *Kummer surface* X is a by definition the minimal resolution $X \to (\mathbb{C}^2/\gamma)/\pm$ of the quotient of a complex torus $\mathbb{C}^2/\gamma$ by the natural involution $x \mapsto -x$. Since there are exactly 16 two-division points, one finds 16 smooth rational curves $C_1, \ldots, C_{16} \subset X$ as exceptional divisors. The Kummer surface can also be obtained by first blowing-up the 16 two-division points and then taking the quotient by the induced involution. It can be shown that any K3 surface X that contains 16 pairwise disjoint smooth rational curves $C_1, \ldots, C_{16} \subset X$ such that $\sum [C_i] \in H^2(X, \mathbb{Z})$ is divisible by two is isomorphic to a Kummer surface. Moreover, if X is a K3 surface such that $(H^{2,0}(X) \oplus H^{0,2}(X)) \cap H^2(X, \mathbb{Z})$ is a sublattice of rank two with $\alpha^2 \equiv 0 \ (4)$ for all its elements α, then X is also a Kummer surface. The latter fact is used to show that Kummer surfaces are dense in the moduli space of all K3 surfaces [16,4].

Example 21.5 If X is a K3 surface, then the general fibre of any non-trivial morphism $\pi : X \to \mathbb{P}^1$ is an elliptic curve. A K3 surface that admits such a morphism is called an *elliptic K3 surface*. Clearly, for the class $\alpha :=$

$\pi^* c_1(\mathcal{O}(1))$ one has $\alpha^2 = 0$. In fact, any K3 surface that admits a class $\alpha \in H^2(X, \mathbb{Z}) \cap H^{1,1}(X)$ with $\alpha^2 = 0$ is an elliptic K3 surface. Note, however, that the class α need not be the fibre class of an elliptic fibration in this case. E.g. if $\pi : X \to \mathbb{P}^1$ is an elliptic fibration with a section C, then $C + \pi^{-1}(t)$ is again a class of self-intersection zero.

21.2   Higher-dimensional Examples

Examples in higher dimensions are extremely hard to construct. E.g. complete intersections of dimension > 2 are never irreducible holomorphic symplectic.

Here is the list of known examples, where manifolds of the same deformation type are not distinguished (for deformations see the next section).

(i) If X is a K3 surface (e.g. the Fermat quartic), then the Hilbert scheme $\mathrm{Hilb}^n(X)$ is an irreducible holomorphic symplectic manifold [12]. Its dimension is $2n$ and for $n > 1$ its second Betti number is $b_2(\mathrm{Hilb}^n(X)) = 23$.

(ii) If X is a complex torus of dimension two, then the *generalized Kummer variety* $\mathrm{K}_n(X)$ is an irreducible holomorphic symplectic manifold [12]. Its dimension is $2n$ and for $n > 2$ its second Betti number is $b_2(\mathrm{K}_n(X)) = 7$.

(iii) O'Grady's 10-dimensional example [169]. The second Betti number is at least 24.

(iv) O'Grady's 6-dimensional example [170]. The second Betti number is 8.

Since the Betti numbers are different in all four examples, one really disposes of two series, (i) and (ii), and two sporadic examples, (iii) and (iv). Note that in any given dimension and for any given second Betti number b_2 one knows at most one real manifold carrying the structure of an irreducible holomorphic symplectic manifold. One would hope that the number of deformation types of irreducible holomorphic symplectic manifolds is finite in any dimension (at least for given b_2). The results presented in §26.5 provide some evidence for this.

There are **other examples** of irreducible holomorphic symplectic manifolds in the literature (Beauville-Donagi [17], Yoshioka [208], O'Grady [168]), but they have all turned out to be deformation equivalent to one of the above. Moduli spaces of stable sheaves on K3 surfaces always admit a holomorphic symplectic structure and frequently they are also compact. But usually it is very difficult to obtain enough global information that shows that the manifold is simply-connected or that the symplectic structure is unique. One way to study global properties of these moduli spaces is to relate them via deformations and birational correspondences to the Hilbert scheme of a K3 surface. The draw-back of this approach is that one often shows at the same time that the moduli space in question is actually deformation equivalent to a Hilbert scheme. Here one uses results on the deformation equivalence of birational irreducible holomorphic symplectic manifolds that will be presented in §27.3.

The construction of O'Grady's examples is rather involved. It uses a fair amount of moduli space theory and Geometric Invariant Theory. Roughly, he shows that certain singular moduli spaces of semi-stable sheaves on a special K3 surface [169] respectively abelian surface [170] admit a resolution which turns out to be holomorphic symplectic. We will only sketch the idea for (i) and (ii).

The Hilbert scheme of a K3 surface: Let X be a K3 surface. The Hilbert scheme $\text{Hilb}^n(X)$ is the moduli space of all zero-dimensional subspaces $Z \subset X$ of length $\ell(\mathcal{O}_Z) = n$. If X is projective, then $\text{Hilb}^n(X)$ can also be viewed as the moduli space of stable rank one sheaves with second Chern number n. Thus, in this case the Hilbert scheme is projective [107]. In fact, for an arbitrary K3 surface the Hilbert scheme is not even a scheme (it should be called Douady space instead), but it is always Kähler due to a result of Varouchas [191]. The existence of a holomorphic symplectic structure can be shown by applying a general observation of Mukai or by using the explicit construction of the Hilbert scheme as a resolution of the symmetric product of the surface. Mukai showed in [162] that the tangent space of the moduli space of simple sheaves at a point that corresponds to a sheaf F is isomorphic to $\text{Ext}^1(F, F)$ and that the natural pairing $\text{Ext}^1(F, F) \times \text{Ext}^1(F, F) \to \text{Ext}^2(F, F) \xrightarrow{\text{tr}} H^2(X, \mathcal{O}_X) = \mathbb{C}$ is non-degenerate.

Let us describe the easiest case $\text{Hilb}^2(X)$ explicitly. For any surface X the Hilbert scheme $\text{Hilb}^2(X)$ is the blow-up $\text{Hilb}^2(X) \to S^2(X)$ of the diagonal $\Delta = \{\{x, x\} \mid x \in X\} \subset S^2(X) = \{\{x, y\} \mid x, y \in X\}$. Equivalently, $\text{Hilb}^2(X)$ is the $\mathbb{Z}/2\mathbb{Z}$-quotient of the blow-up of the diagonal in $X \times X$. Since for a K3 surface there exists only one $\mathbb{Z}/2\mathbb{Z}$-invariant two-form on $X \times X$, the holomorphic symplectic structure on $\text{Hilb}^2(X)$ is unique. For the general case and for the proof that the Hilbert scheme is simply-connected we refer to [12] or [107, Thm. 6.24].

The generalized Kummer varieties: Let T be a complex torus of dimension two. Then $\text{Hilb}^n(T)$ is a holomorphic symplectic manifold, but it is neither simply-connected nor is the symplectic structure unique. Passing to a fibre of $\text{Hilb}^n(T) \to S^n(T) \to T$, $Z \mapsto \{p_i\} \mapsto \sum p_i$ yields an irreducible holomorphic symplectic manifold $K_{n-1}(T)$.

21.3 Birational Maps

Let X and X' be irreducible holomorphic symplectic manifolds and let $f : X' \dashrightarrow X$ be a birational map. Note that if X and X' are not projective one should rather speak of bimeromorphic maps. Although the definition of a bimeromorphic map is slightly more complicated we will nevertheless say "birational" in both cases. Thus, a birational map f is given by a closed analytic subset $Z \subset X \times X'$ such that the two projections $Z \to X$ and $Z \to X'$ are proper and generically injective. Since any compact Kähler manifold that is Moishezon is also projective [156], X is projective if and only if X' is projective.

Proposition 21.6 *Let $f : X' \dashrightarrow X$ be a birational map between irreducible holomorphic symplectic manifolds. Then there exist open subsets $U \subset X$ and $U' \subset X'$ with $\mathrm{codim}(X \setminus U), \mathrm{codim}(X' \setminus U') \geq 2$ such that f induces an isomorphism $U' \cong U$.*

Proof. Let $U' \subset X'$ be the largest open subset where the birational map $f : X' \dashrightarrow X$ is defined. Then $\mathrm{codim}(X' \setminus U') \geq 2$. Let σ and σ' be non-trivial two-forms on X and X', respectively. Since $\mathbb{C} = H^0(X', \Omega^2_{X'}) = H^0(U', \Omega^2_{U'})$, the pull-back $f^* \sigma$ is a non-trivial multiple of σ'. Since σ' is everywhere non-degenerate, f must be quasi-finite on U'. Since it is generically one-to-one, it is an embedding. Thus, if $U' \subset X'$ and $U \subset X$ denote the maximal open subsets where f and f^{-1}, respectively, are defined, then $U \cong U'$ and $\mathrm{codim}(X \setminus U), \mathrm{codim}(X' \setminus U') \geq 2$. $\square$

In fact the proposition holds true for any birational map between manifolds X and X' with nef canonical bundles. In this more general case one shows that for a resolution of the birational map $X \leftarrow \tilde{Z} \to X'$ the set of exceptional divisors is the same for both projections. In particular, this yields $H^2(X, \mathbb{Z}) \cong H^2(X', \mathbb{Z})$.

Example 21.7 Here is a very explicit construction of a birational map due to Mukai [162]. Let X be a complex manifold of dimension $2n > 2$ with an everywhere non-degenerate two-form $\sigma \in H^0(X, \Omega^2_X)$. Furthermore, let $P \subset X$ be a closed submanifold which is isomorphic to a projective space $P \cong \mathbb{P}^n$. Using the symplectic structure one defines the *elementary transformation* (or *Mukai flop*) $X' := \mathrm{elm}_P(X)$ of X along P as follows: Since a projective space $\mathbb{P}^n$ does not admit any regular two-form, the restriction of σ to P is trivial. Thus, the composition of the natural inclusion $\mathcal{T}_P \subset \mathcal{T}_X|_P$, the isomorphism $\sigma|_P : \mathcal{T}_X|_P \cong \Omega_X|_P$, and the projection $\Omega_X|_P \to \Omega_P$ has to be trivial. Hence, σ induces a homomorphism $\mathcal{N}_{P/X} \cong \mathcal{T}_X|_P / \mathcal{T}_P \to \Omega_P$. Since σ is non-degenerate and the dimension of P equals its codimension, this must be an isomorphism.

Let $\pi : Z \to X$ denote the blow-up of X along $P \subset X$ and let $D \subset Z$ be the exceptional divisor. The projection $D \to P$ is isomorphic to the projective bundle $\mathbb{P}(\mathcal{N}_{P/X}) \cong \mathbb{P}(\Omega_P) \to \mathbb{P}^n$. It is well-known that this projective bundle is canonically isomorphic to the incidence variety $\{(x, H) | x \in H\} \subset \mathbb{P}^n \times \mathbb{P}^{n*}$. In particular, there exists the other projection $D \to \mathbb{P}^{n*}$. Since the incidence variety $D \subset \mathbb{P}^n \times \mathbb{P}^{n*}$ is a hypersurface with normal bundle $\mathcal{O}(1) \boxtimes \mathcal{O}(1)$, the adjunction formula yields $(\mathcal{O}(-n) \boxtimes \mathcal{O}(-n))|_D \cong K_D \cong (K_Z \otimes \mathcal{O}_Z(D))|_D \cong (\pi^* K_X \otimes \mathcal{O}_Z(nD))|_D \cong \mathcal{O}_D(nD)$. Thus, $\mathcal{O}_D(D)$ restricts to $\mathcal{O}(-1)$ on every fibre of $D \to \mathbb{P}^{n*}$. By a result of Nakano and Fujiki [61] one concludes that there exists a blow-down $Z \to X'$ to a smooth manifold $X' = \mathrm{elm}_P(X)$

such that $D \subset Z$ is the exceptional divisor and $D \to X'$ is the projection $D \to \mathbb{P}^{n*} \subset X'$.

The symplectic structure on X restricted to the complement of P and then extended again to X' defines a holomorphic two-form σ' on X'. It is certainly everywhere non-degenerate, for the zero locus of $\sigma'^{2n} \in H^0(X', K'_X)$, which is a divisor, would be contained in $\mathbb{P}^{n*} \subset X'$. But the latter has codimension at least two. Hence, X' is again symplectic. Moreover, if σ is unique, i.e. $h^0(X, \Omega^2_X) = 1$, then the same holds true for X'. Furthermore, if X is simply-connected, then also X' is simply-connected. Note that the Mukai flop of a projective or Kähler manifold X need not be projective resp. Kähler again.

Sometimes the Mukai flop of an irreducible holomorphic symplectic manifold X in a projective space $\mathbb{P}^n \subset X$ is isomorphic to X. The easiest example was found by Beauville in [15] (cf. Exercise 21.1). Note, however, that the birational map itself given by a Mukai flop never extends to an isomorphism.

Example 21.8 It is not easy to decide whether two birational irreducible holomorphic symplectiç manifolds are isomorphic. No invariant that could distinguish non-isomorphic birational irreducible holomorphic symplectic manifolds is known and only one example has ever been worked out. This is the following due to Debarre [45]: Let S be a K3 surface such that Pic(S) is generated by $\mathcal{O}(C)$, where $C \subset S$ is a smooth rational curve. Then $X := \text{Hilb}^2(S)$ contains $\mathbb{P}^2 = \text{Hilb}^2(C)$ and the Mukai flop of X in this $\mathbb{P}^2$ yields a complex manifold X' with a holomorphic symplectic structure . Then Debarre argues as follows:

(i) If X' were isomorphic to X, then the birational map $f : X' \dashrightarrow X$ would extend to an automorphism which is absurd . Indeed, if X' were isomorphic to X then one would obtain a birational automorphism of X. It can be shown that under any birational automorphism the exceptional divisor of the contraction $\text{Hilb}^2(S) \to S^2(S)$ is mapped to itself. This follows from the fact that S contains only one curve. A general argument shows that any birational automorphism of a Hilbert scheme that maps the exceptional divisor to itself is induced by a birational map of the underlying K3 surface. But the latter can certainly be extended to an automorphism of the K3 surface and thus also the induced map of the Hilbert scheme can be extended.

(ii) In order to construct examples where X' is an irreducible holomorphic symplectic manifold we have to describe a situation where X' is Kähler. This can be done by choosing S close to a quartic hypersurface S_0 with a (-2)-curve. Then X' is close to the Mukai flop of $\text{Hilb}^2(S_0)$, which is isomorphic to $\text{Hilb}^2(S_0)$ (cf. Exercise 21.1). As we will explain in the next section any small deformation of a Kähler manifold is again Kähler. Thus, we can choose S such that X' is Kähler, i.e. an irreducible holomorphic symplectic manifold.

We do not know any example of two birational and non-isomorphic irreducible holomorphic symplectic manifolds that are projective.

Example 21.9 Assume that X is a projective irreducible holomorphic symplectic manifold with Picard number $\rho(X) = 2$, i.e. $\text{Pic}(X)_{\mathbb{R}}$ is of dimension two. Assume that X contains two disjoint projective spaces $P_1, P_2 \subset X$ which are both contracted by a non-trivial morphism $X \to Y$ onto a projective variety Y. If $\ell_i \subset P_i$ is a line in the projective space P_i then $\text{Pic}(Y)_{\mathbb{R}} \subset \text{Pic}(X)_{\mathbb{R}}$ is a line that is annihilated by both linear forms $x \mapsto (x.\ell_i)$ for $i = 1, 2$. Thus, $\mathbb{R}\ell_1 = \mathbb{R}\ell_2$ in $H^{4n-2}(X, \mathbb{R})$. Let X' be the Mukai flop of X in P_1. We claim that X' cannot be Kähler. Indeed, the birational map $X' \dashrightarrow X$ induces an isomorphism $H^{4n-2}(X, \mathbb{R}) \cong H^{4n-2}(X', \mathbb{R})$ such that $\mathbb{R}\ell_1 = \mathbb{R}\ell_1'$ where ℓ_1' is a line in P_1^*. If ω is a Kähler class on X, then $\omega.\ell_i > 0$ for $i = 1, 2$ and, therefore, $\ell_1 = \lambda\ell_2$ with $\lambda > 0$. Analogously, if there existed a Kähler class on X', then one could again conclude $\ell_1' = \lambda'\ell_2$ with $\lambda' > 0$. On the other hand, for the image $\tilde{\omega}$ of ω under $H^2(X, \mathbb{R}) \cong H^2(X', \mathbb{R})$ one has $\tilde{\omega}.\ell_1' < 0$ and $\tilde{\omega}.\ell_2 = \omega.\ell_2 > 0$. This yields a contradiction. An explicit example of such a situation was described by Yoshioka in [208, 4.3] where the manifold X is a certain moduli space of stable sheaves on an abelian surface. Another example of this type was described by Namikawa in [165, 1.7(ii)]. The manifold X in his example is easier to construct, it is just the Hilbert scheme $\text{Hilb}^2(S)$ of an elliptic K3 surface with two singular fibres each consisting of cycle of three smooth rational curves. He then considers a Mukai flop in all three projectives planes associated to the rational curves of one fibre simultaneously. Note that these examples also shows that the Kähler condition in the definition of an irreducible holomorphic symplectic manifold cannot be dropped.

One might wonder whether there are other birational maps between irreducible symplectic manifolds. In fact, there is a straightforward generalization of the construction described above (cf. Exercise 21.5), but in dimension four Mukai flops in projective planes are the building blocks for all birational maps. The following result has been announced in [203]. Burns, Hu, and Luo [34] had given a proof under the additional assumption that all components of the exceptional locus are normal.

Theorem 21.10 *Let X and X' be two projective irreducible symplectic fourfolds and let $f : X' \dashrightarrow X$ be a birational map. Then there exists a sequence of Mukai flops:*

$$f_1 : X' = X_0 \dashrightarrow X_1, \ldots, f_n : X_{n-1} \dashrightarrow X_n := X \quad \text{with } f = f_n \circ \ldots \circ f_1.$$

Of course, the projectivity should not be essential. In higher dimension the known examples suggest that in codimension two birational maps are (generalized) Mukai flops. A very interesting series of explicit examples has been described by Markman in [147]. The problem of determining the structure of birational morphisms is intimately related to contractions of symplectic manifolds (cf. [42], [202], [165]).

21.4 Fibrations

Due to results of Matsushita [150,151] one knows that irreducible holomorphic symplectic manifolds only allow morphisms of a very special type.

Theorem 21.11 *Let X be an irreducible holomorphic symplectic manifold of dimension $2n$. If $f : X \to B$ is a connected morphism to a normal variety B with $0 < \dim B < \dim X$, then all fibres of f are complex Lagrangians (possibly singular). The general fibre is a complex torus of dimension n.*

A proof of a weaker assertion will be sketched in §24.5. By definition, a complex Lagrangian is a complex subvariety of dimension n such that σ restricts to a trivial two-form on its smooth part. The easiest example is provided by an elliptic K3 surface $S \to \mathbb{P}^1$. The general fibre is an elliptic curve and, for trivial reasons, any fibre is complex Lagrangian. Moreover, the elliptic structure on a K3 surface gives rise to a complex Lagrangian fibration of the Hilbert scheme $\mathrm{Hilb}^n(S) \to \mathrm{Hilb}^n(\mathbb{P}^1) = \mathbb{P}^n$, whose general fibre is isomorphic to the product of n general fibres of $S \to \mathbb{P}^1$.

There are two open questions in this context that are extremely important:

(i) Let X be an irreducible holomorphic symplectic manifold of dimension $2n$ and let $X \to B$ be a non-constant morphism with positive dimensional fibres. Is the base space B always a projective space $\mathbb{P}^n$? Indeed, Matsushita proves in [152] that a smooth base B has the Hodge numbers of a projective space. For $n = 2$ this suffices to conclude that $B \cong \mathbb{P}^2$. This case had been studied earlier by Markushevich in [149]. He could show that the base is a rational surface. But all this is only known for a smooth base. For the general case, i.e. B not necessarily smooth and of arbitrary dimension n, Matsushita showed that B is only mildly singular and has Picard number one.

(ii) Does any non-trivial nef line bundle L on an irreducible holomorphic symplectic manifold X with $q_X(c_1(L)) = 0$ define a complex Lagrangian fibration $f : X \to \mathbb{P}^n$ with $L \cong f^*\mathcal{O}(k)$ for some $k > 0$? Here, q is the Beauville–Bogomolov quadratic form (cf. §23.2). Or, even more generally, is any irreducible holomorphic symplectic manifold X that admits a line bundle L with trivial $q_X(c_1(L))$ (or, equivalently, $c_1(L)^{2n} = 0$) birational to another irreducible holomorphic symplectic manifold that can be fibred? Note that by Verbitsky's result (cf. §24.1) $c_1(L)^n \neq 0$. It can be shown that irreducible holomorphic symplectic manifolds that admit a line bundle with this property are dense in the moduli space. Let us conclude by saying that irreducible holomorphic symplectic manifolds that are fibred by complex Lagrangians are the higher dimensional analogues of elliptic K3 surfaces and play a prominent rôle in mirror symmetry. See Exercise 21.3.

21.5 Exercises

21.1 Let $S \subset \mathbb{P}^3$ be a smooth quartic hypersurface. Note that any length two subscheme $Z \in \mathrm{Hilb}^2(S)$ defines a line $l_Z \subset \mathbb{P}^3$ and the generic line $l \subset$

$\mathbb{P}^3$ intersects S in four points. Describe the rational map $\text{Hilb}^2(S) \dashrightarrow$ $\text{Hilb}^2(S)$, $Z \mapsto (l_Z \cap S) \setminus Z$. Show that the Mukai flop of $X = \text{Hilb}^2(S)$ in $\mathbb{P}^2 = \text{Hilb}^2(\ell)$ is isomorphic to X if S contains exactly one line ℓ.

21.2 Show that any surjective homomorphism $f : X' \to X$ between two irreducible holomorphic symplectic manifolds is an isomorphism.

21.3 Let X be an irreducible holomorphic symplectic manifold of dimension $2n$ and of Picard number one. Assume that X contains two disjoint projective n-spaces $P, P' \subset X$, $P \simeq P' \simeq \mathbb{P}^n$. Show that the Mukai flop of X in P could not be Kähler. (An explicit example for such an easy situation doesn't seem to be known.)

21.4 Show that there is no birational map between tori which does not extend to an isomorphism.

21.5 Generalize Mukai's construction to the case where $P \subset X$ is of codimension m and P is isomorphic to a $\mathbb{P}^m$ bundle $P \cong \mathbb{P}(F) \to Y$.

21.6 Let $C \subset S$ be a smooth curve in a K3 surface. Show that $\text{Hilb}^n(C) \subset \text{Hilb}^n(S)$ is a complex Lagrangian.

22   Deformations of Complex Structures

Deformation theory of complex manifolds is a subject that is not easily accessible. Most of the fundamental results can only be proved by using deep analytic techniques. On the other hand, the main theorems can be formulated without even mentioning any of these. We will only present a few general results and apply those to our situation. The reader who is interested in the general theory, in particular in the proofs of the various existence results, is referred to [131–133].

22.1   General Results from Deformation Theory

Let X be a compact complex manifold. A *deformation* of X consists of a smooth proper morphism $\mathcal{X} \to S$, where $\mathcal{X}$ and S are connected complex spaces, and an isomorphism $X \simeq \mathcal{X}_0$, where $0 \in S$ is a distinguished point. Usually, only the germ of $(S, 0)$ is considered. An *infinitesimal deformation* of X is a deformation with base space $S = \text{Spec}(\mathbb{C}[\varepsilon])$.

Proposition 22.1 *The isomorphism classes of infinitesimal deformations of a compact complex manifold X are parametrized by elements in $H^1(X, \mathcal{T}_X)$.*

Proof. Let $\mathcal{X} \to S = \mathrm{Spec}(\mathbb{C}[\varepsilon])$ be an infinitesimal deformation of $X = \mathcal{X}_0$. The normal bundle sequence of the embedding $X \subset \mathcal{X}$ is of the form $0 \to \mathcal{T}_X \to \mathcal{T}_{\mathcal{X}}|_X \to \mathcal{O}_X \to 0$, for the normal bundle of $X \subset \mathcal{X}$ is the pull-back of the normal bundle of $(0) \subset S$ and thus trivial. The extension class of this short exact sequence is an element in $H^1(X, \mathcal{T}_X)$, the *Kodaira-Spencer class* of $\mathcal{X} \to S$. Conversely, any such class $v \in H^1(X, \mathcal{T}_X)$ can be represented by a Čech-cocycle $v_{ij} \in \gamma(U_i \cap U_j, \mathcal{T}_X)$. Then, $\{v_{ij}\}$ can be used to define a gluing of $\mathcal{O}_{U_i}[\varepsilon]$ and $\mathcal{O}_{U_j}[\varepsilon]$ by $\mathcal{O}_{U_{ij}}[\varepsilon] \cong \mathcal{O}_{U_{ij}}[\varepsilon]$, $f + g\varepsilon \mapsto f + (v(f) + g)\varepsilon$, where $\mathcal{T}_X$ is considered as the sheaf of $\mathbb{C}$-derivations of $\mathcal{O}_X$. This yields a flat $\mathbb{C}[\varepsilon]$-algebra $\mathcal{O}_{\mathcal{X}}$ on X with $\mathcal{O}_{\mathcal{X}}/\varepsilon\mathcal{O}_{\mathcal{X}} \cong \mathcal{O}_X$, i.e. a deformation $\mathcal{X} \to \mathrm{Spec}(\mathbb{C}[\varepsilon])$. That these constructions are inverse to each other is left as an exercise. Another interpretation of the assertion in terms of Dolbeault cohomology classes will be sketched in the proof of Proposition 25.7. $\square$

Recall that a *universal* $\mathcal{X} \to (S, 0)$ of X is a deformation such that any other deformation is obtained by pull-back. For the precise definition see 14.2. If it exists it will be denoted by $\mathcal{X} \to \mathrm{Def}(X)$, where $(\mathrm{Def}(X), 0)$ is again considered as the germ of a complex space. Also recall that due to a deep theorem of Kuranishi (cf. Theorem 14.4) a universal deformation exists for every compact complex manifold without holomorphic vector fields. In this case, the universal deformation is universal for any of its fibres. The theorem readily applies to irreducible holomorphic symplectic manifolds. Indeed, since tangent and cotangent bundle are isomorphic, global holomorphic vector fields define global holomorphic one-forms, which do not exist on a simply-connected compact Kähler manifold. Thus, an irreducible holomorphic symplectic manifold admits a universal deformation.

Proposition 22.2 *Let X be a compact Kähler manifold and let $\mathcal{X} \to S$ be any deformation of X.*

(i) *For $t \in S$ close to $0 \in S$, the fibre is a compact Kähler manifold.*
(ii) *If K_X is trivial, then $K_{\mathcal{X}_t}$ is trivial for t close to 0 and the dimension of $H^1(\mathcal{X}_t, \mathcal{T}_{\mathcal{X}_t})$ is independent of t.*

Proof. (i) An easy proof of this assertion does not seem to be known. Moreover, there exist (large) deformations of compact Kähler manifolds which cease to be Kähler. The rough idea of the argument is as follows. Fix a Kähler form ω_0 on the special fibre X. This $(1,1)$-form can be extended to a real two-form ω on $\mathcal{X}$ which restricts to a $(1,1)$-form ω_t on any nearby fibre $\mathcal{X}_t$ (this is easy by using partition of unity). For t close to 0 the restriction ω_t will again be positive definite and hence defines a hermitian structure on $\mathcal{X}_t$. Thus, it remains to show that ω can be found with $d\omega_t = 0$. (This is counter-intuitive, one would expect that this condition holds for a closed set of points t in the base!) Using ω as above, one considers the space $\mathcal{H}^{1,1}(\mathcal{X}_t, \omega_t)$ of ω_t-harmonic forms on the fibre $\mathcal{X}_t$. These spaces form the fibre of a vector bundle and, in particular, their dimension is constant for small t. Since ω_0 is harmonic with respect to itself, there exists a local section ω' of this vector bundle with $\omega'_0 = \omega_0$. We may in addition assume that ω'_t is real for any t. As before, for t small the restriction ω'_t is again positive definite on $\mathcal{X}_t$ and, since it is harmonic, also d-closed. Thus, one has found a Kähler form on $\mathcal{X}_t$. The problem to make this argument rigorous is the following: Since we don't know that the fibre $\mathcal{X}_t$ is Kähler, the usual Kähler identities do not hold on $\mathcal{X}_t$. In particular, the various Laplace operators are different. Moreover, one has to show, by using results on the upper-semicontinuity of the dimension of the spaces of solutions of a differentiable family of elliptic operators, that the spaces of harmonic forms do form a vector bundle. For the detailed argument we have to refer to [161, Chapt. 4.4].

(ii) The family $\mathcal{X} \to S$ is locally constant as a family of real manifolds. In particular, $H^k(X, \mathbb{C}) \cong H^k(\mathcal{X}_t, \mathbb{C})$ for any $t \in S$ and any k. Hodge decomposition yields $H^k(\mathcal{X}_t, \mathbb{C}) = \bigoplus_{p+q=k} H^{p,q}(\mathcal{X}_t) = \bigoplus_{p+q=k} H^q(\mathcal{X}_t, \Omega^p_{\mathcal{X}_t})$. On the other hand, the numbers $h^q(\mathcal{X}_t, \Omega^p_{\mathcal{X}_t})$ are upper-semicontinuous. Hence, they are constant. In particular, $h^0(\mathcal{X}_t, K_{\mathcal{X}_t}) = h^0(\mathcal{X}_t, \Omega^N_{\mathcal{X}_t}) = 1$ for $N = \dim(X)$. Hence, the canonical bundle $K_{\mathcal{X}_t}$ admits a non-trivial global section for any $t \in S$. Moreover, these sections can be glued to a section s of the relative canonical bundle $K_{\mathcal{X}/S}$ (after shrinking S). The zero set of s does not meet the special fibre X and, since $\mathcal{X} \to S$ is proper, we may assume that s is trivializing on any fibre $\mathcal{X}_t$ for t close to the origin. For the second assertion observe that $H^1(\mathcal{X}_t, \mathcal{T}_{\mathcal{X}_t}) \cong H^1(\mathcal{X}_t, \mathcal{T}_{\mathcal{X}_t} \otimes K_{\mathcal{X}_t}) = H^1(\mathcal{X}_t, \Omega^{N-1}_{\mathcal{X}_t})$. $\square$

Lemma 22.3 *Let $\mathcal{X} \to \mathrm{Def}(X)$ be the universal deformation of a compact complex manifold X with $H^0(X, \mathcal{T}_X) = 0$. For any $t \in \mathrm{Def}(X)$ close to $0 \in \mathrm{Def}(X)$ the Zariski tangent space $T_t \mathrm{Def}(X)$ is naturally isomorphic to $H^1(\mathcal{X}_t, \mathcal{T}_{\mathcal{X}_t})$.*

Proof. By using Theorem 14.4, which in particular says that the universal family is universal for any $\mathcal{X}_t$, we may assume $t = 0$. Any infinitesimal deformation $\mathcal{X}(v) \to \mathrm{Spec}(\mathbb{C}[\varepsilon])$ given by $v \in H^1(X, \mathcal{T}_X)$ is uniquely determined by a morphism $\mathrm{Spec}(\mathbb{C}[\varepsilon]) \to \mathrm{Def}(X)$ with image 0. But the set of such morphisms is by definition the Zariski tangent space of $\mathrm{Def}(X)$. $\square$

Definition 22.4 Let X be a compact complex manifold that admits a universal deformation $\mathcal{X} \to \mathrm{Def}(X)$. We say that the deformations of X are *unobstructed* if $\dim(T_0 \mathrm{Def}(X)) = \dim(\mathrm{Def}(X))$.

In other words, the deformations of X are unobstructed if $\mathrm{Def}(X)$ is smooth or, equivalently, if any infinitesimal deformation can be integrated over a small disk.

Assertion (ii) of the proposition can be applied to the universal deformation $\mathcal{X} \to \mathrm{Def}(X)$ of a Calabi–Yau manifold X. Hence, $\dim(T_t \mathrm{Def}(X)) = \dim(H^1(\mathcal{X}_t, \mathcal{T}_{\mathcal{X}_t}))$ is constant. If $\mathrm{Def}(X)$ is reduced this is enough to conclude that $\mathrm{Def}(X)$ is smooth, i.e. that X has unobstructed deformations.

As the base space $\mathrm{Def}(X)$ of the universal deformation could, *a priori*, be non-reduced one has to argue more carefully to obtain:

Theorem 22.5 *If X is a compact Kähler manifold with trival canonical bundle K_X, then the deformations of X are unobstructed.*

This approach to the unobstructedness of Calabi–Yau manifolds is due to Ran [172] and Kawamata [130]. The fact that $H^1(\mathcal{X}, \mathcal{T}_{\mathcal{X}/S})$ is locally free for any deformation $\mathcal{X} \to S$ of X is called the T^1-lifting property. The unobstructedness goes back to Tian [189] and Todorov [190] and for the special case of holomorphic symplectic manifolds to Bogomolov [21]. Yet another proof for holomorphic symplectic manifolds using a hyperkähler metric on such a manifold was given by Fujiki (cf. [60] and Exercise 25.1). The original proof of Tian and Todorov is completely different. The main lemma in their approach gave rise to a structure of what nowadays is called a dGBV-algebra. This was later used by Barannikov and Kontsevich [3] to construct Frobenius structures in the 'B-model' associated to a Calabi–Yau manifold. For more details on the proof of the unobstructedness theorem see the discussion in §14.1.

22.2 Deforming the Mukai Flop

Deformation theory can be used to study birational geometry of irreducible holomorphic symplectic manifolds. We will explain this in detail for the simple case of a Mukai flop.

Let X be an irreducible holomorphic symplectic manifold of dimension $2n$ and assume that X contains a projective space $P = \mathbb{P}^n$. As we have seen, one can construct a holomorphic symplectic manifold $X' = \mathrm{elm}_P X$ by successively blowing-up $\mathbb{P}^n$ and then blowing-down the exceptional divisor in a different direction.

Proposition 22.6 *There exist deformations $\mathcal{X} \to S$ and $\mathcal{X}' \to S$ of X and X' with S smooth and one-dimensional, such that $\mathcal{X}_{|S\setminus\{0\}}$ and $\mathcal{X}'_{|S\setminus\{0\}}$ are isomorphic families over $S \setminus \{0\}$.*

Proof. Consider a deformation $\mathcal{X} \to S$ over a smooth base space S of dimension one such that its Kodaira-Spencer class in $H^1(X, \mathcal{T}_X) = H^1(X, \Omega_X)$ is a Kähler class. Then one shows that the normal bundle $\mathcal{N}_{P/\mathcal{X}}$ is isomorphic to $\mathcal{O}(-1)^{n+1}$. The same argument as in the construction of the Mukai flop proves that the blow-up of $\mathcal{X}$ in P admits a contraction in a different direction. This yields a family $\mathcal{X}'$. Then one checks that on the special fibre this describes the Mukai flop $X \dashrightarrow X'$. For the details we refer to [102]. □

Thus, although X and X' could be quite different as complex manifolds, each of them can be used as special fibre for the same family over a punctured disk. The same result can be proved for two arbitrary birational irreducible holomorphic symplectic manifolds, but the proof is more involved (cf. §27.2). The advantage of the above proof is that it holds true also in the case that the Mukai flop X' is not Kähler. This phenomenon was observed for the first time in the context of moduli spaces in [71]. In [102] it has been applied to the classification of examples of irreducible holomorphic symplectic manifolds given as moduli spaces of stable sheaves on K3 surfaces.

22.3 Local Torelli Theorem

Analogously to Proposition 22.2 one proves the following

Proposition 22.7 *Let $\mathcal{X} \to S$ be a deformation of an irreducible holomorphic symplectic manifold $X \cong \mathcal{X}_0$. For t close to the origin $0 \in S$ the fibre $\mathcal{X}_t$ is again an irreducible holomorphic symplectic manifold.*

We next want to describe small deformations of an irreducible holomorphic symplectic manifold in terms of its *period*. Let $\mathcal{X} \to S$ be a deformation of an irreducible holomorphic symplectic manifold $X = \mathcal{X}_0$. If S is contractible, then there is a natural isomorphism $H^*(\mathcal{X}_t, \mathbb{Z}) \cong H^*(X, \mathbb{Z})$ for all $t \in S$. For any $t \in S$ the complex structure on the fibre $\mathcal{X}_t$ yields the Hodge decomposition $H^2(X, \mathbb{C}) = H^2(\mathcal{X}_t, \mathbb{C}) = H^{2,0}(\mathcal{X}_t) \oplus H^{1,1}(\mathcal{X}_t) \oplus H^{0,2}(\mathcal{X}_t)$, which is uniquely determined by the line $\mathbb{C}\sigma_t \subset H^2(X, \mathbb{C})$, where σ_t is a non-trivial holomorphic two-form on $\mathcal{X}_t$.

Definition 22.8 Let X be an irreducible holomorphic symplectic manifold and let σ be a holomorphic two-form such that $\int (\sigma\bar{\sigma})^n = 1$. The Beauville–Bogomolov form q_X on $H^2(X, \mathbb{C})$ is defined as follows. If $\alpha = \lambda\sigma + \beta + \mu\bar{\sigma} \in H^2(X, \mathbb{C})$, with $\beta \in H^{1,1}(X)$ we let

$$q_X(\alpha) := \lambda\mu + \frac{n}{2} \int \beta^2 (\sigma\bar{\sigma})^{n-1}.$$

Furthermore, let $Q_X \subset \mathbb{P}(H^2(X, \mathbb{C}))$ be the set

$$Q_X := \{\mathbb{C}\alpha \mid q_X(\alpha) = 0, q_X(\alpha + \bar{\alpha}) > 0\}.$$

Lemma 22.9 *For any t close to $0 \in S$ one has $q_X(\sigma_t) = 0$.*

Proof. Indeed, if σ_t is a $(2,0)$-form on $\mathcal{X}_t$, which is of complex dimension $2n$, then the class $\sigma_t^{n+1} \in H^{2(n+1)}(X, \mathbb{C})$ is trivial. Hence, $\int_X \sigma_t^{n+1} \bar{\sigma}^{n-1} = 0$. Let $\sigma_t = \lambda\sigma + \alpha + \mu\bar{\sigma}$ be the Hodge decomposition of σ_t as a class on X. Then the $(2n, 2)$-component of σ_t^{n+1} on X is $(n+1)\lambda^n\mu\sigma^n\bar{\sigma} + \binom{n+1}{2}\lambda^{n-1}\sigma^{n-1}\alpha^2$. Hence,

$$0 = \int_X \sigma_t^{n+1} \bar{\sigma}^{n-1} = \int_X (\sigma_t^{n+1})_{(2n,2)} \bar{\sigma}^{n-1}$$

$$= (n+1)\lambda^n\mu \int_X (\sigma\bar{\sigma})^n + \binom{n+1}{2}\lambda^{n-1} \int_X (\sigma\bar{\sigma})^{n-1}\alpha^2$$

$$= (n+1)\lambda^{n-1}\left(\lambda\mu + \frac{n}{2} \int_X (\sigma\bar{\sigma})^{n-1}\alpha^2\right) = (n+1)\lambda^{n-1} q_X(\sigma_t).$$

Since $\lambda \neq 0$ for t close to 0, this yields $q_X(\sigma_t) = 0$. $\qquad\qquad\square$

Clearly, $q_X(\sigma_t + \bar\sigma_t) > 0$ and hence one has the following

Definition 22.10 Let $\mathcal{X} \to S$ be a deformation of $X = X_0$ as above. The map $\mathcal{P}_X : S \longrightarrow Q_X \subset \mathbb{P}(H^2(X,\mathbb{C}))$, $t \mapsto [\sigma_t]$ is called the *local period map*.

The period map $\mathcal{P}_X$ is holomorphic. This is due to the fact that $\pi_* \Omega^2_{\mathcal{X}/S}$ is a holomorphic subbundle of $R^2\pi_*\mathbb{C} \otimes_{\mathbb{C}} \mathcal{O}_S$. Since $q_X(\sigma_t) = 0$, the image of $\mathcal{P}_X$ is contained in Q_X ([39]). Period maps of Calabi–Yau 3-folds are discussed in §16.2.

Proposition 22.11 (Local Torelli, [12]) *Consider the universal family $\mathcal{X} \to \mathrm{Def}(X)$. The induced period map $\mathcal{P}_X : \mathrm{Def}(X) \to Q_X$ is a local isomorphism.*

Proof. As in [39], the differential $d\mathcal{P}_X$ of the period map at the point $0 \in \mathrm{Def}(X)$ is the contraction homomorphism

$$T_0\mathrm{Def}(X) = H^1(X,\mathcal{T}_X) \to \mathrm{Hom}(H^{2,0}(X), H^2(X,\mathbb{C})/H^{2,0}(X)).$$

Since $H^{2,0}(X) = \mathbb{C}\sigma$ and the contraction $\sigma : H^1(X,\mathcal{T}_X) \to H^1(X,\Omega_X) = H^{1,1}(X) \subset H^2(X,\mathbb{C})/H^{2,0}(X)$ is induced by the isomorphism $\sigma : \mathcal{T}_X \cong \Omega_X$, the differential $d\mathcal{P}_X$ is injective. Both spaces $\mathrm{Def}(X)$ and Q_X are of dimension $h^{1,1}(X)$ and the image of $\mathcal{P}_X$ is contained in Q_X. Hence, $\mathcal{P}_X : \mathrm{Def}(X) \to Q_X$ is a local isomorphism in $0 \in \mathrm{Def}(X)$. $\qquad\square$

Later we will see that Q_X is smooth. Using the unobstructedness of irreducible holomorphic symplectic manifolds we can already conclude that Q_X is smooth in the image of $\mathcal{P}_X$.

22.4 Exercises

22.1 Compute the dimension of $\mathrm{Def}(X)$ for $X = \mathrm{Hilb}^2(K3)$.

22.2 Prove Proposition 22.7.

22.3 Generalize Proposition 22.6 in the sense of Exercise 21.5.

23 The Beauville–Bogomolov Form

Here we will introduce an integral quadratic form q_X on $H^2(X,\mathbb{Z})$ of a compact hyperkähler manifold X. By means of this quadratic form we will define the global period map and the period domain in §25.2.

The approach we present here is similar to the classical one for the Hodge–Riemann pairing, which is first studied on each tangent space. Restricting to harmonic forms yields results (e.g. the Lefschetz decomposition) on the level of cohomology.

Usually, the quadratic form q_X on $H^2(X)$ of a compact hyperkähler manifold X is studied by using the deformation theory of X and in particular its unobstructedness due to Bogomolov, Tian, and Todorov. The approach presented here stays on a fixed X, but uses the existence of a hyperkähler metric for each given Kähler class. That it is possible to construct and describe the form q_X either in terms of the deformation theory of X or by using the Kähler cone, could be seen, if one wants to, as an effect of mirror symmetry for hyperkähler manifolds.

23.1   Compact Hyperkähler Manifolds

In these lectures compact hyperkähler manifolds are defined as follows.

Definition 23.1 A compact Riemannian manifold (M, g) of dimension $4n$ is called *hyperkähler* if the holonomy group of g equals $\mathrm{Sp}(n)$. In this case g is called a hyperkähler metric.

Remark 23.2 If g is a hyperkähler metric, then there exist three complex structures I, J, and K on M, such that g is Kähler with respect to all three of them and such that $K = I \circ J = -J \circ I$. Thus, I is orthogonal with respect to g and the Kähler form $\omega_I := g(I(\ ), \ )$ is closed (similarly for J and K). Often, this is taken as a definition of a hyperkähler metric. Note that our condition is stronger, as we not only want the holonomy be contained in $\mathrm{Sp}(n)$, but be equal to it. Here holonomy groups are discussed in Part I.

Sometimes, we use the names irreducible holomorphic symplectic manifold and compact hyperkähler manifold interchangeably. They refer to different aspects of such manifolds. Whereas "holomorphic symplectic" refers to a complex manifold with certain properties, "hyperkähler" is a property of a Riemannian metric on the underlying real manifold. Let us make this more precise.

Proposition 23.3 ([12]) *If (M, g) is a compact hyperkähler manifold, then the complex manifolds (M, I), (M, J), and (M, K) are irreducible holomorphic symplectic manifolds.*

Proof. It suffices to prove the assertion for $X := (M, I)$. Clearly, X is a compact Kähler manifold. It remains to show that X is simply-connected and that $H^0(X, \Omega_X^2)$ is generated by an everywhere non-degenerated two-form. Any form in $H^0(X, \Omega_X^2)$ is parallel. Since the holonomy group is $\mathrm{Sp}(n)$, there is only one parallel section of $\Lambda^{2,0} T^*_{(M,I)}$ and this one defines a holomorphic symplectic structure σ. Analogously, one shows that $H^0(X, \Omega_X^k)$ is trivial for

k odd and generated by $\sigma^{k/2}$ for k even. Therefore, $\chi(X, \mathcal{O}_X) = n + 1$. Since $h^1(X, \mathcal{O}_X) = 0$, the decomposition theorem, Theorem 14.15, shows that the fundamental group of X is finite. Let $\pi : X' \to X$ be a finite unramified cover. The holonomy of the pull-back metric is again $\mathrm{Sp}(n)$ and, therefore, $\chi(X', \mathcal{O}_{X'}) = n + 1$. On the other hand, $\chi(X', \mathcal{O}_{X'}) = \deg(\pi) \cdot \chi(X, \mathcal{O}_X)$. Hence, $\deg(\pi) = 1$. Thus, X is simply-connected. $\square$

Thus any compact hyperkähler manifold is an irreducible holomorphic symplectic manifold in many ways. The converse is also true, but much harder to prove.

Definition 23.4 Let X be a compact Kähler manifold. The *Kähler cone* $\mathcal{K}_X \subset H^{1,1}(X, \mathbb{R})$ is the open convex cone of all Kähler classes on X, i.e. classes that can be represented by some Kähler form.

For the openness of the Kähler cone see Proposition 14.14. The most important single result on irreducible holomorphic symplectic manifolds is the following consequence of Yau's proof of the Calabi Conjecture [206], taken from Theorem 5.11:

Theorem 23.5 *Let X be an irreducible holomorphic symplectic manifold. Then for any $\alpha \in \mathcal{K}_X$ there exists a unique hyperkähler metric on the underlying real manifold M, such that $X = (M, I)$ and $\alpha = [\omega_I]$.*

For comments on the proof, see §5.

23.2 Quaternionic Hermitian Vector Spaces

Before we are going to use the existence of this special metric in order to define the quadratic form q_X we will study the induced structure on each tangent space.

Let V be a real vector space with an $\mathbb{H}$-action, i.e. there exist three complex structures I, J, and K on V such that $I \circ J = -J \circ I = K$. Any linear combination $\lambda = aI + bJ + cK$ with $a^2 + b^2 + c^2 = 1$ also defines a complex structure on V. The resulting complex vector spaces are denoted by (V, I), (V, J), (V, K), and (V, λ), respectively. Note that the real dimension of V is divisible by four and we write $\dim_{\mathbb{R}}(V) = 4n$. For any of these complex structures the exterior algebra $\Lambda^* V_{\mathbb{C}}^*$ allows a decomposition with respect to the bidegree. E.g. we write $\Lambda^* V_{\mathbb{C}}^* = \bigoplus_{p,q} \Lambda_I^{p,q} V^*$ for the decomposition with respect to I and we say that α is a $(p, q)_I$-form if $\alpha \in \Lambda_I^{p,q} V^*$.

Let, in addition, V be endowed with a scalar product $\langle \, , \, \rangle$, i.e. $(V, \langle \, , \, \rangle)$ is an euclidian vector space. Furthermore, we assume that $\langle \, , \, \rangle$ is compatible with I, J, and K, i.e. all three are orthogonal. The associated Kähler forms are the real forms $\omega_I = \langle I(\ ), \ \rangle \in \Lambda_I^{1,1} V^*$, $\omega_J = \langle J(\ ), \ \rangle \in \Lambda_J^{1,1} V^*$, and $\omega_K = \langle K(\ ), \ \rangle \in \Lambda_K^{1,1} V^*$

Next, we define "holomorphic" two-forms on the complex vector spaces (V, I), (V, J), and (V, K) by $\sigma_I := \omega_J + i\omega_K \in \Lambda_I^{2,0} V^*$, $\sigma_J := \omega_K + i\omega_I \in$

$\Lambda_J^{2,0}V^*$, and $\sigma_K := \omega_I + i\omega_J \in \Lambda_K^{2,0}V^*$. That these are really forms of the asserted types is shown by an explicit calculation. (Choose an orthonormal basis (e_j, f_j, g_j, h_j) of V^* with $I(e_j) = f_j$, $J(e_j) = g_j$, and $K(e_j) = h_j$. Then, $\omega_J + i\omega_K = \sum((e_j \wedge g_j - f_j \wedge h_j) + i(e_j \wedge h_j + f_j \wedge g_j)) = \sum(e_j + if_j) \wedge (g_j + ih_j)$.)

For any complex structure $\lambda = aI + bJ + cK$ the associated Kähler form ω_λ defines a Hodge–Riemann pairing on the space of two-forms

$$\begin{array}{ccc} \Lambda^2 V^* \times \Lambda^2 V^* \to & \Lambda^{4n} V^* & \cong \mathbb{R} \\ (\alpha \quad , \quad \beta) \mapsto & \alpha \wedge \beta \wedge \omega_\lambda^{2n-2}. \end{array}$$

For the last isomorphism we use the orientation given by any of the complex structures λ. Let us try to clarify the relation between all these different Hodge–Riemann pairings.

Proposition 23.6 *For $n > 1$ one has*

$$\binom{2n-2}{n}(\sigma_J^n \bar\sigma_J^{n-2} + \sigma_J^{n-2} \bar\sigma_J^n) = 2^{2n-3}(\omega_K^{2n-2} - \omega_I^{2n-2}).$$

Proof. The defining equation for σ_J yields $\sigma_J + \bar\sigma_J = 2\omega_K$ and $\sigma_J - \bar\sigma_J = 2i\omega_I$. Hence,

$$2^{2n-2}(\omega_K^{2n-2} - \omega_I^{2n-2}) = (\sigma_J + \bar\sigma_J)^{2n-2} - (-1)^{n-1}(\sigma_J - \bar\sigma_J)^{2n-2}.$$

Since $\sigma_J^k = \bar\sigma_J^k = 0$ for $k > n$, one finds

$$(\sigma_J + \bar\sigma_J)^{2n-2} = \binom{2n-2}{n}(\sigma_J^n \bar\sigma_J^{n-2} + \sigma_J^{n-2} \bar\sigma_J^n) + \binom{2n-2}{n-1}(\sigma_J \bar\sigma_J)^{n-1}$$

and

$$(i\sigma_J - i\bar\sigma_J)^{2n-2} = -\binom{2n-2}{n}(\sigma_J^n \bar\sigma_J^{n-2} + \sigma_J^{n-2} \bar\sigma_J^n) + \binom{2n-2}{n-1}(\sigma_J \bar\sigma_J)^{n-1}.$$

Therefore,

$$2^{2n-2}(\omega_K^{2n-2} - \omega_I^{2n-2}) = 2\binom{2n-2}{n}(\sigma_J^n \bar\sigma_J^{n-2} + \sigma_J^{n-2} \bar\sigma_J^n).$$

$\square$

Proposition 23.7 *If α is a $(1,1)_I$-form, then*

$$\alpha\omega_J^{2n-2} = \alpha\omega_K^{2n-2} = 2^{2-2n}\binom{2n-2}{n-1}\alpha(\sigma_I \bar\sigma_I)^{n-1}.$$

Proof. This time we use $\sigma_I = \omega_J + i\omega_K$ to conclude $2\omega_K = i(\bar\sigma_I - \sigma_I)$. Hence,

$$2^{2n-2}\omega_K^{2n-2} = (-1)^{n-1}\sum_{p=0}^{2n-2}(-1)^p\binom{2n-2}{p}\sigma_I^p \bar\sigma_I^{2n-2-p}.$$

If α is a $(1,1)_I$-form, then $\alpha\sigma_I^p = \alpha\bar\sigma_I^p = 0$ for $p \geq n$ and hence $\alpha\sigma_I^p \bar\sigma_I^{2n-2-p} = 0$ for $p \neq n-1$. Thus, $2^{2n-2}\omega_K^{2n-2}\alpha = \binom{2n-2}{n-1}(\sigma_I \bar\sigma_I)^{n-1}\alpha$.

$\square$

Corollary 23.8 *If α is a $(1,1)_{I,J}$-form, i.e. it is of type $(1,1)$ with respect to both complex structures I and J, then*

$$\alpha\omega_I^{2n-2} = \alpha\omega_J^{2n-2} = \alpha\omega_K^{2n-2} = 2^{2-2n}\binom{2n-2}{n-1}\alpha(\sigma_I\bar\sigma_I)^{n-1}.$$

Proof. If α is a $(1,1)_J$-form, then $\alpha(\sigma_J^n\bar\sigma_J^{n-2} + \sigma_J^{n-2}\bar\sigma_J^n) = 0$. By Proposition 23.6 we have $\omega_K^{2n-2}\alpha = \omega_I^{2n-2}\alpha$ and combined with Proposition 23.7 this yields the assertion.   □

Notice, that a $(1,1)_{I,J}$-form is of type $(1,1)$ with respect to any λ. The above calculations show that on the space of two-forms which are of type $(1,1)$ with respect to I and J (hence to K and, in fact, to all λ) the Hodge–Riemann pairing is independent of the complex structure. On the other hand, ω_I itself is $(1,1)_I$ and $(2,0)_J + (0,2)_J$. For $\alpha = \omega_I$ the Hodge–Riemann pairings compare as follows:

Corollary 23.9 *Under the same assumptions as above one has*

$$\omega_I\omega_I^{2n-2} = (2n-1)\omega_I\omega_J^{2n-2} = (2n-1)\omega_I\omega_K^{2n-2} = 2^{1-2n}n\binom{2n}{n}\omega_I(\sigma_I\bar\sigma_I)^{n-1}.$$

Proof. Applying the standard formula (cf. [201, p. 23]), that compares the action of the Hodge $*$-operator with the action of the Lefschetz operator, to the ω_I-primitive form $1 \in \Lambda^0 V^*$ yields $*\omega_I = (1/(2n-1)!)\omega_I^{2n-1}$. Applied to the ω_K-primitive form ω_I it gives $*\omega_I = (1/(2n-2)!)\omega_K^{2n-2}\omega_I$. (The argument that any $(1,1)_I$-form is ω_K-primitive is given in the proof of Proposition 23.10.) The assertion follows from a comparison of these two equalities and the equality $\binom{2n-2}{n-1} = \frac{n}{2(2n-1)}\binom{2n}{n}$.   □

Note that already for $n = 2$ the space of $(1,1)_I$-forms which are primitive with respect to ω_I is not spanned by $(1,1)_{I,J}$-forms.

23.3   The Quadratic Form

We now wish to come back to the **geometric situation**. Let M be a compact hyperkähler manifold with a fixed hyperkähler metric g. As above, I, J, and K denote the three associated anticommuting complex structures on M. The Kähler forms and the holomorphic symplectic structures are $[\omega_I] \in H^{1,1}((M,I))$, $[\omega_J] \in H^{1,1}((M,J))$, $[\omega_K] \in H^{1,1}((M,K))$ and $\sigma_I \in H^{2,0}((M,I))$, $\sigma_J \in H^{2,0}((M,J))$, $\sigma_K \in H^{2,0}((M,K))$, respectively. Recall that the holomorphic two-form σ_I on (M,I) is unique, i.e. $H^{2,0}((M,I)) = \mathbb{C}\sigma_I$. This has the following consequence.

Proposition 23.10 *Let α be a real harmonic $(1,1)$-form on (M,I) which is primitive with respect to ω_I. Then α is of type $(1,1)$ with respect to any complex structures $\lambda = aI + bJ + cK$.*

Proof. It suffices to show that α is of type $(1,1)$ on (M, J). Let $\alpha = \alpha^{2,0} + \alpha^{0,2} + \alpha^{1,1}$ be its bidegree decomposition with respect to J. Thus, $\alpha^{1,1}$ is a real $(1,1)_J$-form and $\alpha^{2,0} + \alpha^{0,2}$ is a real $((2,0)_J + (0,2)_J)$-form. First note that any $(1,1)_J$-form is automatically ω_I-primitive. Indeed, $\alpha^{1,1}\omega_I^{2n-1} = (1/2i)^{2n-1}\alpha^{1,1}(\sigma_J - \bar{\sigma}_J)^{2n-1} = 0$, since $\alpha^{1,1}\sigma_J^p = \alpha^{1,1}\bar{\sigma}_J^p = 0$ for $p \geq n$. Hence, α is ω_I-primitive if and only if $\alpha^{2,0} + \alpha^{0,2}$ is ω_I-primitive.

Next, Hodge theory says that with α also $\alpha^{2,0}$, $\alpha^{0,2}$, and $\alpha^{1,1}$ are harmonic. The real harmonic $(2,0)+(0,2)$-form $\alpha^{2,0}+\alpha^{0,2}$ on (M, J) is of the form $\lambda\omega_I + \mu\omega_K = (\lambda/2i)(\sigma_J - \bar{\sigma}_J) + (\mu/2)(\sigma_J + \bar{\sigma}_J)$ with $\lambda, \mu \in \mathbb{R}$, since $H^{2,0}((M, J))$ is one-dimensional. Since $\alpha^{2,0} + \alpha^{0,2}$ is ω_I-primitive, one has $\lambda = 0$. Since $\alpha\omega_K^{2n-1} = (1/2i)^{2n-1}\alpha(\sigma_I - \bar{\sigma}_I)^{2n-1} = 0$ and $\alpha^{1,1}\omega_K^{2n-1} = (1/2)^{2n-1}\alpha^{1,1}(\sigma_J - \bar{\sigma}_J)^{2n-1} = 0$, one obtains $0 = (\alpha^{2,0} + \alpha^{0,2})\omega_K^{2n-1} = \mu\omega_K^{2n}$, i.e. $\mu = 0$. Thus, $\alpha^{2,0} + \alpha^{0,2} = 0$, i.e. α is a $(1,1)_J$-form. $\square$

Let X be an irreducible holomorphic symplectic manifold of complex dimension $2n$ and let σ be a holomorphic two-form on X such that $\int_X(\sigma\bar{\sigma})^n = 1$. Recall that the *Beauville–Bogomolov form* q_X on $H^2(X, \mathbb{R})$ was defined by (see Definition 22.8)

$$q_X(\alpha) = (n/2)\int_X \alpha^2(\sigma\bar{\sigma})^{n-1} + (1-n)\left(\int_X \alpha\sigma^{n-1}\bar{\sigma}^n\right)\left(\int_X \alpha\sigma^n\bar{\sigma}^{n-1}\right).$$

Later we shall denote by q_X a scalar multiple of the above quadratic form which is primitive and integral. That such a rescaling is possible will be shown in Proposition 23.14. On the complex manifold (X, I), the quadratic form q_X does not depend on the choice of a hyperkähler metric g or even on the choice of a Kähler class. Moreover, the decomposition

$$(H^{2,0} \oplus H^{0,2})(X) \oplus H^{1,1}(X) = H^2(X, \mathbb{C})$$

is orthogonal and on $H^{1,1}(X)$ it is of the form $\int \alpha^2(\sigma\bar{\sigma})^{n-1}$ (up to a scalar factor).

Let us first compute the index of q_X. This is an immediate consequence of the last proposition.

Corollary 23.11 *The Beauville–Bogomolov quadratic form q_X on the second cohomology $H^2(X, \mathbb{R})$ of an irreducible holomorphic symplectic manifold X has index $(3, b_2(X) - 3)$. More precisely, if $[\omega] \in H^{1,1}(X)$ is a Kähler class, q_X is positive definite on $\mathbb{R}[\omega] \oplus (H^{2,0} \oplus H^{0,2})(X)_{\mathbb{R}}$ and negative definite on the primitive $(1,1)$-part $H^{1,1}(X)_\omega$.*

Proof. For given $[\omega]$ there is a unique hyperkähler metric such that the induced Kähler form ω_I associated to it on the complex manifold X given by the complex structure I represents $[\omega]$. Once this hyperkähler structure is chosen we will write ω_I instead of ω and σ_I instead of σ.

If $0 \neq \alpha \in H^{1,1}(X, \mathbb{R})_\omega$, then α is of type $(1,1)$ with respect to J. Hence, by Corollary 23.8 and the usual Hodge–Riemann bilinear relations (cf. [77, p. 123])

$$2^{2-2n} \binom{2n-2}{n-1} \alpha^2 (\sigma \bar{\sigma})^{n-1} = \alpha^2 \omega_I^{2n-2} < 0.$$

Thus, q_X is negative definite on $H^{1,1}(X, \mathbb{R})_{\omega_I}$. Clearly, q_X is positive definite on $(H^{2,0} \oplus H^{0,2})(X)_\mathbb{R} = \{\mu\sigma + \bar{\mu}\bar{\sigma} | \mu \in \mathbb{C}\}$ and by Corollary 23.9 we also have $q_X(\omega_I) > 0$. Obviously, $(H^{2,0} \oplus H^{0,2})(X)$ and $H^{1,1}(X)$ are orthogonal with respect to q_X and $[\omega_I]$ is orthogonal to any ω_I-primitive form (cf. Corollary 23.9). $\qquad\square$

The essential idea of the proof is the following. On the primitive forms in $\Lambda_I^{1,1} V^*$ which are $(1,1)_J$ the form $\alpha^2 (\sigma_I \bar{\sigma}_I)^{n-1}$ is negative definite, since there it coincides with the usual Hodge–Riemann pairing. For a compact hyperkähler manifold every harmonic (with respect to a hyperkähler metric) primitive $(1,1)_I$-form is a $(1,1)_J$-form. Thus, the above approach to compute the index of q_X allows to refine the statement. Roughly, one can say that on the space of harmonic two-forms $\mathcal{H}^2(X)$ the quadratic form q_X is pointwise of index $(3, b_2(X) - 3)$. The reason why this statement is not absolutely correct is that the space of harmonic two-forms does usually not embed into the space of two-forms at a given point, i.e. there will always be some harmonic forms that vanish in a given point.

Definition 23.12 Let X be an irreducible holomorphic symplectic manifold. The *positive cone* $\mathcal{C}_X \subset H^{1,1}(X, \mathbb{R})$ is the connected component of the set $\{\alpha \mid q_X(\alpha) > 0\}$ that contains the Kähler cone $\mathcal{K}_X$.

Note that the quadratic form q_X has signature $(1, b_2(X) - 3)$ on $H^{1,1}(X, \mathbb{R})$. Therefore, the positive cone is one of the two connected components, i.e. $\{\alpha \mid q_X(\alpha) > 0\} = \mathcal{C}_X \sqcup (-\mathcal{C}_X)$.

Definition 23.13 Let α be a Kähler class on the irreducible holomorphic symplectic manifold X, then $F(\alpha)$ is the three-dimensional space spanned by α, $\mathrm{Re}(\sigma)$, and $\mathrm{Im}(\sigma)$. If g is a hyperkähler metric on the underlying real manifold, then we denote by $F(g)$ the space $F(\omega_I)$, where I is one of the complex structures associated to g.

As we have seen, the Beauville–Bogomolov quadratic form is positive definite on the three-space $F(g)$ and negative definite on its orthogonal complement $H^{1,1}(X, \mathbb{R})_{\omega_I}$. Sometimes, we will also associate a positive three-space $F(\alpha)$ to an arbitrary class $\alpha \in \mathcal{C}_X$ in the same way.

So far q_X is a quadratic form on a real vector space. One way to show that q_X is actually integral on $H^2(X, \mathbb{Z})$ (up to a scalar factor) is by comparing q_X with the top intersection form. This will be done next.

23.4   Beauville–Fujiki Relation

Proposition 23.14 *Let X be an irreducible holomorphic symplectic manifold. Then there exists a positive constant $c \in \mathbb{R}$ such that $q_X(\alpha)^n = c \int_X \alpha^{2n}$ for all $\alpha \in H^2(X)$. In particular, q_X can be renormalized such that q_X is a primitive integral quadratic form on $H^2(X, \mathbb{Z})$. [12,60,54]*

Proof. Let us fix a hyperkähler metric g on X and let us assume that the induced holomorphic two-form $\sigma_I = \omega_J + i\omega_K$ satisfies $\int (\sigma_I \bar\sigma_I)^n = 1$, i.e. that the assumption for the definition of q_X in terms of σ_I holds true. Note that again we use σ_I rather than σ whenever a hyperkähler structure is chosen. By Corollary 23.9 we have $\omega_I^2(\sigma_I \bar\sigma_I)^{n-1} = 2^{2n-2} \frac{2}{n} \binom{2n}{n}^{-1} \omega_I^{2n}$. Hence

$$q_X(\alpha) = \frac{n}{2} \int_X \alpha^2 (\sigma_I \bar\sigma_I)^{n-1} = 2^{2n-2} \binom{2n}{n}^{-1} \int_X \alpha^{2n}$$

for $\alpha = [\omega_I]$. On the other hand, $*1 = \frac{1}{(2n)!} \omega_I^{2n}$, but also

$$*1 = \frac{1}{(2n)!} \omega_J^{2n} = \frac{1}{(2n)!} 2^{-2n} (\sigma_I + \bar\sigma_I)^{2n} = \frac{\binom{2n}{n} 2^{-2n}}{(2n)!} (\sigma_I \bar\sigma_I)^n.$$

Thus, $\int \omega_I^{2n} = 2^{-2n} \binom{2n}{n} \int (\sigma_I \bar\sigma_I)^n = 2^{-2n} \binom{2n}{n}$.

Now, any Kähler class $\beta \in \mathcal{K}_X$ is of the form $\beta = \mu\alpha$ with $\mu \in \mathbb{R}_{>0}$, such that α is represented by a hyperkähler form ω_I, whose associated holomorphic two-form σ satisfies $\int (\sigma\bar\sigma)^n = 1$. Therefore,

$$
\begin{aligned}
q_X(\beta)^n &= (\tfrac{n}{2} \int \beta^2 (\sigma_I \bar\sigma_I)^{n-1})^n = \mu^{2n} q_X(\alpha)^n \\
&= (2^{2n-2} \mu^2 \binom{2n}{n}^{-1})^n (\int \alpha^{2n})^n = (2^{2-2n} \binom{2n}{n})^{-n} \mu^{2n} (\int \alpha^{2n})(\int \alpha^{2n})^{n-1} \\
&= (2^{2-2n} \binom{2n}{n})^{-n} (\int \beta^{2n})(2^{-2n} \binom{2n}{n})^{n-1} = \binom{2n}{n}^{-1} \int \beta^{2n}.
\end{aligned}
$$

Since the Kähler cone $\mathcal{K}_X \subset H^{1,1}(X,\mathbb{R})$ is open, the equation holds for all $\beta \in H^{1,1}(X,\mathbb{R})$. Now we use (i) of the remark following this proof, which only depends on what has been said so far, to conclude that the asserted equality holds for all $\beta \in H^2(X,\mathbb{R})$. Indeed, the Kähler cones $\mathcal{K}_{(X,\lambda)}$ for all complex structures $\lambda = aI + bJ + cK$ sweep out an open subset of $H^2(X,\mathbb{R})$. Therefore, $q_X(\beta)^n = \binom{2n}{n}^{-1} \int \beta^{2n}$ for all $\beta \in H^2(X,\mathbb{R})$. More directly, one also sees that for $\beta = \mu\sigma + \bar\mu\bar\sigma$ one has $q_X(\beta) = \mu\bar\mu$, $\int \beta^{2n} = \binom{2n}{n}(\mu\bar\mu)^n$, and, therefore, $q_X(\beta)^n = \binom{2n}{n}^{-1} \int \beta^{2n}$ for all $\beta \in (H^{2,0} \oplus H^{0,2})(X)$, as well.

If now $q_X(\alpha)^n = c \int_X \alpha^{2n}$ for all $\alpha \in H^2(X,\mathbb{R})$ and some $c > 0$, then the quadratic form $\tilde{q}_X := c^{-1/n} q_X$, for a positive real n-th root $c^{1/n}$ of c, is integral. This is a general fact, which can be checked by a straightforward argument. Indeed, since $\tilde{q}_X(\alpha)^n = \int_X \alpha^{2n}$ is integral, the form $\tilde{q}_X$ is integral up to an n-th root of unity. Since on the other hand $\tilde{q}_X(\alpha) \in \mathbb{R}$, this root must be real, i.e. equal ± 1. Moreover, $\tilde{q}_X$ may be further rescaled to make it primitive. $\qquad\square$

Remark 23.15 (i) Recall that q_X was introduced on a fixed complex manifold $X = (M, I)$ without using the hyperkähler metric, but rather by means of the holomorphic two-form σ_I. By Corollary 23.8 we know $q_X(\alpha) = q_{(M,\lambda)}(\alpha)$ for all primitive $(1,1)$-classes α on X and an arbitrary complex structure $\lambda = aI + bJ + cK$. Together with Corollary 23.9 and the calculation $\int \omega_I^{2n} = 2^{-2n}\binom{2n}{n}$ presented in the proof above, one finds that q_X does in fact not depend on the complex structure I either, i.e. the corresponding quadratic forms on (M, J) and (M, K) defined in terms of σ_J and σ_K, respectively, coincide with q_X.

(ii) Let q_X be the above primitive integral quadratic form on $H^2(X, \mathbb{Z})$ as above and let p be another such quadratic form which also satisfies $p(\alpha)^n = c \int \alpha^{2n}$ for some $c \in \mathbb{R}_{>0}$ and all $\alpha \in H^2(X, \mathbb{Z})$. Then $p = \pm q$. Indeed, since $p^n = cq_X^n$ for some $c \in \mathbb{Q}_{>0}$, one has $p = c^{1/n}q_X$ for some n-th root of c. Since p and q are integral, this yields $c^{1/n} = \pm 1$. Note that for $b_2 > 3$ the minus sign cannot occur if the signature of p is also $(3, b_2(X) - 3)$. Thus, the primitive Beauville–Bogomolov form is uniquely determined, up to a possible factor -1 in the case $b_2 = 6$, by its property that $q_X(\alpha)^n = c \int \alpha^{2n}$ for some $c \in \mathbb{R}_{>0}$ and all $\alpha \in H^2(X, \mathbb{Z})$.

(iii) Later we will argue that the normalization of the Beauville–Bogomolov form to a primitive form may not, after all, be the best choice. In fact, in the examples the primitive form is not unimodular, so classification theory of unimodular lattices cannot be applied. Moreover, using a natural quadratic class on the manifold gives rise to a quadratic form that is a scalar multiple of the above, but is not primitive a priori.

The independence of q_X of the complex structure is used to prove the following two results:

Corollary 23.16 *For any $\alpha \in H^{1,1}(X)$ and $0 \leq p \leq n$ one has*

$$q_X(\alpha)^p = c_p \int \alpha^{2p}(\sigma_I \bar{\sigma}_I)^{n-p},$$

where q_X is defined as in Definition 22.8 and $c_p = c\binom{2n}{2p}\binom{2n-2p}{n-p}\binom{n}{p}^{-1}$.

Proof. Use $q_X(t\alpha + \sigma_I + \bar{\sigma}_I)^n = c \int (t\alpha + \sigma_I + \bar{\sigma}_I)^{2n}$ and compare the coefficients of t^{2p}. $\qquad\square$

The following result, although not quite in this generality, was proved by Fujiki [60] and more recently by Lunts and Looijenga [145] using results that will be desribed later and which are based on [195]. The proof we will give follows [23].

Corollary 23.17 *Assume $\beta \in H^{4j}(X, \mathbb{R})$ is of type $(2j, 2j)$ on all small deformations of X (in particular on X itself). Then there exists a constant $c \in \mathbb{R}$ depending on β such that*

$$\int_X \beta \alpha^{2(n-j)} = c \cdot q_X(\alpha)^{n-j}$$

for all $\alpha \in H^2(X, \mathbb{R})$. The case in Proposition 23.14 is the special case $\beta = [X] \in H^0(X, \mathbb{R})$.

Proof. By the Local Torelli Theorem (cf. Proposition 22.11) the local period map $\mathcal{P} : \mathrm{Def}(X) \to Q_X$ is a local isomorphism. The image $\mathcal{P}(t)$ is given by the $(2,0)$-form σ_t on the fibre $\mathcal{X}_t$. Since β is of type $(2j, 2j)$ on $\mathcal{X}_t$, the integral $\int \beta \sigma_t^{2(n-j)}$ vanishes. Thus, the homogeneous polynomial $g_\beta : \alpha \mapsto \int \beta \alpha^{2(n-j)}$ vanishes on the open subset $\mathcal{P}(\mathrm{Def}(X))$ of the irreducible quadric hypersurface Q_X. This already suffices to conclude that q_X divides g_β.

We have seen that q_X is non-degenerate. Thus, it defines an irreducible hypersurface $Q_X \subset \mathbb{P}(H^2(X, \mathbb{C}))$. By the Local Torelli Theorem the period map defines an isomorphism $\mathcal{P} : \mathrm{Def}(X) \cong U$, where U is an open subset of Q_X. If $\alpha = \mathcal{P}(t)$, then α represents the $(2,0)$-form σ_t on the fibre $\mathcal{X}_t$. Since β is of type $(2j, 2j)$ on $\mathcal{X}_t$ by assumption and $\mathcal{X}_t$ is of complex dimension $2n$, the classes $\beta \alpha^i \in H^{4j+2i}(X, \mathbb{C})$ vanish for $2i + 2j > 2n$, i.e. for $i > n - j$. Thus, for $i > n - j$ the complete intersection defined by $\beta(.)^i = 0$ contains U and hence Q_X.

Let us now come to the proof of the assertion. The integral $\int \beta \alpha^{2(n-j)}$ defines a homogeneous polynomial g_β of degree $2(n-j)$ on the projective space $\mathbb{P}(H^2(X, \mathbb{C}))$. The associated hypersurface is denoted (g_β). If $g_\beta = 0$, then set $c = 0$. If $g_\beta \neq 0$ and $\mathrm{supp}(g_\beta) \subset Q_X$, then there exists a constant c such that $g_\beta = c \cdot q_X^{n-j}$, i.e. $\int \beta \alpha^{2(n-j)} = c q_X(\alpha)^{n-j}$. Let now $g_\beta \neq 0$ and $\mathrm{supp}(g_\beta) \not\subset Q_X$. Then pick $\alpha \in (g_\beta)$ such that $\alpha \notin Q_X$. For the generic $\gamma \in \mathbb{P}(H^2(X, \mathbb{C})) \setminus (g_\beta)$ the line $\overline{\alpha\gamma}$ meets Q_X in two distinct points δ_0, δ_1. Denoting representatives of all the classes in $H^2(X, \mathbb{C})$ by the same symbols, one can write $\gamma = a_0 \delta_0 + a_1 \delta_1$ and $\alpha = b_0 \delta_0 + b_1 \delta_1$. Since $\alpha \notin Q_X$, both coefficients b_0 and b_1 are non-zero. Now use $\gamma \notin (g_\beta)$ and $\beta \delta_0^i = \beta \delta_1^i = 0$ for $i > n - j$ to prove

$$0 \neq \int \beta \gamma^{2(n-j)} = \int \beta (a_0 \delta_0 + a_1 \delta_1)^{2(n-j)}$$

$$= \binom{2(n-j)}{n-j}(a_0 a_1)^{n-j} \int \beta(\delta_0 \delta_1)^{n-j}$$

$$= \binom{2(n-j)}{n-j}\left(\frac{a_0 a_1}{b_1}\right)^{n-j} \int \beta \delta_0^{n-j}(\alpha - b_0 \delta_0)^{n-j}$$

$$= \binom{2(n-j)}{n-j}\left(\frac{a_0 a_1}{b_1}\right)^{n-j} \int \beta(\delta_0 \alpha)^{n-j}.$$

Hence, $\int \beta(\delta_0 \alpha)^{n-j} \neq 0$. On the other hand, using $\alpha \in (g_\beta)$ a similar calculation yields

$$0 = \int \beta \alpha^{2(n-j)} = \binom{2(n-j)}{n-j} b_0^{n-j} \int \beta(\delta_0 \beta)^{n-j}.$$

Since $b_0 \neq 0$, this contradicts $\int \beta(\delta_0 \alpha)^{n-j} \neq 0$. $\square$

A typical example of a cohomology class β that is of type $(2j, 2j)$ on all deformations of X is the Chern class $c_{2j}(X)$ or, more generally, any class contained in the subring generated by the Chern classes of X.

Corollary 23.18 *Let X be an irreducible holomorphic symplectic manifold. Then there exist constants $b_i \in \mathbb{Q}$ such that for any line bundle L on X one has:*

$$\chi(L) = \sum_{i=0}^{n} b_i q_X(c_1(L))^i.$$

Proof. Indeed, by the Hirzebruch–Riemann–Roch formula one has

$$\chi(L) = \sum \int_X \mathrm{ch}_i(L) \mathrm{td}_{2n-i}(X) = \sum \int_X \frac{c_1(L)^i}{i!} \mathrm{td}_{2n-i}(X).$$

The i-th Todd polynomial $\mathrm{td}_i(X)$ can be expressed as a polynomial in the Chern classes $c_j(X)$. Since $c_j(X) = c_j(\mathcal{T}_X) = c_j(\Omega_X) = (-1)^j c_j(\mathcal{T}_X)$, the odd Chern classes of X are trivial. Hence, $\mathrm{td}_j(X) = 0$ for $j \equiv 1(2)$. For the even Todd polynomials one applies Corollary 23.17 in order to get

$$\int_X c_1(L)^{2i} \mathrm{td}_{2n-2i}(X) = a_i q_X(c_1(L))^i,$$

where a_i is the constant c in the notation of Corollary 23.17 corresponding to the class $\mathrm{td}_{2n-2i}(X)$. Then set $b_i = a_i/(2i)!$.   □

Example 23.19 Let X be the Hilbert scheme $\mathrm{Hilb}^n(S)$ of a K3 surface S. For $n > 1$ the primitive Beauville–Bogomolov form q_X on X has the form

$$(H^2(X, \mathbb{Z}), q_X) \cong (H^2(S, \mathbb{Z}), \cup) \oplus (-2(n - 1))\mathbb{Z},$$

where $\cup$ is the intersection pairing on $H^2(S, \mathbb{Z})$ and $(-2(n-1))\mathbb{Z}$ is the lattice of rank one generated by a class δ with $q_X(\delta) = -2(n-1)$. Moreover, 2δ is the class of the exceptional divisor E of the natural projection $\mathrm{Hilb}^n(S) \to S^n(S)$.

If L is a line bundle on S, then $L \boxtimes \ldots \boxtimes L$ is an S_n-invariant line bundle on $S \times \ldots \times S$ and thus descends to a line bundle $S^n L$ on the symmetric product $S^n(S)$. The pull-back of it to the Hilbert scheme $\mathrm{Hilb}^n(S)$ is denoted by L_n. Since for an ample line bundle L the induced line bundle L_n on the Hilbert scheme is big and nef, one deduces

$$\chi(X, L_n) = h^0(X, L_n) = h^0(S^n(S), S^n L)$$

$$= \binom{\chi(L) + n - 1}{n} = \binom{L^2/2 + n + 1}{n} = \binom{q_X(c_1(L_n))/2 + n + 1}{n}.$$

This then suffices to conclude the same formula for any line bundle M on X. By Corollary 23.18 one knows that $\chi(X, M)$ is a polynomial in $q_X(c_1(M))$

with coefficients that only depend on X. Consider for any $k \in \mathbb{Z}_{>0}$ the line bundle $M = (L^{\otimes k})_n$ with L ample. The above formula determines $\chi(X, M)$, which is a polynomial in $k^2 q_X(c_1(L_n))$ with the same coefficients. Since $q_X(c_1(L_n)) \neq 0$, this determines the polynomial uniquely. Thus, we obtain

$$\chi(\text{Hilb}^n(S), L) = \binom{q_X(c_1(L))/2 + n + 1}{n}$$

for any line bundle L on $\text{Hilb}^n(S)$ (cf. [52]).

Example 23.20 For the second series of examples one can prove a similar formula for the Euler characteristic of a line bundle. The proof, due to Britze, is geometrically more interesting (cf. [29]). First, let us note that for the generalized Kummer variety $X = K_{n-1}(T)$ and $n > 2$ one has

$$(H^2(X, \mathbb{Z}), q_X) = (H^2(T, \mathbb{Z}), \cup) \oplus (-2n)\mathbb{Z}.$$

This was observed in [12]. Then one finds

$$\chi(X, L) = n \binom{q_X(c_1(L))/2 + n - 1}{n - 1}.$$

Remark 23.21 The above formulae for the Beauville–Bogomolov form on the two standard series might look mysterious. Here are some more explanations. Let v be the vector $(1, 0, 1 - n) \in H^0(S, \mathbb{Z}) \oplus H^2(S, \mathbb{Z}) \oplus H^4(S, \mathbb{Z})$. In fact, v is the Mukai vector of the ideal sheaf I_Z of any $Z \in \text{Hilb}^n(S)$, i.e. $v = \text{ch}(I_Z)\sqrt{\text{td}(S)}$. A natural quadratic form on $H^*(S, \mathbb{Z})$ is defined by $(\alpha_0, \alpha_1, \alpha_2)^2 = \alpha_1^2 - 2\alpha_0\alpha_2$. This yields the lattice $\tilde{H}(S, \mathbb{Z})$. In particular, $v^2 = 2(n - 1)$. Thus, v is isotropic only if $n = 1$. For $n > 1$ the orthogonal complement $v^\perp$ of $v \in \tilde{H}(S, \mathbb{Z})$ is $H^2(S, \mathbb{Z}) \oplus \mathbb{Z}\delta$, where $\delta = (1, 0, n-1)$. Then, the above description of $(H^2(X, \mathbb{Z}), q_X)$ has to be read as an isomorphism $(H^2(X, \mathbb{Z}), q_X) \cong v^\perp$ (cf. [12,163]). Indeed, $\delta^2 = -2(n - 1)$. This description was later successfully generalized by O'Grady in [168] to the case of moduli spaces of stable sheaves of rank at least two (see also [207] for further generalizations). The same explanation for the Beauville–Bogomolov form on the generalized Kummer varieties and higher rank analogues were given by Yoshioka in [208].

Remark 23.22 Let us come back to the general form of the Hirzebruch–Riemann–Roch formula of Corollary 23.18. In [167] Nieper-Wißkirchen computed these coefficients in terms of characteristic numbers of the manifold X. More precisely, if q_X is the Beauville–Bogomolov form before normalization as in Definition 22.8, then the coefficients b_i in $\chi(L) = \sum b_i q_X(c_1(L))^i$ are given as $b_i = f_i(X)(\int c_2(X)(\sigma\bar{\sigma})^{n-1})^{-i}$, where $f_i(X)$ is a certain linear combination of Chern numbers. In fact, Britze and Nieper-Wißkirchen were able to use this result and the explicit calculation for the generalized Kummer varieties to compute all Chern numbers in dimension ten.

23.5   Exercises

23.1 Let X be a compact hyperkähler manifold. Prove the Hodge-Index theorem: If α is a Kähler class and $\beta \in H^{1,1}(X)$ is q_X-orthogonal to α then $\beta = 0$ or $q_X(\beta) < 0$. Show that $q_X(\alpha, \beta) > 0$ if β is in the positive cone, i.e. in the connected component of $\{\beta \in H^{1,1}(X, \mathbb{R}) \mid q_X(\beta) > 0\}$ that contains a Kähler class, if β is the cohomology class of an effective divisor $Y \subset X$.

23.2 Show that for all $\alpha, \beta \in H^2(X)$ one has

$$q_X(\alpha, \beta) \int \alpha^{2n} = 2q_X(\alpha) \left(\int \alpha^{2n-1} \beta \right).$$

Deduce that α^{2n-1} and $(\sigma\bar\sigma)^{n-1}\alpha$ are linearly dependent for all $\alpha \in H^{1,1}(X)$.

23.3 Let $f : X \to B$ be a morphism with $\dim B < \dim X$. Show that $q_X(f^*\alpha) = 0$ for all $\alpha \in H^2(B)$.

23.4 Let X be a compact hyperkähler manifold and ω a Kähler class. Prove the following assertions for a class $\alpha \in H^{1,1}(X, \mathbb{R})$ with $\int \alpha^{2n} = 0$ (cf. [150]):

(i) If $\int \alpha\omega^{2n-1} = 0$, then $\alpha = 0$.
(ii) If $\int \alpha\omega^{2n-1} > 0$, then $q(\alpha, \omega) > 0$ and $\int \alpha^m \omega^{2n-m} = 0$ for m bigger than n and positive otherwise.
(Hint: Compute $q_X(t\omega + \alpha)$ and use Proposition 23.14.)

23.5 Find a new proof for the fact that the index of $q_X|_{H^{1,1}(X,\mathbb{R})}$ is $(1, b_2(X) - 3)$ by just using Proposition 23.14 and $q_X(\omega) > 0$ for a Kähler class ω. (Hint: Compute $q_X(\omega + t\alpha)$ for a class α orthogonal to ω.)

23.6 Use Proposition 23.14 and Exercise 23.2 to show that for any $\alpha, \beta \in H^2(X, \mathbb{R})$ the following holds:

$$\left(\int \alpha^{2n} \right)^2 q_X(\beta) = q_X(\alpha) \left((2n-1) \int \alpha^{2n-2}\beta^2 - (2n-2) \left(\int \beta\alpha^{2n-1} \right)^2 \right).$$

(This formula was first proved by Beauville in [12]. He then used it to deduce Proposition 23.14. So, the above approach reverses the order of the arguments.)

24   Cohomology of Compact Hyperkähler Manifolds

The cohomology ring of an arbitrary compact hyperkähler manifold is a great mystery. Even in the case of the Hilbert scheme of a K3 surface a complete description of the ring structure was only given recently by Lehn and Sorger [142]. The Betti and Hodge numbers of the two standard series, however, have been known for quite some time by the work of Göttsche and Soergel [70].

24.1   The Subring Generated by H^2

For a general compact hyperkähler manifold Verbitsky has shown that the subalgebra generated by all cohomology classes of degree two can be explicitly described. We follow Bogomolov's exposition in [23].

Proposition 24.1 *Let X be an irreducible holomorphic symplectic manifold of dimension $2n$ and let $\mathrm{SH}^2(X,\mathbb{C}) \subset H^*(X,\mathbb{C})$ be the subalgebra generated by $H^2(X,\mathbb{C})$. Then $\mathrm{SH}^2(X,\mathbb{C}) = S^*H^2(X,\mathbb{C})/\langle \alpha^{n+1} \mid q_X(\alpha) = 0 \rangle$.*

Proof. If $\sigma_t \in H^{2,0}(\mathcal{X}_t) \subset H^2(X,\mathbb{C})$, where $\mathcal{X}_t$ is a small deformation of X, then $\sigma_t^{n+1} = 0$. By the Local Torelli theorem 22.11 the local period map $\mathrm{Def}(X) \to Q_X$ is an open embedding and hence $\alpha^{n+1} = 0$ for an open subset of the quadric $\{\alpha \mid q_X(\alpha) = 0\}$. Thus, the canonical homomorphism $S^*H^2(X,\mathbb{C}) \to \mathrm{SH}^2(X,\mathbb{C})$ factorizes through the quotient by the ideal $\langle \alpha^{n+1} \mid q_X(\alpha) = 0 \rangle$.

There is a general fact from representation theory that says that if V is a complex vector space with a non-degenerate quadratic form q, then the algebra $A = S^*V / \langle \alpha^{n+1} \mid q(\alpha) = 0 \rangle$ has the following properties: (i) $A_{2n} = \mathbb{C}$ and (ii) $A_i \times A_{2n-i} \to A_{2n} = \mathbb{C}$ is non-degenerate.

Then one concludes as follows: If the map

$$A := S^*H^2(X,\mathbb{C})/\langle \alpha^{n+1} \mid q_X(\alpha) = 0 \rangle \to \mathrm{SH}^2(X;\mathbb{C}) \subset H^*(X,\mathbb{C})$$

were not injective, then by (ii) the kernel would be an ideal which contains the generator of A_{2n}. But $(\sigma + \bar{\sigma})^{2n} \in H^{4n}(X,\mathbb{C})$ is certainly not trivial. $\square$

The proposition says nothing about the cohomology in odd degree (which definitely is not trivial in general) nor about the even degree cohomology classes not contained in SH^2. In fact, almost nothing is known about those in general.

Let X be a compact Kähler manifold of dimension N. To any Kähler class $\alpha \in H^{1,1}(X)$ one associates a natural sl_2-representation

$$\varphi_\alpha : \mathrm{sl}_2(\mathbb{R}) \to \mathrm{End}(H^*(X,\mathbb{R}))$$

by $L_\alpha := \varphi_\alpha \left(\begin{smallmatrix} 0 & 1 \\ 0 & 0 \end{smallmatrix}\right) = $ multiplicaton by α, $\Lambda_\alpha := \varphi_\alpha \left(\begin{smallmatrix} 0 & 0 \\ 1 & 0 \end{smallmatrix}\right) = $ (the dual Lefschetz operator), and $H := \varphi_\alpha \left(\begin{smallmatrix} 1 & 0 \\ 0 & -1 \end{smallmatrix}\right)$ with $H|_{H^k(X,\mathbb{C})} = (N - k) \cdot id$.

In order to produce a bigger Lie algebra that acts on the cohomology one could try two things. Firstly, one could just change the Kähler class α on the complex manifold X. Then all the induced Lie algebras $\varphi_\alpha(\mathrm{sl}_2)$ generate the so called Kähler Lie algebra. Secondly, one could change the complex structure, i.e. one considers a deformation $\mathcal{X} \to S$ of $X = \mathcal{X}_0$ and Kähler classes α_t on any fibre $\mathcal{X}_t$. Since the cohomology $H^*(\mathcal{X})$ is locally constant, this yields a family of Lie algebras $\varphi_{\alpha_t}(\mathrm{sl}_2)$. However, in general there will be no canonical choice for α_t on the nearby fibres, as we can deform the Kähler structure in many different ways. In both cases one would wish to describe the resulting Lie algebra.

Until recently this question had never been studied. In [145] Looijenga and Lunts show that these Lie algebras can be classified under the additional assumption that the dual Lefschetz operators Λ_α, $\Lambda_{\alpha'}$ commute for different classes α and α'.

For hyperkähler manifolds all this works very nicely. It had partially been studied before by Fujiki and Verbitsky. Some of the results in this direction admit an explanation from the point of view of supersymmetry (cf. [55]).

24.2   The so(4, 1)-action

Here, we will discuss the action of the Lie algebra $\mathrm{so}(4,1)$ on the cohomology $H^*(X, \mathbb{R})$ of a compact hyperkähler manifold which is associated to any particular choice of a hyperkähler structure or, more precisely, to the induced positive three space $\langle \omega_I, \omega_J, \omega_K \rangle \subset H^2(X, \mathbb{R})$. This fits in the setting explained above. Indeed, the Lie algebra generated by $\varphi_{\omega_I, \omega_J, \omega_K}(\mathrm{sl}_2)$ is the Lie algebra induced by the twistor space $\mathcal{X} \to \mathbb{P}^1$ (cf. §25.3) endowed with the canonical choice of a Kähler class ω_λ on the fibre $\mathcal{X}_\lambda$ over any $\lambda = aI + bJ + cK \in S^2 = \mathbb{P}^1$.

Let V be a euclidian vector space with an orthogonal $\mathbb{H}$-action as in §23.2. In particular, one has the three complex structures I, J, K and the three real two-forms $\omega_I, \omega_J, \omega_K \in \Lambda^2 V^*$. By definition the Lefschetz operators are given by $L_\lambda(\alpha) = \omega_\lambda \wedge \alpha$ for $\lambda = I, J, K$ and its adjoint operators are $\Lambda_\lambda = *^{-1} L_\lambda *$. To any $\lambda = I, J, K$ one has the natural sl_2-algebra generated by $\langle L_\lambda, \Lambda_\lambda, H \rangle$. The following proposition is the quaternionic version of the isomorphism $\mathrm{sl}_2(\mathbb{R}) = \mathrm{so}(2,1)$.

Proposition 24.2 *Let* $\mathfrak{g} \subset \mathrm{End}(\Lambda^* V^*)$ *be the Lie algebra generated by the operators* $\{L_\lambda, \Lambda_\lambda, H\}_{\lambda=I,J,K}$. *Then* $\mathfrak{g}$ *is isomorphic to* $\mathrm{so}(4,1)$.

Proof. The quaternionic vector space V can be written as the orthogonal direct sum of four-dimensional ones, i.e. $V = \bigoplus V_i$ with $V_i = \mathbb{H}$. The assertion thus reduces to the case $n = 1$. The representation $\Lambda^* \mathbb{H}$ splits into $\Lambda^{\mathrm{odd}} \mathbb{H}$, $W := \mathbb{R} \oplus \langle \omega_I, \omega_J, \omega_K \rangle \oplus \mathbb{R}$, and the trivial representation $(\Lambda^2 \mathbb{H})_{\mathrm{prim}}$ of all two-forms annihilated by all L_λ. The vector space W is endowed with a quadratic form of index $(4,1)$ as follows: On $\mathbb{R} \oplus \mathbb{R}$ one takes the hyperbolic form and on $\langle \omega_I, \omega_J, \omega_K \rangle$ one declares the basis $\omega_I, \omega_J, \omega_K$ to be orthonormal. By applying L_λ to $\alpha = 1, \omega_\mu$, and vol, respectively, one easily checks that $L_\lambda \in so(W)$. Since Λ_λ is the adjoint, this shows that the image of $\mathfrak{g} \to \text{End}(W)$ is contained in $so(W) = so(4,1)$. The Lie algebra $so(W)$ is ten-dimensional (e.g. use the description $so(W) = \Lambda^2 W$). Consider the elements $H, L_I, L_J, L_K, \Lambda_I, \Lambda_J, \Lambda_K, K_{IJ} := [L_I, \Lambda_J], K_{IK} := [L_I, \Lambda_K]$, and $K_{JK} := [L_J, \Lambda_K]$ of $\mathfrak{g}$. Note that L_λ is of degree 2, Λ_λ is of degree -2, and $K_{\lambda\mu}$ is of degree zero with e.g. $K_{IJ}(\omega_I) = -2\omega_J$, $K_{IJ}(\omega_J) = 2\omega_I$, $K_{IJ}(\omega_K) = 0$. Using this, one easily verifies that these ten operators are linearly independent on W. Thus, $\mathfrak{g} \to so(W)$ is surjective. Eventually, one checks that $\mathfrak{g}$ is also generated by these operators by showing in addition to the standard commutator relations for L and Λ that e.g. $[L_J, K_{JK}] = 2L_K$. This yields the injectivity of the map $\mathfrak{g} \to so(W)$. □

Corollary 24.3 *Let $[\omega_I], [\omega_J]$, and $[\omega_K]$ be the three Kähler classes induced by a hyperkähler metric g on a irreducible holomorphic symplectic manifold X. If $\mathfrak{g}_g \subset \text{End}(H^*(X, \mathbb{R}))$ is the Lie algebra that is generated by the three associated sl_2-representations, then $\mathfrak{g}_g$ is isomorphic to $so(4,1)$. [192]*

Proof. The crucial point here is, that all operators $L_\lambda, \Lambda_\lambda$ preserve the space of harmonic forms. This is due to the fact that they are all associated with the same metric. Identifying the space of harmonic forms with cohomology yields that $so(4,1)$ acts on $H^*(X, \mathbb{R})$. Since the Lie algebra cannot be smaller then $so(4,1)$ this proves the assertion. □

Remark 24.4 Note that the same trick will not apply to Lefschetz operators L_ω, Λ_ω and $L_{\omega'}, \Lambda_{\omega'}$ associated with different metrics, e.g. L_ω does not respect the space of forms which are harmonic with respect to ω'.

The following corollary is due to Looijenga and Lunts [145].

Corollary 24.5 *The Hodge structure on the cohomology $H^*(X)$ of an irreducible holomorphic symplectic manifold X is uniquely determined by the Hodge structure on $H^2(X)$ and the action of $H^2(X)$ on $H^*(X)$.*

Proof. Choose a hyperkähler metric g on X and let I, J, K be the associated complex structures. We may assume that I is the complex structure that gives the complex manifold X. Then we use the operator $K_{JK} = [L_J, \Lambda_K]$ introduced in the proof of the proposition. Since $\sigma_I = \omega_J + i\omega_K$, this operator only depends on the Hodge structure on $H^2(X)$, and the action of $H^2(X)$ on $H^*(X)$. Moreover, again by reducing to the one-dimensional case, one finds that $K_{JK} = i \sum (p - q)\Pi_{pq}$, where Π_{pq} is the projection to the space of $(p,q)_I$-forms. Thus, the operator K_{JK} detects the Hodge structure on $H^*(X)$. □

24.3 The Action of the Total Lie Algebra

If one not only wants to consider one hyperkähler metric on the underlying real manifold X but allows all possible ones, then one obtains an sl_2-action for any class α in an open subset in $H^2(X, \mathbb{R})$. The Lie algebra generated by all these representations generates the *total Lie algebra* $\mathfrak{g}_{tot}$ of X. This Lie algebra was described by Verbitsky [195,193] and Looijenga-Lunts [145]:

Proposition 24.6 *The total Lie algebra of a compact hyperkähler manifold X is naturally isomorphic to $\mathrm{so}((H^2(X, \mathbb{R}), q_X) \oplus U)$, where U is the hyperbolic plane.*

Proof. The crucial point here is to show that the dual Lefschetz operators Λ_i, $i = 1, 2$ for two different Kähler classes α_i commute, i.e. $[\Lambda_1, \Lambda_2] = 0$. In fact, the Kähler classes α_i need not be of type $(1, 1)$ on the same complex manifold. We have seen however, that the dual Lefschetz operators associated to Kähler classes contained in the positive three-space $F(g)$ of a hyperkähler metric g do commute (even on the level of forms). Now we use the following easy observation: The set of positive three-spaces $F(g)$ associated to a hyperkähler metric g forms an open subset of the set of all positive three-spaces $F \subset H^2(X, \mathbb{R})$ (use the Local Torelli theorem and the openness of the Kähler cone). Since the set of all positive three-spaces is connected, this proves that for a given α_1 the commutator $[\Lambda_1, \Lambda_2] = 0$ vanishes for all α_2 in the open subset of all classes for which there exists a positive three-space that contains this class together with α_1. Thus, the commutator always vanishes, as the vanishing of the commutator is an algebraic condition. The vanishing of the commutator implies that the total Lie algebra contains only elements of degree $-2, 0$, and 2. The rest of the proof is purely algebraic and we refer the reader to [145]. □

Remark 24.7 It is remarkable that in the above proof the commutativity of the dual Lefschetz operators was not obtained by proving it on each tangent space and then passing to cohomology. The commutativity holds on the level of forms, but passing to cohomology is problematic, as the space of harmonic forms definitely changes when changing the hyperkähler metric (see [106]).

24.4 Further Results on the Cohomology

What else do we know about the cohomology of a compact hyperkähler manifold? Here we just wish to present a list of known results without giving complete proofs.

1. Wakakuwa [198]: The odd Betti numbers of a compact hyperkähler manifold are all divisible by four. This result is the hyperkähler version of the well-known fact that the odd Betti numbers of a compact Kähler manifold are even. We refer the reader to [60]. Note that b_1 is always zero, but in general b_3 is not, e.g. $b_3(K_2(T)) = 8$.

2. Salamon [179]: If X is a compact hyperkähler manifold of dimension $2n$ then

$$\sum_{i=0}^{4n}(-1)^i(6i^2 - n(6n+1))b_i = 0$$

or equivalently

$$\sum_{i=0}^{4n}(-1)^i 6i^2 b_i = n(6n+1)e(X).$$

Using Wakakuwa's results this shows in particular that $ne(X)$ is divisible by 24. Note that for a K3 surface the Euler number is 24. (The divisibility by 24 was also obtained by Gritsenko and Hirzebruch [79].)

3. Beauville: In dimension four one knows $b_2 \leq 23$. Indeed, for $n = 2$ Salamon's results yields $b_3 + b_4 = 46 + 10b_2$. On the other hand, Verbitsky's result (Proposition 24.1) implies $b_4 \geq b_2(b_2 + 1)/2$. Thus, $(23 - b_2)(b_2 + 4) \geq 2b_3 \geq 0$. Recall that $b_2(\mathrm{Hilb}^n(K3)) = 23$ for $n \geq 2$.

4. Guan: Using similar techniques as Beauville and Verbitsky he shows that in dimension four either $3 \leq b_2 \leq 8$ or $b_2 = 23$ (see [89]). He also obtains further bounds on b_3.

24.5   Application: Proof of Matsushita's Theorem

In the following we will prove a version of Matsushita's result Theorem 21.11. On the one hand, we will allow more generally non-projective hyperkähler manifolds, but on the other hand will assume that the base of the fibration is smooth. Conjecturally, the latter assumption always holds. The proof follows closely the arguments of Matsushita [150–152].

Proposition 24.8 *Let X be an irreducible holomorphic symplectic manifold of dimension $2n$ and let $\pi : X \to B$ be a non-constant morphism of positive fibre dimension onto a Kähler manifold B. Then*

(i) B is of dimension n, projective and satisfies $b_2(B) = \rho(B) = 1$. Moreover, K_B^* is ample, i.e. B is Fano.

(ii) Every fibre is complex Lagrangian and, in particular, of dimension n.

(iii) Every smooth fibre is an n-dimensional complex torus.

Proof. Let ω and ω_B be Kähler classes on X and B, respectively.

(i) Since $\dim(B) < 2n$, one has $(\pi^*\omega_B)^{2n} = 0$. Thus, $q_X(\pi^*\omega_B) = 0$. Using Proposition 24.1 this proves $(\pi^*\omega_B)^{n+1} = 0$ and, therefore, $\dim(B) \leq n$. On the other hand, since $\pi^*\omega_B \neq 0$ also $(\pi^*\omega_B)^n \neq 0$ by Proposition 24.1. Thus, $\dim(B) = n$.

If $\alpha \in H^2(X, \mathbb{R})$ with $\int_X \alpha^{2n} = 0$ and $\int_X \alpha^n \pi^* \omega_B^n = 0$, then $\alpha \in \mathbb{R}\pi^*\omega_B$. Indeed, $\int_X (\alpha - \lambda\pi^*\omega_B)^{2n} = \binom{2n}{n}\lambda^n \int_X \alpha^n \pi^* \omega_B^n = 0$, since $\int_X \alpha^{2n} = 0$ implies $\alpha^{n+1} = 0$ by Proposition 24.1. Now choose λ such that $\int_X (\alpha -$

$\lambda \pi^* \omega_B) \omega^{2n-1} = 0$, which is possible as $\int_X \pi^* \omega_B \omega^{2n-1} \neq 0$. Using Exercise 23.4 one concludes that $\alpha - \lambda \pi^* \omega_B = 0$.

If $\alpha = \pi^* \beta$ with $\beta \in H^2(B, \mathbb{R})$, then the above yields $\beta \in \mathbb{R} \omega_B$. Thus, $b_2(B) = 1$. Hence, B admits an integral Kähler class and is thus projective (cf. Theorem 26.1). Since $H^1(B, \mathcal{O}_B) = 0$, the canonical bundle K_B must be a multiple of the ample generator H of $\mathrm{Pic}(B)$. On the other hand, its pull-back $\pi^* K_B$ is a subsheaf of the semistable degree zero vector bundle Ω_X^n and, thus, $K_B = mH$ with $m \leq 0$. Here we use that any hyperkähler metric on X defines a Hermite-Einstein metric on the cotangent bundle und hence also on Ω_X^n. This is enough to conclude that the bundle is polystable (and hence semistable) of degree zero. If $m = 0$, then the pull-back $\pi^*(s)$ of a trivializing section s of K_B would be a non-trivial section of Ω_X^n. For n odd such a section cannot exist. For n even it must be a multiple of $\sigma^{n/2}$. Since $\pi^*(s) \wedge \pi^*(s) = 0$ and $\sigma^{n/2} \wedge \sigma^{n/2} = \sigma^n \neq 0$, also for n even $\pi^*(s) = 0$. Hence, K_B is negative.

(ii) Let F be a smooth fibre of X. In order to show that $\sigma|_F = 0$, it suffices to show that $\int_F (\sigma \bar\sigma) \omega^{2n-2} = 0$ (use that $\sigma \bar\sigma$ is a weakly positive form). The latter equals (up to a non-zero scalar) the integral $\int_X (\sigma \bar\sigma) \omega^{n-2} \pi^* \omega_B^n$. Now use $\int_X (\sigma + \bar\sigma + s\omega + t\pi^* \omega_B)^{2n} = c q_X (\sigma + \bar\sigma + s\omega + t\pi^* \omega_B)^n = c(q_X(\sigma + \bar\sigma) + s^2 q_X(\omega) + 2stq_X(\omega, \pi^* \omega_B))^n$ for some positive scalar c. This yields $\int_X (\sigma \bar\sigma) \omega^{n-2} \pi^* \omega_B^n = 0$, as the coefficient of $s^{n-2} t^n$ on the right hand side is trivial.

In order to show that any irreducible component F of a fibre $\pi^{-1}(t)$ is complex Lagrangian one uses the result of Kollár and Saito [178] that says that $R^2 \pi_* \omega_X$ is torsion free. Since in our case $\omega_X \cong \mathcal{O}_X$, this shows that $R^2 \pi_* \mathcal{O}_X$ is torsion free. Consider $\bar\sigma \in H^2(X, \mathcal{O}_X)$ and its image $\rho \in H^0(B, R^2 \pi_* \mathcal{O}_X)$. As the general fibre is complex Lagrangian, ρ must be torsion and hence zero. On the other hand, if $\tilde F \to F$ is a resolution of F, then the image of $\bar\sigma$ in $H^2(\tilde F, \mathcal{O}_{\tilde F})$ is contained in the image of $R^2 \pi_* \mathcal{O}_X \otimes k(t) \to H^2(F, \mathcal{O}_F) \to H^2(\tilde F, \mathcal{O}_{\tilde F})$ and hence trivial. This implies that the image of σ in $H^0(\tilde F, \Omega_{\tilde F}^2)$ is trivial, i.e. $\sigma|_F = 0$. Since $\dim F \geq n$, this shows that F is complex Lagrangian.

(iii) is a consequence of a well-known (in the real situation this is called the Liouville-Arnold theorem) result. The morphism $\pi : X \to B$ is a completely integrable system and the general fibre is thus isomorphic to a complex torus (cf. [13,148]).   □

24.6 Exercises

24.1 Show that $H^*(X, \mathbb{R}) = \mathrm{SH}^2(X, \mathbb{R}) \oplus \mathrm{SH}^2(X, \mathbb{R})^\perp$ with respect to the intersection pairing.

24.2 Use the classical result of Chen-Ogiue or the Bogomolov inequality for stable vector bundles to deduce that the SH^2-component of $c_2 \in$

$H^4(X, \mathbb{R})$ of an irreducible holomorphic symplectic manifold is not trivial.

24.3 Compare the cohomology of a complex torus with the cohomology of a compact hyperkähler manifold.

24.4 Prove that the base space B in Proposition 24.8 for $n = 2$ is isomorphic to the projective plane.

24.5 Let $S \to \mathbb{P}^1$ be an elliptic K3 surface. Show without using Proposition 24.8 that the fibres of $\text{Hilb}^n(S) \to \mathbb{P}^n$ are complex Lagrangians (cf. §21.4 and use the description of the symplectic structure in §21.2).

25   Twistor Space and Moduli Space

25.1   The Twistor Space

We have learned that on an irreducible holomorphic symplectic manifold there always exists a hyperkähler metric g. Actually, any Kähler class $\alpha \in \mathcal{K}_X$ can be represented by a hyperkähler form. In particular, the given complex structure I comes along with two other complex structures J and K, with respect to which g is Kähler and which satisfy the relation $K = IJ = -JI$. In fact, any $\lambda = aI + bJ + cK$ with $a^2 + b^2 + c^2 = 1$ defines a complex structure on the underlying real manifold X with respect to which g is a Kähler metric. The associated Kähler form $g(\lambda\, ,\,)$ is denoted ω_λ. Thus, a hyperkähler metric g defines a family of complex Kähler manifolds $(X, \lambda, \omega_\lambda)$, where $\lambda \in S^2 \cong \mathbb{P}^1$. This can be turned into the concept of the twistor space of a hyperkähler manifold: Consider the product $X \times \mathbb{P}^1$ and define

$$\mathbb{I} : T_m X \oplus T_\lambda \mathbb{P}^1 \to T_m X \oplus T_\lambda \mathbb{P}^1 \quad , \quad (v, w) \mapsto (\lambda(v), I_{\mathbb{P}^1}(w)),$$

where $\lambda \in \mathbb{P}^1 \cong S^2$ as above. Obviously, $\mathbb{I}$ is an almost complex structure, but in fact (cf. [97])

Lemma 25.1 *The almost complex structure $\mathbb{I}$ on $X \times \mathbb{P}^1$ is integrable.*

Thus, we obtain a complex manifold $\mathcal{X}(\alpha) := (X \times \mathbb{P}^1, \mathbb{I})$ for any Kähler class $\alpha \in H^{1,1}((X, I))$. Equivalently, one could say that $\mathcal{X}(\alpha)$ is associated with the hyperkähler metric g. The class α is the Kähler class of the Kähler form ω_I. By construction, the projection $\pi : \mathcal{X}(\alpha) \to \mathbb{P}^1 =: T(\alpha)$ is holomorphic and the fibre over $\lambda \in \mathbb{P}^1$ is isomorphic to (X, λ). The complex manifold $\mathcal{X}(\alpha)$ together with this projection is called the *twistor space* of X with respect to α.

Remark 25.2 Although any fibre of $\pi : \mathcal{X}(\alpha) \to T(\alpha)$ has a natural Kähler structure, the twistor space $\mathcal{X}(\alpha)$ is not Kähler. Indeed, if ω were a Kähler form on the complex manifold $\mathcal{X}(\alpha)$, then the restriction to any fibre (X, λ) would again be Kähler. Since the underlying real manifold of $\mathcal{X}(\alpha)$ is $X \times \mathbb{P}^1$, the associated cohomology class $\gamma := [\omega|_{(X,\lambda)}] \in H^{1,1}((X, \lambda)) \subset H^2(X)$ does not depend on λ. If $\omega_\lambda := g(\lambda ,)$ is the canonical Kähler form on (X, λ) induced by the hyperkähler metric g, then $\int_X \gamma^{2n-1} \omega_\lambda > 0$. But, since $\omega_{-I} = -\omega_I$, the inequality cannot hold for $-I, I \in \mathbb{P}^1$ at the same time. This yields a contradiction.

25.2 Global Period Map

To any irreducible holomorphic symplectic manifold X one associates the lattice $(H^2(X, \mathbb{Z}), q_X)$ together with the Hodge-decomposition of $H^2(X, \mathbb{C})$. The latter is given by $\mathbb{C}\sigma \subset H^2(X, \mathbb{Z}) \otimes \mathbb{C}$, since $H^{2,0}(X)$ and $H^{0,2}(X)$ are both one-dimensional. This datum is called the "period" of X.

Definition 25.3 For a lattice (γ, q_γ) of index $(3, b-3)$ one defines the *period domain* $Q_\gamma \subset \mathbb{P}(\gamma \otimes \mathbb{C})$ as the set

$$Q_\gamma := \{x \in \mathbb{P}(\gamma \otimes \mathbb{C})| \ q_\gamma(x) = 0, \ q_\gamma(x + \bar{x}) > 0\}.$$

So Q_γ is an open subset of a smooth quadric hypersurface in $\mathbb{P}(\gamma \otimes \mathbb{C})$. In §22 we have considered the period map

$$\mathcal{P}_X : \mathrm{Def}(X) \longrightarrow Q_X \subset \mathbb{P}(H^2(X, \mathbb{C})), \ t \mapsto [\sigma_t],$$

where $\mathcal{X} \to \mathrm{Def}(X)$ is the universal deformation of $X = \mathcal{X}_0$. In order to make the target independent of X, one has to introduce the moduli space of marked manifolds.

Definition 25.4 A marking of an irreducible holomorphic symplectic manifold X is a lattice isomorphism $\varphi : (H^2(X, \mathbb{Z}), q_X) \cong (\gamma, q_\gamma)$. The pair (X, φ) is called a *marked manifold*. The *moduli space* of marked irreducible holomorphic symplectic manifolds $\mathfrak{M}_\gamma$ is the set $\{(X, \varphi)\}/ \sim$, where (X, φ) is a marked irreducible holomorphic symplectic manifold of dimension $2n$ and $(X, \varphi) \sim (X', \varphi')$ if and only if there exists an isomorphism $f : X \cong X'$ such that $f^* = \varphi^{-1} \circ \varphi'$.

The composition of the local period map $\mathcal{P}_X$ (cf. Definition 22.10) and the isomorphism $\mathbb{P}(H^2(X, \mathbb{C})) \cong \mathbb{P}(\gamma \otimes \mathbb{C})$ induced by a marking φ gives rise to the period map $\mathcal{P} : \mathrm{Def}(X) \to \mathbb{P}(\gamma \otimes \mathbb{C})$ for any marked manifold (X, φ). The image is contained in Q_γ and by the Local Torelli Theorem 22.11 the map $\mathcal{P} : \mathrm{Def}(X) \to Q_\gamma$ is a local isomorphism. This allows one to prove that the deformation spaces $\mathrm{Def}(X)$ glue and thus define on $\mathfrak{M}_\gamma$ the structure of a smooth non-separated (i.e. non-Hausdorff) complex space.

Remark 25.5 The set $\mathfrak{M}_\gamma$ is a moduli space of marked irreducible holomorphic symplectic manifolds in the following sense: For any family $(\mathcal{X}, \varphi) \to S$ of marked irreducible holomorphic symplectic manifolds there exists a uniquely defined holomorphic map $S \to \mathfrak{M}_\gamma$ such that $t \in S$ is mapped to $(\mathcal{X}_t, \varphi_t) \in \mathfrak{M}_\gamma$ and such that locally $\mathcal{X} \to S$ is the pull-back of the universal family $\mathcal{X} \to \mathrm{Def}(X)$ under the holomorphic map $S \to \mathrm{Def}(X) \subset \mathfrak{M}_\gamma$, where $\mathrm{Def}(X) \subset \mathfrak{M}_\gamma$ is defined by φ_0. But, there does not exist a universal family over $\mathfrak{M}_\gamma$! In order to construct a universal family over $\mathfrak{M}_\gamma$ one would have to glue the universal families over $\mathrm{Def}(X), \mathrm{Def}(X') \subset \mathfrak{M}_\gamma$ over the intersection $\mathrm{Def}(X) \cap \mathrm{Def}(X')$, but there might be different ways to do this and one usually runs into problems of compatibility for three of them. Indeed, an irreducible holomorphic symplectic manifold might have automorphisms that act trivially on the second cohomology. An example for such a phenomenon was given by Beauville in [11]. Note that in dimension two, i.e. for K3 surfaces, $\mathfrak{M}_\gamma$ is fine, i.e. there exists a universal family (the Global Torelli holds!).

Definition 25.6 The *global period map* is the holomorphic map $\mathcal{P} : \mathfrak{M}_\gamma \to Q_\gamma \subset \mathbb{P}(\gamma \otimes \mathbb{C})$ obtained by glueing the local period maps $\mathrm{Def}(X) \to Q_X \cong Q_\gamma \subset \mathbb{P}(\gamma \otimes \mathbb{C})$.

25.3   Twistor Lines in the Moduli Space

As we explained above, the choice of a Kähler class $\alpha \in \mathcal{K}_X$ on a compact hyperkähler manifold gives rise to a global family, the twistor space $\mathcal{X}(\alpha) \to \mathbb{P}^1$. Moreover, as the total space $\mathcal{X}$ is naturally diffeomorphic to the product $X \times \mathbb{P}^1$, any marking of X induces a marking of all twistor fibres $\mathcal{X}_t$. The classifying map $\mathbb{P}^1 \to \mathfrak{M}_\gamma$ composed with the period map yields a morphism $\mathbb{P}^1 \to Q_\gamma$. The following proposition describes this map infinitesimally and globally.

Proposition 25.7 (i) *The image of the Kodaira–Spencer map*

$$T_{\{I\}}\mathbb{P}^1 \to H^1(X, \mathcal{T}_X)$$

given by the twistor space $\mathcal{X}(\alpha) \to T(\alpha) = \mathbb{P}^1$ is the one-dimensional subspace spanned by the image of α under the natural isomorphism $H^1(X, \Omega_X) \cong H^1(X, \mathcal{T}_X)$. [60].

(ii) *Let $F(\alpha) := \langle \omega_I, \omega_J, \omega_K \rangle \subset H^2(X, \mathbb{R})$ be the positive three-space associated with the hyperkähler metric g uniquely determined by the Kähler class α. Then the period map $\mathcal{P}_X$ identifies the base $\mathbb{P}^1$ of the twistor space $\mathcal{X}(\alpha) \to \mathbb{P}^1$ with the quadric $Q_X \cap \mathbb{P}(F(\alpha)_{\mathbb{C}})$.*

Proof. (i) Let us first recall the definition of the Kodaira–Spencer class in slightly different terms than in §22. Let $I(t)$ be a family of complex structures with $I(0) = I$ and let $\phi(t) : T_I^{0,1} \to T_I^{1,0}$ be the vector bundle homomorphism with $v + \phi(t)(v) \in T^{0,1}_{I(t)}$ for all $v \in T^{0,1}_I$ and all t. Thus, $\phi(t) \in \mathcal{A}^{0,1}(\mathcal{T}_X)$. Let $\phi(t) = \phi_0 + \phi_1 t + \phi_2 t^2 + \dots$ be its power series expansion. Then $\phi_0 = 0$ and the Dolbeault cohomology class $[\phi_1] \in H^1(X, \mathcal{T}_X)$ of the linear coefficient is the Kodaira-Spencer class. Note that $\bar{\partial}\phi_1 = 0$ follows from the integrability condition $[T^{0,1}_{I(t)}, T^{0,1}_{I(t)}] \subset T^{0,1}_{I(t)}$.

In our situation let $I(t) = \lambda_t \in \mathbb{P}^1$. If $\lambda_t = I + \lambda_1 t + \lambda_2 t^2 + \dots$, we may assume $\lambda_1 = J$, as the tangent space we want to describe is complex one-dimensional. Comparing the linear coefficients in

$$\lambda_t(v + \phi(t)(v)) = -i(v + \phi(t)(v))$$

yields $\lambda_1(v) + I(\phi_1(v)) = -i\phi_1(v)$. Since $\phi_1(v) \in T^{1,0}_I$, one finds $\lambda_1(v) = -2i\phi_1(v)$. Now apply $-\omega_K : T \to T^*$ to both sides. This gives $-g(K(J(v))) = 2i\omega_K(\phi_1(v))$. The left hand side is $\omega_I(v)$ and the right hand side equals $2i\omega_K(\phi_1(v)) = (\sigma + \bar{\sigma})(\phi_1(v)) = \sigma(\phi_1(v))$, as $\sigma = \omega_J + i\omega_K$ and $\phi_1(v) \in T^{1,0}_I$. Hence, $\omega_I(v) = \sigma(\phi_1(v))$ for all $v \in T^{0,1}_I$. This proves (i). There is also a more algebraic proof of this assertion (cf. [103]).

(ii) We know by Proposition 22.11 that the period map takes value in Q_X. On the other hand, for any complex structure $\lambda \in \mathbb{P}^1 = T(\alpha)$ the holomorphic two-form σ_λ on the fibre (X, λ) is contained in $F(\alpha)_{\mathbb{C}}$. Thus, $\mathcal{P}(\mathbb{P}^1) \subset Q_X \cap \mathbb{P}(F(\alpha)_{\mathbb{C}})$. Since $Q_X \cap \mathbb{P}(F(\alpha)_{\mathbb{C}})$ is a non-degenerate quadric, this yields a holomorphic map $\mathbb{P}^1 \to Q_X \cap \mathbb{P}(F(\alpha)_{\mathbb{C}}) \cong \mathbb{P}^1$. Moreover, by (i) its differential is non-zero at every point λ. Hence, $\mathcal{P}$ defines an isomorphism $T(\alpha) \cong Q_X \cap \mathbb{P}(F(\alpha)_{\mathbb{C}})$. $\qquad\square$

25.4 Surjectivity of the Period Map

Any Calabi–Yau manifold has unobstructed deformations (cf. Theorem 22.5), but only for hyperkähler manifolds some of them naturally globalize to complete families. This is due to the existence of the twistor space $\mathcal{X}(\alpha) \to T(\alpha) = \mathbb{P}^1$ for any Kähler class $\alpha \in \mathcal{K}_X$ on an irreducible holomorphic symplectic manifold X. This will be crucial in the proof of the surjectivity of the period map $\mathcal{P} : \mathfrak{M}_\gamma \to Q_\gamma$. As we have seen above, the period map identifies the base space $T(\alpha)$ of the twistor space with the quadric $Q_\gamma \cap \mathbb{P}(\varphi(F(\alpha)_{\mathbb{C}}))$, where φ is a marking of X. The essential idea is to show that any two points $x, x' \in Q_\gamma$ can be connected by a sequence of curves of this form.

We first study this aspect of the period domain separately from the period map. Most of what will be said about Q_γ is a straightforward generalization of well-known results in the case of K3 surfaces.

Proposition 25.8 *The map $x \mapsto P(x) := \langle \mathrm{Re}(x), \mathrm{Im}(x) \rangle$ defines a bijection between the period domain Q_γ and the space $\mathrm{Gr}^{po}_2(\gamma_{\mathbb{R}})$ of oriented positive planes in the real vector space $\gamma_{\mathbb{R}}$.*

Proof. For $x \in Q_\gamma$ one has $q_\gamma(x) = 0$ and $q_\gamma(x + \bar{x}) > 0$. Hence, $\mathrm{Re}(x)$ and $\mathrm{Im}(x)$ form an oriented orthogonal base of the positive space $P(x)$. For the inverse map choose such a basis (v_1, v_2) of $P \in \mathrm{Gr}_2^{\mathrm{po}}(\gamma_\mathbb{R})$ and define $x := v_1 + iv_2$. As a point in $\mathbb{P}(\gamma_\mathbb{C})$ the definition does not depend on the choice of the basis (v_1, v_2) . $\qquad\square$

Corollary 25.9 *The period domain Q_γ is naturally isomorphic to the homogeneous space $\mathrm{O}(\gamma_\mathbb{R})/(\mathrm{SO}(2) \times \mathrm{O}(1, b - 3))$. In particular, Q_γ is simply-connected.*

Proof. Fix a plane $P_0 \in \mathrm{Gr}_2^{\mathrm{po}}(\gamma_\mathbb{R})$. Any other plane $P \in \mathrm{Gr}_2^{\mathrm{po}}(\gamma_\mathbb{R})$ can be obtained by an orthogonal transformation $\psi \in \mathrm{O}(\gamma_\mathbb{R})$. The stabilizer of P_0 is naturally isomorphic to $\mathrm{SO}(P_0) \times \mathrm{O}(P_0^\perp)$. Then use the following classical results on the homotopy groups of $\mathrm{O}(3, b - 3)$, $\mathrm{SO}(2)$, and $\mathrm{O}(1, b - 3)$ (see e.g. [141]): The orthogonal groups $\mathrm{O}(3, b - 3)$ and $\mathrm{O}(1, b - 3)$ both have four connected components, which correspond to each other via the natural inclusion. Thus, the quotient is connected. Let $\mathrm{SO}(r, s)^o$ denote the connected component of the identity of $\mathrm{O}(r, s)$. Thus, $\mathrm{O}(\gamma_\mathbb{R})/(\mathrm{SO}(2) \times \mathrm{O}(1, b - 3)) = \mathrm{SO}(\gamma_\mathbb{R})^o/(\mathrm{SO}(2) \times \mathrm{SO}(1, b-3)^o)$. Then one uses $\pi_1(\mathrm{SO}(r, s)^o) = \pi_1(\mathrm{SO}(r)) \times \pi_1(\mathrm{SO}(s))$ and $\pi_1(\mathrm{SO}(r)) = \mathbb{Z}/2\mathbb{Z}$ for $r \geq 3$. Thus, $\pi_1(\mathrm{O}(\gamma_\mathbb{R})) = \mathbb{Z}/2\mathbb{Z} \times \mathbb{Z}/2\mathbb{Z}$ and $\pi_1(\mathrm{SO}(2) \times \mathrm{O}(1, b-3)) = \mathbb{Z} \times \mathbb{Z}/2\mathbb{Z}$. The map $\pi_1(\mathrm{SO}(2) \times \mathrm{O}(1, b-3)) \to \pi_1(\mathrm{O}(\gamma_\mathbb{R}))$ induced by the inclusion is the natural one and thus surjective. This shows that the period domain is simply-connected. $\qquad\square$

Definition 25.10 Two points $x, x' \in Q_\gamma$ are called equivalent if there exists a sequence $x = x_1, \ldots, x_k = x' \in Q_\gamma$ such that the spaces $\langle P(x_i), P(x_{i+1}) \rangle \subset \gamma_\mathbb{R}$ are positive three-spaces.

In other words, two points are equivalent if they can be connected by a chain of curves of the form $Q_\gamma \cap \mathbb{P}(F_\mathbb{C})$ with $F \subset \gamma_\mathbb{R}$ a positive three-space.

Proposition 25.11 *Any two points in Q_γ are equivalent.*

Proof. It suffices to show that any equivalence class is open, for Q_γ is connected. Let $x = x_1 + ix_2 \in Q_\gamma$. Choose $z \in \gamma_\mathbb{R}$ such that $\langle x_1, x_2, z \rangle \subset \gamma_\mathbb{R}$ is a positive three-space. Then there exists an open neighbourhood U of $x \in Q_\gamma$ such that for any $y = y_1 + iy_2 \in U \subset Q_\gamma$ the two spaces $\langle y_1, y_2, z \rangle$ and $\langle x_1, y_2, z \rangle$ are both positive and three-dimensional. Then, $x \sim (x_1 + iz) \sim (y_2 + iz) \sim y$. $\qquad\square$

Let us now come back to the period map $\mathcal{P} : \mathfrak{M}_\gamma \to Q_\gamma$. We know that twistor spaces $\mathcal{X}(\alpha) \to T(\alpha) = \mathbb{P}^1$ induce positive three-spaces $F(\alpha)$ and we have seen that two points in the period domain can be connected by a sequence of curves of the form $Q_\gamma \cap \mathbb{P}(F_\mathbb{C})$ with $F \subset \gamma_\mathbb{R}$ a positive three-space. But, not every positive three-space used to connect two points necessarily comes from a twistor space. So we have to modify this approach slightly. The other result that is used is the description of the Kähler cone of a very general irreducible holomorphic symplectic manifold which will be discussed in §26.3.

Proposition 25.12 *Let $\mathfrak{M}_\gamma^\circ$ be a connected component of the moduli space $\mathfrak{M}_\gamma$. Then the period map $\mathcal{P} : \mathfrak{M}_\gamma^\circ \to Q_\gamma$ is surjective.*

Proof. It suffices to prove that for any two points $x, x' \in Q_\gamma$ such that $\langle P(x), P(x') \rangle \subset \gamma_\mathbb{R}$ is a positive three-space $x \in \mathcal{P}(\mathfrak{M}_\gamma^\circ)$ if and only if $x' \in \mathcal{P}(\mathfrak{M}_\gamma^\circ)$. Let $x = \mathcal{P}(X, \varphi)$. For any deformation $(\mathcal{X}_t, \varphi)$ of (X, φ) with $t \in \mathrm{Def}(X)$ close to $0 \in \mathrm{Def}(X)$ the space $\langle P(\mathcal{P}(\mathcal{X}_t, \varphi)), P(x') \rangle$ is still positive and three-dimensional. Thus, $\langle P(\mathcal{P}(\mathcal{X}_t, \varphi)), P(x') \rangle = \langle P(\mathcal{P}(\mathcal{X}_t, \varphi)), \alpha \rangle$ for some $\alpha \in \mathcal{C}_{\mathcal{X}_t}$. If t is chosen very general, then $\mathcal{K}_{\mathcal{X}_t} = \mathcal{C}_{\mathcal{X}_t}$ by Proposition 26.12. Hence, $\langle P(\mathcal{P}(\mathcal{X}_t, \varphi)), P(x') \rangle = F(\alpha)$ for some Kähler class α. Hence, $x' \in \mathcal{P}(F(\alpha))$. $\qquad\square$

25.5 Global Torelli Theorem??

After having seen that the period map $\mathcal{P} : \mathfrak{M}_\gamma \to Q_\gamma$ is surjective and locally an isomorphism, one would like to go on and study the fibres $\mathcal{P}^{-1}(x)$ for $x \in X$. Due to the work of many people (see [16] or [4]) one knows that the Global Torelli theorem holds for K3 surfaces.

Theorem 25.13 *(Global Torelli) Let X and X' be two K3 surfaces and let $\psi : H^2(X', \mathbb{Z}) \cong H^2(X, \mathbb{Z})$ be an isomorphism which respects Hodge structure and intersection pairing, and maps a Kähler class on X' to a Kähler class on X. Then there exists a unique isomorphism $f : X \cong X'$ with $f^* = \psi$.*

What does the Global Torelli theorem say for the fibres of the period map $\mathcal{P} : \mathfrak{M}_\gamma \to Q_\gamma$, where γ is the K3 lattice $2(-E_8) \oplus 3U$? First of all, if $\mathcal{P}(X, \varphi) = \mathcal{P}(X', \varphi')$, then X and X' are isomorphic K3 surfaces. Indeed, $\varphi^{-1} \circ \varphi'$ defines an isomorphism of the second cohomology that respects the Hodge structure. A priori, this isomorphism might not be compatible with the Kähler cones. However, it is well-known for K3 surfaces, and will be discussed in greater generality later, that the general K3 surface X has maximal Kähler cone, i.e. $\mathcal{K}_X = \mathcal{C}_X$. Thus, if $\mathcal{P}(X, \varphi) = \mathcal{P}(X', \varphi')$ is general, then $(\varphi^{-1} \circ \varphi')(\mathcal{K}_{X'}) = \pm \mathcal{C}_X = \pm \mathcal{K}_X$. For arbitrary K3 surfaces one can always find an additional automorphism of $H^2(X, \mathbb{Z})$, such that the composition with $\varphi^{-1} \circ \varphi'$ maps any Kähler class to a Kähler class. These extra isomorphisms relate points in one fibre, which are all non-separated from each other. Concretely, the Weil group $W \subset \mathrm{Aut}(H^2(X, \mathbb{Z}))$, which is generated by reflections in hyperplanes orthogonal to (-2)-classes of type $(1, 1)$, acts freely and transitively on the fibres. It turns out that $\mathfrak{M}_\gamma$ has two connected components $\mathfrak{M} = \mathfrak{M}^0 \cup$

$\mathfrak{M}^1$ which are identified by $(X, \varphi) \mapsto (X, -\varphi)$. Hence, $\mathcal{P} : \mathfrak{M}^i_\gamma \to Q_\gamma$ is a generically injective étale map.

As the Local Torelli theorem and the surjectivity of the period map generalize nicely to higher dimensions, one would hope that some kind of Global Torelli theorem also holds in arbitrary dimension. In [11] Beauville showed that non-trivial automorphisms may act trivially on the second cohomology. Thus, the uniqueness does not generalize. Moreover, already in 1984 Debarre observed that irreducible holomorphic symplectic manifolds with isomorphic periods need not be isomorphic. Recall, that Debarre had constructed (cf. Example 21.8) two birational irreducible holomorphic symplectic manifolds which are not isomorphic. Thus, a counterexample to the Global Torelli theorem in higher dimension is obtained by applying the following proposition, a proof of which, together with more general results, will be discussed in §27.1.

Proposition 25.14 *Any birational correspondence between two irreducible holomorphic symplectic manifolds X and X' defines an isomorphism of lattices $(H^2(X, \mathbb{Z}), q_X) \cong (H^2(X', \mathbb{Z}), q_{X'})$ which is compatible with the Hodge structures.*

Remark 25.15 ((i) We will see that often there exist other isomorphisms of the second cohomology groups of birational irreducible holomorphic symplectic manifolds, which are given by the birational correspondence composed with a higher dimensional analogue of the reflections defined by (-2)-classes. (ii) Note that the birational correspondence never sends a Kähler class to a Kähler class, except when the birational map extends to an isomorphism ([58] and Proposition 27.7). In this sense, the above example is still not a counterexample to the Global Torelli theorem as formulated in Theorem 25.13.

For quite some time it was felt that the correct Global Torelli theorem in higher dimension should say that irreducible holomorphic symplectic manifolds with isomorphic periods are birational. Only very recently, actually after these lectures had been delivered, Namikawa [164] showed that also this version cannot hold true. His example is surprisingly simple and very nice. So we cannot resist to outline it here.

Example 25.16 The basic idea to construct isomorphic periods goes back to Shioda [182]. Let T and T^* be a complex two-dimensional torus and its dual. We may assume that they are not isomorphic. Then there exists an isomorphism $H^2(T, \mathbb{Z}) \cong H^2(T^*, \mathbb{Z})$ which respects intersection pairing and Hodge decomposition. Indeed, the map is given by $H^2(T, \mathbb{Z}) = \Lambda^2 H^1(T, \mathbb{Z}) \cong \Lambda^2 H^1(T^*, \mathbb{Z})^* = H^2(T^*, \mathbb{Z})^* \cong H^2(T^*, \mathbb{Z})$, where the last isomorphism is given by Poincaré duality. Now, let $X := K_2(T)$ and $X' := K_2(T^*)$. Then

$$(H^2(X, \mathbb{Z}), q_X) = (H^2(T, \mathbb{Z}), \cup) \oplus \delta \mathbb{Z}$$

and similarly

$$(H^2(X', \mathbb{Z}), q_{X'}) = (H^2(T^*, \mathbb{Z}), \cup) \oplus \delta' \mathbb{Z}$$

with $\delta^2 = \delta'^2 = -6$ (cf. Example 23.20). Hence, X and X' have isomorphic periods. It remains to show that for appropriate T the two irreducible holomorphic symplectic manifolds X and X' cannot be birational. Let us suppose that there exists a birational map $f : X \dashrightarrow X'$. By E and E' we denote the exceptional divisors for $K_2(T) \to S^3(T)$ and $K_2(T^*) \to S^3(T^*)$, respectively. If $\operatorname{Pic}(T) = 0$, then $\operatorname{Pic}(X) \otimes \mathbb{Q}$ is generated by $\mathcal{O}(E)$. Hence, $f^*\mathcal{O}(E')$ is a multiple of $\mathcal{O}(E)$. Since E is exceptional and f induces an isomorphism in codimension two (cf. Proposition 21.6), we obtain a birational map $f : E \dashrightarrow E'$. On the other hand, the Albanese is a birational invariant. Thus, $T \cong \operatorname{Alb}(E) \cong \operatorname{Alb}(E') \cong T^*$. This yields a contradiction.

Remark 25.17 Let us come back to (ii) of Remark 25.15. In the above example an isomorphism of the period was constructed without paying attention to the Kähler cones. In fact, Namikawa also produced an example of projective non-birational generalized Kummer varieties associated to an abelian surface and its dual together with an isomorphism of their periods that maps an ample class to an ample class. So, eventually one can conclude that Theorem 25.13 does not hold already in dimension four. One could also use results which will be discussed in §27.2 in order to construct to any example of non-birational irreducible holomorphic symplectic manifolds with isomorphic periods another example of non-birational irreducible holomorphic symplectic manifolds whose periods admit an isomorphism that maps a Kähler class to a Kähler class.

25.6   Exercises

25.1 Use the existence of the twistor space to give an alternative proof of the unobstructedness of compact hyperkähler manifolds.

25.2 Verify that the isomorphism $H^2(T, \mathbb{Z}) \cong H^2(T^*, \mathbb{Z})$ in Example 25.16 is indeed compatible with the Hodge structures.

25.3 Prove the connectivity of the period domain explicitly (cf. [16]).

26   Projectivity of Hyperkähler Manifolds

Not all compact hyperkähler manifolds are projective. In fact, the very general hyperkähler manifold is not projective. (Any point in the complement of a countable union of proper analytic subsets is called very general.) On the other hand, as we will see, the projective ones are dense in the moduli space. As projective hyperkähler manifolds can be studied by means of higher dimensional algebraic geometry, it is worthwhile to study in detail which hyperkähler manifolds are projective and what the subset of projective ones looks like.

26.1   Projectivity after Small Deformations

Let X be a compact Kähler manifold. Mapping a line bundle L to its first Chern class $c_1(L)$ defines a homomorphism

$$\text{Pic}(X) \to H^{1,1}(X) \cap H^2(X, \mathbb{Z}) =: H^{1,1}(X, \mathbb{Z}).$$

Due to the Lefschetz theorem on $(1,1)$-classes this map is surjective (use the exponential sequence). If $H^1(X, \mathbb{Z}) = 0$ it defines an isomorphism $\text{Pic}(X) \cong H^{1,1}(X, \mathbb{Z})$.

Let $\mathcal{K}_X \subset H^{1,1}(X) \cap H^2(X, \mathbb{R})$ be the Kähler cone of X. If X is projective and L is an ample line bundle on X, then the restriction of the Fubini-Study metric endows L with a hermitian product whose curvature gives back the Fubini-Study Kähler form on X. Hence, $c_1(L)$ can be represented by a Kähler form, i.e. $c_1(L) \in \mathcal{K}_X$. Conversely, the Kodaira embedding theorem says:

Theorem 26.1 *The manifold X is projective if and only if $\mathcal{K}_X$ contains an integral point, i.e. $\mathcal{K}_X \cap H^2(X, \mathbb{Z}) \neq \emptyset$ (cf. [77]).*

Actually, the intersection $\mathcal{K}_X \cap H^2(X, \mathbb{Z})$ is the cone of ample classes, i.e. the set of all $c_1(L)$ where L is an ample line bundle on X. As we have seen, deforming the complex structure on X deforms the Hodge structure $H^2(X, \mathbb{C}) = H^{2,0}(X) \oplus H^{1,1}(X) \oplus H^{0,2}(X)$. In particular, the size of the intersection $H^{1,1}(X) \cap H^2(X, \mathbb{Z})$ depends on the complex structure on X. Therefore, it may occur that for a family $\mathcal{X} \to S$ of compact Kähler manifolds one has $H^{1,1}(\mathcal{X}_0) \cap H^2(\mathcal{X}_0, \mathbb{Z}) = 0$, but $H^{1,1}(\mathcal{X}_t) \cap H^2(\mathcal{X}_t, \mathbb{Z}) \neq 0$ for some point $t \in S$ arbitrary close to 0. The same applies to the intersection $\mathcal{K}_X \cap H^2(X, \mathbb{Z})$. Thus, it could happen that in the family $\mathcal{X} \to S$ the special fibre $X := \mathcal{X}_0$ is not projective, but some nearby fibres $\mathcal{X}_t$ are.

There is a general conjecture saying that any compact Kähler manifold can be deformed to a projective one. Evidence for this conjecture is provided solely by the fact that so far no restriction on the topology (e.g. the fundamental group) of a projective manifold has been found, which is not satisfied by an arbitrary Kähler manifold (cf. [2]). We will shortly present a result saying that for hyperkähler manifolds the conjecture is indeed true.

Before stating the proposition we take the opportunity to introduce the locus of manifolds on which a given class is of type $(1,1)$ and to prove a number of easy facts about it.

Definition 26.2 Let (X, φ) be a marked irreducible holomorphic symplectic manifold. For any $\alpha \in H^2(X, \mathbb{R})$ let

$$S_\alpha := \{t \in \text{Def}(X) | q_\gamma(\varphi(\alpha), \mathcal{P}(t)) = 0\},$$

i.e. S_α is the pull-back of the hyperplane in $\mathbb{P}(\gamma_\mathbb{C})$ defined by $q_\gamma(\varphi(\alpha), \,.\,)$.

Sometimes, we will also denote the hyperplane section of $Q_\gamma \subset \mathbb{P}(\gamma_\mathbb{C})$ by S_α. Using the natural identification of the cohomology $H^2(\mathcal{X}_t, \mathbb{R})$ of a fibre $\mathcal{X}_t$ for $t \in \text{Def}(X)$ with $H^2(X, \mathbb{R})$ one has

Lemma 26.3 *The set S_α for $\alpha \in H^2(X, \mathbb{R})$ is the set of points $t \in \mathrm{Def}(X)$ such that α is of type $(1,1)$ on $\mathcal{X}_t$.*

Proof. Indeed, since $\alpha \in H^2(\mathcal{X}_t, \mathbb{R}) = H^2(X, \mathbb{R})$ is real and orthogonal to $\sigma_t \in H^{2,0}(\mathcal{X}_t)$ for $t \in S_\alpha$, it has to be of type $(1,1)$ on $\mathcal{X}_t$.   □

Lemma 26.4 *Let $\alpha \in H^{1,1}(X, \mathbb{R})$, i.e. $0 \in S_\alpha \subset \mathrm{Def}(X)$. Then the tangent space of S_α at 0 is the kernel of the surjective contraction map*

$$\tilde{\alpha} : H^1(X, \mathcal{T}_X) \to H^2(X, \mathcal{O}_X).$$

Proof. The first assertion is a general fact: If $\beta \in H^{p,p}(X, \mathbb{R})$ then β stays of type (p,p) in a first order deformation of X given by $v \in H^1(X, \mathcal{T}_X)$ if and only if the contraction $\tilde{\beta}(v) \in H^{p-1,p+1}(X)$ is trivial. Indeed, if we write locally β as $\sum \beta_{IJ} dz_I \wedge d\bar{z}_J$ and $v = \sum v_{ij}(\partial/\partial z_i) d\bar{z}_j$, then $\tilde{\beta}(v) = \sum \beta_{IJ} v_{ij} dz_{I\setminus i} \wedge d\bar{z}_j \wedge d\bar{z}_J$. On the other hand, the infinitesimal coordinate change is given by $dz_i + \varepsilon \sum v_{ik} d\bar{z}_k$. Thus, β stays of type (p,p) to the first order if and only if $\sum \beta_{IJ}(\prod_{i\in I}(dz_i + \varepsilon \sum v_{ik} d\bar{z}_k)) \wedge (\prod_{j\in J}(d\bar{z}_j + \varepsilon \sum \bar{v}_{j\ell} dz_\ell)) = \beta$, i.e. $\sum \beta_{IJ} v_{ik} dz_{I\setminus i} \wedge d\bar{z}_k \wedge d\bar{z}_J = 0$ (use that β is real).

If $q_\gamma(\varphi(\alpha)) \neq 0$, then q_γ restricted to $\varphi(\alpha)^\perp$ is non-degenerate and thus $S_\alpha \subset Q_\gamma$ is smooth. If $q_\gamma(\varphi(\alpha)) = 0$, then the quadric defined by q_γ on $\varphi(\alpha)^\perp$ is singular in exactly one point, namely in $\varphi(\alpha)$. But this point is not contained in Q_γ for α real. Thus, also in this case $S_\alpha \subset Q_\gamma$ is smooth. This proves in particular, that the contraction map is surjective.   □

The arguments of the proof in particular show that the pairing

$$H^1(X, \mathcal{T}_X) \times H^1(X, \Omega_X) \to H^2(X, \mathcal{O}_X) = \mathbb{C}\bar{\sigma}$$

is non-degenerate. This readily yields:

Lemma 26.5 *If $\alpha, \beta \in H^2(X, \mathbb{R})$ are linearly independent, then S_α and S_β intersect transversally.*

The following result was proved by Fujiki [59], Campana [35], and later again by Verbitsky [194].

Proposition 26.6 *Let X be an irreducible holomorphic symplectic manifold, let $0 \in S \subset \mathrm{Def}(X)$ be an analytic subset of positive dimension and let $\mathcal{X} \to S$ denote the restriction of the Kuranishi family to S. Then any open neighbourhood $U \subset S$ of $0 \in S$ contains a point $t \neq 0$ such that the fibre $\mathcal{X}_t$ over t is projective.*

Proof. By shrinking S and passing to arbitrarily close points of 0 we may assume that S is one-dimensional and smooth. Then, its tangent space at zero $T_0 S$ is a line in $T_0 \mathrm{Def}(X) = H^1(X, \mathcal{T}_X)$ spanned by, say, $0 \neq v \in H^1(X, \mathcal{T}_X)$.

The induced map $\tilde{v} : H^1(X, \Omega_X) \to H^2(X, \mathcal{O}_X) \cong \mathbb{C}$ is surjective. Since the Kähler cone $\mathcal{K}_X$ is an open subset of $H^{1,1}(X, \mathbb{R}) = H^{1,1}(X) \cap H^2(X, \mathbb{R})$ and $H^{1,1}(X, \mathbb{R})$ spans $H^1(X, \Omega_X)$ as a complex vector space, there exists a Kähler form ω on X such that $\tilde{v}([\omega]) \neq 0$. Consider the corresponding hypersurface $S_{[\omega]} \subset \mathrm{Def}(X)$. Since $\tilde{v}([\omega]) \neq 0$, one has $T_0 S_{[\omega]} \cap T_0 S = 0$, i.e. $S_{[\omega]}$ and S intersect transversally in $0 \in \mathrm{Def}(X)$. Shrinking S we may assume that $S_{[\omega]} \cap S = \{0\}$.

Next, pick classes $\alpha_i \in H^2(X, \mathbb{Q})$ converging to $[\omega]$ and consider the associated hypersurfaces $S_{\alpha_i} \subset \mathrm{Def}(X)$. Then the hypersurfaces S_{α_i} converge to $S_{[\omega]}$ and therefore $S_{\alpha_i} \cap S \neq \emptyset$ for $i \gg 0$. Moreover, if we choose α_i such that they are not of type $(1,1)$ on X, then $0 \notin S_{\alpha_i} \cap S$. Hence, there exist points $t_i \in (S_{\alpha_i} \cap S) \setminus \{0\}$ converging to 0. Using the isomorphisms $H^2(\mathcal{X}_t, \mathbb{Z}) \cong H^2(X, \mathbb{Z})$, the classes α_i can be considered as rational classes of type $(1,1)$ on $\mathcal{X}_{t_i}$. Intuitively, the classes α_i on $\mathcal{X}_{t_i}$ converge to the Kähler class $[\omega]$ on $X = \mathcal{X}_0$ and thus should be Kähler for $i \gg 0$. This can be made rigorous as follows: Fix a diffeomorphism $\mathcal{X} \cong X \times S$ (compatible with the projections to S). By a result of Kodaira and Spencer (Thm. 3.1 in [132], Thm. 15 in [133]) there exists a real two-form $\tilde{\omega}$ on $X \times S$ such that the restriction $\tilde{\omega}_t$ of $\tilde{\omega}$ to $\mathcal{X}_t = X \times \{t\}$ is a Kähler form for all $t \in S$ and $\tilde{\omega}_0 = \omega$. One also finds real two-forms $\tilde{\alpha}_i$ on $X \times S$ such that $(\tilde{\alpha}_i)_t$ is harmonic with respect to $\tilde{\omega}_t$ and $[(\tilde{\alpha}_i)_t] \equiv \alpha_i$ for all t (§2 in [132]). In particular, $(\tilde{\alpha}_i)_{t_i}$ is a harmonic $(1,1)$-form on $\mathcal{X}_{t_i}$ representing α_i. Since α_i converges to $[\alpha]$, also $(\tilde{\alpha}_i)_{t_i}$ converges to ω. Hence, the harmonic $(1,1)$-form $(\tilde{\alpha}_i)_{t_i}$ is a Kähler form on $\mathcal{X}_{t_i}$ for $i \gg 0$. Hence, α_i is a Hodge class on $\mathcal{X}_{t_i}$ for $i \gg 0$. In particular, $\mathcal{X}_{t_i}$ is projective for $i \gg 0$. $\square$

26.2   Nakai–Moishezon and Demailly–Paun

The aim is to establish an explicit projectivity criterion for hyperkähler manifolds which allows one to decide whether a given compact hyperkähler manifold is projective just from the knowing its period. We need to recall the following classical result, the so called Nakai–Moishezon criterion.

Theorem 26.7 *Let X be a projective variety and L a line bundle on X. Then L is ample if and only if for any subvariety $Y \subset X$ of dimension d the intersection $\int_Y c_1(L)^d$ is positive.*

The proof of this result can be found in [90, Thm 5.1, Ch. I]. Note that $Y = X$ is among the subvarieties that have to be tested. There is another version of the Nakai–Moishezon criterion that says that a line bundle is ample if and only if the induced linear form on $H^{N-1,N-1}(X)$ is strictly positive on the closure of the cone generated by all effective curves.

A remarkable recent theorem of Demailly and Paun shows that with the necessary modification the Nakai–Moishezon criterion also holds for compact Kähler manifolds.

Theorem 26.8 ([12]) *Let X be a compact Kähler manifold. Then the Kähler cone $\mathcal{K}_X$ of X is a connected component of the set $\mathcal{P}_X$ of all classes $\alpha \in H^{1,1}(X, \mathbb{R})$ such that $\int_Y \alpha^d > 0$ for any irreducible analytic subset $Y \subset X$ of dimension d.*

The proof of this theorem is difficult. It uses Demailly's regularization of currents [48]. The techniques of [49] also provide an alternative proof of a result originally due to Bonavero and Ji–Shiffman which we will recall in Theorem 26.10. Before stating it let us summarize a few basic concepts from the theory of positive currents. This will also be needed in the next sections.

Definition 26.9 Let X be a compact Kähler manifold of dimension N. A real $(1,1)$-current is an $\mathbb{R}$-linear continuous map $T : \mathcal{A}^{2N-2}(X) \to \mathbb{R}$ whose $\mathbb{C}$-linear extension is trivial on $\mathcal{A}^{p,q}(X)$ for $(p, q) \neq (N-1, N-1)$. Here, the topology on the space of forms is the natural one used for the definition of distributions, i.e. a series is convergent if all partial derivatives converge uniformly. The current T is closed if $T(d\alpha) = 0$ for all $\alpha \in \mathcal{A}^{2N-3}(X)$. The current T is positive if $T(\alpha) \geq 0$ for any $\alpha \in \mathcal{A}^{N-1,N-1}(X)$ which locally is of the form $(v_1 \wedge i\bar{v}_1) \wedge \ldots \wedge (v_{N-1} \wedge i\bar{v}_{N-1})$.

Clearly, any $(1,1)$-form ω defines a real $(1,1)$-current by $\alpha \mapsto \int_X \alpha \wedge \omega$. This current is closed if and only if ω is closed. If ω is a Kähler form, then this current is positive. Any closed real $(1,1)$-current T defines a de Rham cohomology class $[T] \in H^{1,1}(X, \mathbb{R}) = \mathrm{Hom}(H^{N-1,N-1}(X, \mathbb{R}), \mathbb{R})$ by $[\alpha] \mapsto T(\alpha)$.

As was noted before, if $c_1(L)$ of a line bundle $L \in \mathrm{Pic}(X)$ can be represented by a Kähler form, i.e. by a real closed strictly-positive $(1,1)$-form, then X is projective. Something weaker is still true if $c_1(L)$ can be represented by a positive current:

Theorem 26.10 *Let X be a compact complex manifold and let L be a line bundle such that $c_1(L)$ can be represented by a closed positive current T such that $T - \omega$ is positive (as a current) for some small positive $(1,1)$-form ω. Then X is a Moishezon manifold, i.e. $\mathrm{trdeg}K(X) = \dim(X)$. [24],[108]*

This result will be applied in conjunction with the following result due to Moishezon [156].

Theorem 26.11 *A compact Kähler manifold X is projective if and only if X is Moishezon.*

26.3 Very General Hyperkähler Manifolds

The following result, which is proved in [105, Cor. 1], can be seen as a special case of a more general statement that the Kähler cone of a compact hyperkähler manifold that does not contain any rational curve is maximal, but the proof of this more general result uses the special case and will be explained later. Recall that the positive cone $\mathcal{C}_X$ is the component of the open set of all classes $\alpha \in H^{1,1}(X, \mathbb{R})$ with $q_X(\alpha) > 0$ that contains the Kähler cone.

Proposition 26.12 *Let X be a very general compact hyperkähler manifold. Then $\mathcal{K}_X = \mathcal{C}_X$.*

Proof. We apply the Demailly–Paun result, Theorem 26.8. Since $\mathcal{C}_X$ is connected and $\mathcal{K}_X$ is a connected component of $\mathcal{P}_X$, it suffices to prove $\mathcal{C}_X \subset \mathcal{P}_X$. Let $Y \subset X$ be an irreducible analytic subset of X.

Claim: If X is very general then Y is of even codimension $2p$ and $[Y] \in H^{2p,2p}(X)$ is of type $(2p, 2p)$ on all small deformations of X.

If this holds true then one can apply Corollary 23.17: There exists a constant $c_{[Y]} \in \mathbb{R}$ such that $\int_Y \alpha^{2(n-p)} = c_{[Y]} q_X(\alpha)^{n-p}$ for all $\alpha \in H^2(X)$. The constant $c_{[Y]} \in \mathbb{R}$ has to be positive, since both $\int_Y \alpha^{2(n-p)}$ and $q_X(\alpha)$ are positive for a Kähler class α. Hence, $\int_Y \alpha^{2(n-p)}$ is positive for any $\alpha \in \mathcal{C}_X$.

It remains to prove the claim: Let us first show that the general compact hyperkähler manifold does not admit any odd-dimensional subvariety (cf. [60, Prop. 5.11]). In fact, we will show that for a fixed hyperkähler metric g the very general fibre $\mathcal{X}_t$ in the associated twistor family does not contain any odd-dimensional subvariety. Indeed, if $Y \subset X$ is any subvariety which is analytic with respect to all complex structures parametrized by the twistor base $\mathbb{P}^1$, then $\int_Y \omega_\lambda^{\dim(Y)} > 0$ for all $\lambda \in \mathbb{P}^1$. On the other hand, $\omega_{-I} = -\omega_I$. Thus, if the dimension of Y is odd this yields a contradiction. Eventually, one uses the fact that the Douady space of cycles Y in the fibres of the twistor space $\mathcal{X} \to \mathbb{P}^1$ with fixed cohomology class $[Y] \in H^*(X, \mathbb{Z})$ is proper over $\mathbb{P}^1$. The general point in $\mathbb{P}^1$ therefore is not contained in the countable union of the images of the Douady spaces associated to odd degree integral cohomology classes. To finish the proof of the claim one has to show that for any cohomology class $\beta \in H^{4p}(X, \mathbb{R})$ the set $S_\beta := \{t \in \mathrm{Def}(X) \mid \beta \in H^{2p,2p}(\mathcal{X}_t, \mathbb{R})\}$ is a closed analytic subset. This is again a general fact. The locus can be described as the zero set of the induced section of the holomorphic vector bundle that is given as the quotient of the inclusion $\mathcal{F}^{2p} \subset R^{4p}\pi_* \mathbb{C} \otimes \mathcal{O}$, where $\mathcal{F}^r$ is fibrewise $\bigoplus_{k \geq r} H^{k,4p-k}(\mathcal{X}_t)$ (cf. [39]). Eventually, one finds that a point t in the complement of the union of all proper S_β with $\beta \in H^{4p}(X, \mathbb{Z})$ gives a compact hyperkähler manifold on which any integral $(2p, 2p)$-class stays of type $(2p, 2p)$ under any small deformation. □

Note that it is a priori not clear whether the general hyperkähler manifold contains any non-trivial subvarieties at all. If there are none, then the last proposition is an immediate consequence of the Demailly–Paun result. For the general deformation of the Hilbert scheme Verbitsky has shown that there are no subvarieties [196]. For the general deformation of a generalized Kummer variety the situation is different. In [126] is was shown that the general deformation of the generalized Kummer variety does contain non-trivial subvarieties (this corrects [127]).

The structure of the Kähler cone of an arbitrary hyperkähler manifold will be taken up in §28.

Proposition 26.12 also shows that on a very general hyperkähler manifold X any class $\alpha \in \mathcal{C}_X$ can be represented by a closed positive $(1,1)$-current T such that $T - \omega$ is still positive for some small Kähler form ω. This weaker statement generalizes.

26.4   The Projectivity Criterion

We will next show how the Demailly–Paun–Nakai–Moishezon theorem can be used to prove the following projectivity criterion for irreducible holomorphic symplectic manifolds.

Proposition 26.13 *Let X be an irreducible holomorphic symplectic manifold. Then X is projective if and only if $\mathcal{C}_X \cap H^2(X, \mathbb{Z}) \neq \emptyset$.*

Proof. One direction is obvious. If X is projective and L is an ample line bundle on X then $c_1(L)$ is a Kähler class and thus $c_1(L) \in \mathcal{C}_X$. For the other direction one uses the result on the structure of the Kähler cone of a very general hyperkähler manifold (Proposition 26.12).

Let now X be an arbitrary compact hyperkähler manifold and L be a line bundle with $c_1(L) \in \mathcal{C}_X$. We claim that any class $\alpha \in \mathcal{C}_X$ can be represented by a real closed positive $(1,1)$-current T such that $T - \omega$ is positive for some small Kähler form ω. This applied to $\alpha = c_1(L) \in \mathcal{C}_X$ and using Theorems 26.10 and 26.11 proves that X is projective.

Since the set of very general $t \in \mathrm{Def}(X)$ is dense, any class $\alpha \in \mathcal{C}_X \subset H^2(X, \mathbb{R})$ can be approximated by a sequence $\alpha_{t_i} \in H^2(X, \mathbb{R})$ such that the points t_i converge to $0 \in \mathrm{Def}(X)$ and α_{t_i} are Kähler classes on $\mathcal{X}_{t_i}$. Then one copies an argument of Demailly in [48, Prop. 6.1]: Fix Kähler classes ω_t on $\mathcal{X}_t$ depending continuously on $t \in \mathrm{Def}(X)$. Then the mass of α_{t_i}, which is $\int_X \omega_{t_i}^{2n-1} \alpha_{t_i}$, converges to $\int_X \omega_0^{2n-1} \alpha$. Hence the sequence of forms (α_{t_i}) is weakly bounded and thus weakly compact. In particular, it contains a weakly convergent subsequence. As the α_{t_i} are closed positive $(1,1)$-forms on $\mathcal{X}_{t_i}$, the limit current is closed and positive of bidegree $(1,1)$ on X. As $\mathcal{C}_X$ is open, we may repeat this argument for the class $\alpha - \omega$ for some small Kähler class ω. This proves the claim.   □

26.5   Application: Finiteness of Hyperkähler Manifolds

The failure of the Global Torelli theorem (cf. Example 25.16) shows that the second cohomology $H^2(X, \mathbb{Z})$ of an irreducible holomorphic symplectic manifold X together with its Beauville–Bogomolov form and the Hodge structure does not determine its birational isomorphism type. However, there is evidence that it does determine the deformation type at least up to finitely many choices. As an application of the previous two sections we will outline the main arguments of [100], which goes in this direction.

In order to illustrate what is meant by "determined up to finitely many choices" we recall the following classical result of Sullivan [185].

Theorem 26.14 *Let $H^* = \bigoplus_{i=0}^{2n} H^i$ be a graded ring and let $c_i \in H^{2i} \otimes \mathbb{Q}$ be fixed. Then there exist finitely many differentiable manifolds $M_1, \ldots, M_m$ with the following property: If X is a simply-connected Kähler manifold that admits an isomorphism $\psi : H^*(X, \mathbb{Z}) \cong H^*$ of graded algebras with $\psi_{\mathbb{Q}}(c_i(X)) = c_i$, then X is diffeomorphic to M_k for some $k \in \{1, \ldots, m\}$.*

We will present a similar result for compact hyperkähler manifolds with the only difference that much less of the structure needs to be fixed in order to determine the manifold up to finite ambiguity.

Whereas Sullivan's finiteness comes from topological considerations, our result will make use of the following result which is an immediate consequence of the highly non-trivial Kollár-Matsusaka theorem [134].

Theorem 26.15 *There exists only a finite number of deformation types of projective manifolds X of dimension N with $c_1(X) = 0$ that admit an ample line bundle L such that $c_1(L)^N$ is bounded by a fixed number.*

Remark 26.16 Comparing Theorems 26.15 (or rather its general version in [134]) with Theorem 26.14 one finds more evidence for the general philosophy that compact Kähler manifolds behave very much like projective manifolds.

Corollary 26.17 *There exists only a finite number of deformation types, and hence diffeomorphism types, of irreducible holomorphic symplectic manifolds X of dimension $2n$ that admit a class $\alpha \in H^2(X, \mathbb{Z})$ with $q_X(\alpha) > 0$ and $\int_X \alpha^{2n}$ bounded. (Using Proposition 23.14, we can drop the condition on the sign of $q_X(\alpha)$ in case n is odd.)*

Proof. Let X be an irreducible holomorphic symplectic manifold that satisfies the assumption. We will show that X can be deformed to a projective irreducible holomorphic symplectic manifold Y such that $c_1(L) = \alpha$, where L is an ample line bundle on Y. Applying Theorem 26.15 then yields the assertion.

The orthogonal complement of α in $H^2(X, \mathbb{R})$ contains a positive plane. Thus, there exists a point y in the period domain $Q_{(H^2(X,\mathbb{Z}), q_X)}$, such that α is orthogonal to y. In fact, we have seen before that the set of these points is the hyperplane section S_α of the period domain. Using the surjectivity of the period map (Proposition 25.12) we find a marked irreducible holomorphic symplectic manifold (Y, ψ) in the same connected component of $\mathfrak{M}_{(H^2(X,\mathbb{Z}), q_X)}$ as (X, id) with $\mathcal{P}(Y, \psi) = y$. By Lemma 26.3 $\alpha = \psi(c_1(L))$ for some line bundle

L on Y. Using the projectivity criterion Proposition 26.13 one concludes that Y is projective, but a priori L might not be ample. However, if we choose $y \in S_\alpha$ very general, then $\mathrm{Pic}(Y)_\mathbb{Q} = \mathbb{Q}\alpha$. Indeed, $\bigcup S_\beta \subset \mathrm{Def}(X)$, where β runs through the set of all elements in $H^2(X, \mathbb{Z})$ linearly independent of α, is a countable union of hyperplane sections transversal to S_α (use Lemma 26.5). This, union cannot cover S_α. Hence, we can choose $y \in S_\alpha$ such that $\alpha = c_1(L)$ and thus L is ample. $\square$

This result is not optimal. Neither does it seem to be an appropriate version of Sullivan's result for hyperkähler manifolds nor does it make use of the Beauville–Bogomolov form. Using Rozansky-Witten invariants one can actually apply the same techniques to obtain

Proposition 26.18 *Let H be a $\mathbb{Z}$-module and let $c : H \to \mathbb{Z}$ be an homogeneous polynomial of degree $2n - 2$. Then there exists only a finite number of deformation types of irreducible holomorphic symplectic manifolds X such that $(H^2(X, \mathbb{Z}) \to \mathbb{Z}, \alpha \mapsto \int \alpha^{2n-2}c_2(X))$ is isomorphic to $(c : H \to \mathbb{Z})$.*

Proof. The idea of the proof is to reduce the assertion to Corollary 26.17 by bounding the top intersection product in terms of the polynomial defined by the second Chern class. We use the following formula due to Hitchin and Sawon [98]:

$$\frac{\|R\|^{2n}}{(192\pi^2 n)^n} = \int_X \sqrt{\hat{A}(X)} \cdot (\mathrm{vol}(X))^{n-1}.$$

Here, $\|R\|$ is the norm of the curvature, which is positive and can also be expressed as $\frac{8\pi^2}{(2n-2)!} \int_X c_2(X)\omega^{2n-2}$ and is certainly positive. Moreover, the $\hat{A}$-genus of the manifold in our case nothing is but the Todd class $\mathrm{td}(X)$. Thus,

$$\left(\int_X c_2(X)\omega^{2n-2}\right)^n = c \int_X \sqrt{\mathrm{td}(X)} \cdot \left(\int_X \omega^{2n}\right)^{n-1}$$

with $c = (24n)^n/((2n)!)^{n-1}$. Since the Kähler cone is open in $H^{1,1}(X, \mathbb{R})$ and the union of all Kähler cones in a twistor family is open in $H^2(X, \mathbb{R})$, the same equation holds for any class $\alpha \in H^2(X, \mathbb{R})$ instead of ω. Hitchin and Sawon further observed that this equality implies that the topological factor $\int_X \sqrt{\mathrm{td}(X)}$ on the right hand side is positive. In fact, since the square root of the Todd class is a rational cohomology class whose denominator can be bounded from above by a constant c_n depending only on n, this yields more precisely $\int_X \sqrt{\mathrm{td}(X)} > c_n^{-1}$. Hence, for any class $\alpha \in H^2(X, \mathbb{Z})$ on an $2n$-dimensional irreducible holomorphic symplectic manifold X one has $\int_X \alpha^{2n} < (\int_X c_2(X)\alpha^{2n-2})^{n/(n-1)}(c_n/c)^{1/(n-1)}$.

Thus, if $\alpha \mapsto \int_X c_2(X)\alpha^{2n-2}$ of an irreducible holomorphic symplectic manifold X is fixed, then $\alpha \mapsto \int_X \alpha^{2n}$ is bounded and we may apply Corollary 26.17. $\square$

Note that the finiteness in this form really is analogous to Sullivan's result. Moreover, it is in the spirit of many other boundedness results, e.g. for surfaces of general type and stable vector bundles on an algebraic variety, where only the first two Chern classes need to be fixed in order to obtain boundedness of the family.

In order to reduce further to the quadratic form we need the following generalization of the Hitchin–Sawon formula due to Nieper-Wißkirchen [167]

$$\left(\int_X \sqrt{\mathrm{td}(X)}\alpha^{2k}\right)^n = \left((2k)!\binom{n}{k}\right)^n \left(\frac{1}{(2n)!}\int_X \alpha^{2n}\right)^k \cdot \left(\int_X \sqrt{\mathrm{td}(X)}\right)^{n-k}.$$

The most interesting cases are $k = n - 1$, which gives back the Hitchin–Sawon formula, and $k = 1$. Of course, due to Corollary 23.17 we already know that the two quadratic forms given by $\int_X \alpha^2 \sqrt{\mathrm{td}(X)}$ and q_X only differ by a scalar. The virtue of the above formula is that we see that this scalar is not trivial and, in fact, positive due to the observation made by Hitchin–Sawon.

Usually, the Beauville–Bogomolov form is normalized so that it becomes integral and primitive. On the other hand, the primitivity has never been used, as the form, even after making it primitive, is in general not unimodular. Thus, we may equally well take the above formula as a definition.

Definition 26.19 The *unnormalized Beauville–Bogomolov form* $\tilde{q}_X$ on an irreducible holomorphic symplectic manifold is

$$\tilde{q}_X(\alpha) = c_n \int_X \alpha^2 \sqrt{\mathrm{td}(X)},$$

where c_n is some universal positive constant which only depends on n.

The coefficient is inserted in order to make the right hand side integral. The advantage of this definition becomes apparent by the following

Proposition 26.20 *Let (γ, q_γ) be a lattice. Then there exist only finitely many deformation types of irreducible holomorphic symplectic manifolds X of fixed dimension $2n$, such that $(H^2(X, \mathbb{Z}), \tilde{q}_X)$ is isomorphic to (γ, q_γ).*

Proof. This time we use the Nieper-Wißkirchen formula to bound the top intersection product of an integral class in terms of the unnormalized Beauville–Bogomolov form. The rest of the proof is analogous to the proof of Proposition 26.18. □

26.6   Exercises

26.1 Show that any compact Kähler manifold X with $H^{2,0}(X) = 0$ is projective.

26.2 Find an example of a compact Kähler manifold for which $\mathcal{P}$ as in Theorem 26.8 has more than one connected component.

26.3 Describe the Kähler cone of a complex torus.

26.4 What is the codimension of the subset of projective compact hyperkähler manifolds inside the moduli space of all compact hyperkähler manifolds? Compare this to the codimension of the moduli space of abelian varieties inside the moduli space of all complex tori.

26.5 Use the projectivity criterion to give an alternative proof of Proposition 26.6.

27   Birational Hyperkähler Manifolds

We have seen very explicit birational maps between irreducible holomorphic symplectic manifolds, but a complete classification of all birational maps seems not in reach. In this section we will approach this problem from a different point of view. We will prove, as a generalization of Proposition 22.6, that any two birational irreducible holomorphic symplectic manifolds are deformation equivalent. It will turn out that the existence of non-trivial birational maps is related to the non-Hausdorffness of the moduli space.

27.1   Periods of Birational Hyperkähler Manifolds

Let us first prove that two birational irreducible holomorphic symplectic manifolds have isomorpic periods. This was asserted in Proposition 25.14. Let

$$
\begin{array}{ccc}
 & Z & \\
\pi' \swarrow & & \searrow \pi \\
f : X' & \dashrightarrow & X
\end{array}
$$

be a resolution of the birational map f. We may choose it to be minimal, i.e. over the maximal open subsets where the birational map is defined the two morphisms $X \leftarrow Z \rightarrow X'$ are bijective. We can also assume that $Z \rightarrow X$ is a sequence of blow-ups with smooth center and we denote the exceptional divisors by $E_1, \ldots, E_\ell$. By Proposition 21.6 the induced map $[Z]_* : H^2(X, \mathbb{Z}) \rightarrow H^2(X', \mathbb{Z})$ is bijective. Clearly, this isomorphism respects the Hodge structures. In particular, we may assume that under this isomorphism the holomorphic two-forms σ and σ' on X and X', respectively, correspond to each other. It remains to verify that the isomorphism respects the Beauville–Bogomolov forms q_X and $q_{X'}$. The isomorphism

$H^2(X,\mathbb{Z}) \cong H^2(X',\mathbb{Z})$ can also be described by the two direct sum decompositions

$$H^2(Z,\mathbb{Z}) = H^2(X,\mathbb{Z}) \oplus \bigoplus \mathbb{Z} \cdot [E_i]$$
$$= H^2(X',\mathbb{Z}) \oplus \bigoplus \mathbb{Z} \cdot [E_i].$$

Using Definition 22.8 of the Beauville–Bogomolov form and $\pi^*(\sigma) = \pi'^*(\sigma')$ one finds that it suffices to show that $\pi^* \sigma^{n-1}|_{E_i} = 0$. If the image of $E_i \to X$ is of codimension at least three, then this is trivial. If there exists an exceptional divisor E_i such that its image in X is of codimension two, then there are two cases. Either the second contraction $E_i \to X'$ is different or there exists another exceptional divisor E_j with the same image in X as E_i and for which the two contractions $X \leftarrow E_j \to X'$ are different. Indeed, if such an E_j did not exist then the birational map would be defined on a non-empty subset of the image of $E_i \to X$, which we avoided by assuming Z be minimal. In both cases one concludes that σ restricted to the image of $E_i \to X$ has rank at most $2n-3$ (Here we use $\pi^*(\sigma)|_{E_i} = \pi'^*(\sigma')|_{E_i}$). Hence, $\pi^*(\sigma^{n-1})|_{E_i} = 0$. $\square$

The birational correspondence induces more generally a homomorphism $[Z]_* : H^*(X) \to H^*(X')$. One would like to know under which circumstances this map is bijective or, even stronger, a ring isomorphism. In the next two examples we will treat the case of Mukai's elementary transformation.

Example 27.1 Let X be an irreducible holomorphic symplectic manifold of dimension $2n$ containing a projective space $P := \mathbb{P}^n$ and let $X \leftarrow Z \to X'$ be the Mukai flop of X in P. By P' we denote the dual projective space in X'. We wish to study the action of $[Z]_*$ and $[\gamma]_* = [Z]_* + [P \times P']_*$ on the cohomology and in particular on the class $[P]$. Firstly, $[P \times P']_*[P] = (\int_X [P]^2) \cdot [P']$. Since $N_{P/X} \cong \Omega_P$, this yields

$$[P \times P']_*[P] = \left(\int_{\mathbb{P}^n} c_n(\Omega_{\mathbb{P}^n}) \right) \cdot [P'] = (-1)^n (n+1)[P'].$$

The exceptional locus in Z is of the form $\mathbb{P}(\Omega_P) = \mathbb{P}(\Omega_{P'})$. Its normal bundle on each fibre of the projection of $\mathbb{P}(\Omega_{P'}) \to P'$ is $\mathcal{O}(-1)$. On the other hand, the normal bundle of $P \times X'$ in $X \times X'$ is Ω_P, whose restriction to the fibres of $\mathbb{P}(\Omega_{P'}) \to P'$ is $\Omega_{\mathbb{P}^n}|_{\mathbb{P}^{n-1}} = \mathcal{O}(-1) \oplus \Omega_{\mathbb{P}^{n-1}}$. Hence (cf. [63, Prop. 6.7])

$$[Z]_*[P] = \left(\int_{\mathbb{P}^{n-1}} c_{n-1}(\Omega_{\mathbb{P}^{n-1}}) \right) \cdot [P'] = (-1)^{n-1} n[P'].$$

The extra factor indicates that we should not expect $[Z]_*$ to define an isomorphism of the integral(!) cohomology. On the other hand, $[\gamma]_*[P] = [Z]_*[P] + [P \times P']_*[P] = (-1)^n[P']$, which suggests that $[\gamma]_*$ is better behaved on the integral cohomology. In fact, by Proposition 22.6 we already know that $[\gamma]_*$ defines a ring isomorphism $H^*(X, \mathbb{Z}) \cong H^*(X', \mathbb{Z})$, as γ is the limit cycle of a sequence of isomorphisms. This will be generalized by Corollary 27.9.

Example 27.2 If we assume that in the previous example the two manifolds X and X' are projective, we can consider the two Fourier-Mukai functors

$$\mathrm{FM}_{\mathcal{O}_Z} : D^b(X) \to D^b(X') \ , \quad F \mapsto \pi'_*(\pi^*(F) \otimes \mathcal{O}_Z)$$
$$\mathrm{FM}_{\mathcal{O}_\Gamma} : D^b(X) \to D^b(X') \ , \quad F \mapsto \pi'_*(\pi^*(F) \otimes \mathcal{O}_\Gamma)$$

between their derived categories of coherent sheaves. In the light of the previous example, the second functor should be considered as the more natural one, as it descends to an isomorphism of the integral cohomology. Indeed Namikawa has shown in [166] that $\mathrm{FM}_{\mathcal{O}_\Gamma}$ defines an equivalence, whereas $\mathrm{FM}_{\mathcal{O}_Z}$ does not.

27.2 The Positive Cone

Let us recall that for a compact hyperkähler manifold X the positive cone $\mathcal{C}_X$ is the connected component of the subset $\{\alpha \mid q_X(\alpha) > 0\} \subset H^{1,1}(X, \mathbb{R})$ that contains the Kähler cone $\mathcal{K}_X$ of all Kähler classes on X.

Clearly, $\mathcal{C}_X$ is a non-empty open convex subset of $H^{1,1}(X, \mathbb{R})$ and $\mathcal{K}_X \subset \mathcal{C}_X$ is an open subset. Note that the positive cone $\mathcal{C}_X$ is by definition not the cone of all positive, i.e. Kähler classes, but rather of those classes α that have positive square $q_X(\alpha)$. The goal of this section is to show that any *general* class α in the positive cone can be viewed as a Kähler class on some compact hyperkähler manifold X' birational to X.

Definition 27.3 A class $\alpha \in \mathcal{C}_X$ is *general* if it is not orthogonal to any integral class $\beta \in H^{1,1}(X) \cap H^2(X, \mathbb{Z})$.

Proposition 27.4 *Let $(X, \varphi) \in \mathfrak{M}_\gamma$ be a marked compact hyperkähler manifold and let $\alpha \in \mathcal{C}_X$ be general. Then there exists a point $(X', \varphi') \in \mathfrak{M}_\gamma$, which cannot be separated from (X, φ) such that $(\varphi'^{-1} \circ \varphi)(\alpha) \in H^2(X', \mathbb{R})$ is a Kähler class.*

Proof. Let $F(\alpha)_{\mathbb{C}} := \mathbb{C}\alpha \oplus H^{2,0}(X) \oplus H^{0,2}(X)$ and let $T(\alpha) := \mathrm{Def}(X) \cap \mathcal{P}^{-1}(\varphi(\mathbb{P}(F(\alpha)_{\mathbb{C}})))$. By $\mathcal{X}(\alpha) \to T(\alpha)$ we denote the restriction of the universal family $\mathcal{X} \to \mathrm{Def}(X)$ to $T(\alpha)$. If α is a Kähler class, then $\mathcal{X}(\alpha) \to T(\alpha)$ is the associated twistor space or rather a germ of it. Analogously to the way we have defined the Kähler forms ω_t on the fibres $\mathcal{X}_t$ of the twistor space, one introduces cohomology classes $\alpha_t \in H^{1,1}(\mathcal{X}(\alpha)_t)$ by $\alpha_t = a\alpha + b\mathrm{Re}(\sigma) + c\mathrm{Im}(\sigma)$, where $t = (a, b, c) \in T(\alpha) \subset \mathbb{P}^1 = S^2$ and the origin 0 corresponds to $(1, 0, 0)$. Since $\alpha \in \mathcal{C}_X$, also $\alpha_t \in \mathcal{C}_{\mathcal{X}(\alpha)_t}$.

We claim that for α and $t \in T(\alpha)$ very general the fibre $\mathcal{X}(\alpha)_t$ is very general in the sense of Proposition 26.12, i.e. $\mathcal{C}_{\mathcal{X}(\alpha)_t} = \mathcal{K}_{\mathcal{X}(\alpha)_t}$. Inspecting the proof of Proposition 26.12 we find that it suffices to ensure that the fibre over t is not contained in a countable union of closed analytic proper subset of $\mathrm{Def}(X)$. But moving α in an open subset of $\mathcal{C}_X$ the base spaces $T(\alpha)$ cover a subset which cannot be contained in such a subset.

Let α and $t \in T(\alpha)$ be very general in this sense, then $\beta = \alpha_t$ is a Kähler class on the fibre $Y := \mathcal{X}(\alpha)_t$. In particular, we can associate to it its twistor space $\mathcal{Y}(\beta) \to \mathbb{P}^1$. The marking φ of X induces a marking φ' of Y. Hence, $\mathcal{Y}(\beta)$ is a family of marked manifolds and we can consider the period map $\mathcal{P} : \mathbb{P}^1 \to \mathbb{P}(\gamma_\mathbb{C})$. By Proposition 25.7 one has $\mathcal{P}(\mathbb{P}^1) = \varphi'(\mathbb{P}(F(\beta)_\mathbb{C})) = \varphi(\mathbb{P}(F(\alpha)_\mathbb{C}))$. In particular, the period map $\mathcal{P}$ identifies $T(\alpha)$ with an open subset of the base of the twistor space $\mathcal{Y}(\beta) \to \mathbb{P}^1$. Thus, one obtains two families $\mathcal{X}(\alpha) \to T(\alpha)$ and $\mathcal{Y}(\beta) \to T(\alpha)$ over the same base with isomorphic fibres $\mathcal{X}(\alpha)_t \cong Y$ over $t \in T(\alpha)$ and identical periods.

By the Local Torelli theorem (cf. Proposition 22.11) there exists a maximal open subset $t \in U \subset T(\alpha)$ over which the two families $\mathcal{X}(\alpha)$ and $\mathcal{Y}(\beta)$ are isomorphic. The two points (X, φ) and $(X', \varphi') := (\mathcal{Y}(\beta)_0, \varphi')$ are non-separated if the origin $0 \in T(\alpha)$ is contained in its closure.

Let $t_0 \in \partial U$ be any point in the boundary of U and let $t_i \in U$ be a sequence converging to t_0. Then the graph γ_i of the isomorphism $\mathcal{X}(\alpha)_{t_i} \cong \mathcal{Y}(\beta)_{t_i}$ degenerates to a cycle

$$\gamma = Z + \sum Y_i \subset \mathcal{X}(\alpha)_{t_0} \times \mathcal{Y}(\beta)_{t_0}$$

such that $\mathcal{X}(\alpha)_{t_0} \leftarrow Z \to \mathcal{Y}(\beta)_{t_0}$ is a birational correspondence, the projections $Y_i \to \mathcal{X}(\alpha)_{t_0}$ and $Y_i \to \mathcal{Y}(\beta)_{t_0}$ have positive fibre dimension, and $[\gamma]_* = {\varphi'}^{-1} \circ \varphi$. (Here, one first has to show that the volume of γ_i with respect to some family of Kähler forms is bounded. In a second step one shows that γ has the claimed special form. We refer to [104] for the details.)

Let us assume that in addition $\mathcal{Y}(\beta)_{t_0}$ neither contains non-trivial (rational) curves nor effective divisors, e.g. $\mathrm{Pic}(\mathcal{Y}(\beta)_{t_0}) = 0$. Then, $\mathcal{X}(\alpha)_{t_0} \cong Z \cong \mathcal{Y}(\beta)_{t_0}$ and $[Y_i]_* : H^2(\mathcal{X}(\alpha)_{t_0}, \mathbb{Z}) \to H^2(\mathcal{Y}(\beta)_{t_0}, \mathbb{Z})$ is trivial. Indeed, the fibres of the two projections from Z are covered by rational curves and must, therefore, be contracted in both directions. The map $[Y_i]_*$ on $H^2(\mathcal{X}(\alpha)_{t_0})$ can only be non-trivial if the image of $Y_i \to \mathcal{Y}(\beta)_{t_0}$ is a divisor. Thus, $\beta_{t_0} = ({\varphi'}^{-1} \circ \varphi)(\alpha_{t_0}) = [\gamma]_*(\alpha_{t_0}) = [Z]_*(\alpha_{t_0})$. Hence, α_{t_0} is a Kähler class on $\mathcal{X}(\alpha)_{t_0}$. By computing the volume of γ with respect to $\alpha_{t_0} \times \beta_{t_0}$ and using that γ is the limit cycle of a sequence of isomorphisms, one proves that $\gamma = Z$ (see [104] for details of this calculation). Thus, if t_0 is in the boundary of U, then $\mathcal{Y}(\beta)_{t_0}$ contains curves or divisors. On the other hand, the set of those points $t_0 \in T(\alpha) \subset \mathbb{P}^1$ is countable. Hence, ∂U is countable and, therefore, $\overline{U} = T(\alpha)$.

This proves the assertion for very general α. The assertion for general α is reduced to this case as follows. If α is not orthogonal to any integral class β, then $T(\alpha)$ is not contained in S_β, i.e. for general $t \in T(\alpha)$ the Picard group of $\mathcal{X}(\alpha)_t$ is trivial. This implies $\mathcal{C}_{\mathcal{X}(\alpha)_t} = \mathcal{K}_{\mathcal{X}(\alpha)_t}$. Indeed if β is a very general class in the positive cone of $\mathcal{X}(\alpha)_t$, then the first part of the proof applied to β on $\mathcal{X}(\alpha)_t$ yields a cycle $\gamma = Z + \sum Y_i \subset \mathcal{X}(\alpha)_t \times Y$ as above. In particular, $[\gamma]_*(\beta) \in \mathcal{K}_Y$. By using $\mathrm{Pic}(\mathcal{X}(\alpha)_t) = 0$, we have seen before that this implies that β is a Kähler class. This suffices to conclude that the Kähler cone of $\mathcal{X}(\alpha)_t$ is maximal. The final step consists of repeating part one of the proposition. $\qquad\square$

If two marked compact hyperkähler manifolds (X, φ) and (X', φ') define two non-separated points in the moduli space, then X and X' are birational. The exact relation between X and X' is spelled out in the following corollary. Some of the arguments were already sketched in the proof of the proposition.

Corollary 27.5 *Let X be a compact hyperkähler manifold of dimension $2n$ and let $\alpha \in \mathcal{C}_X$ be a general class. Then there exists a compact hyperkähler manifold X' and an effective cycle $\gamma = Z + \sum Y_i \subset X \times X'$ of dimension $2n$ with the following properties:*

- *The correspondence $X \leftarrow Z \rightarrow X'$ defines a birational map $X' \dashrightarrow X$ and the general fibre of $Y_i \rightarrow X$ (resp. $Y_i \rightarrow X'$) is of positive dimension and is covered by rational curves.*
- *The map $[\gamma]_* : H^*(X, \mathbb{Z}) \cong H^*(X', \mathbb{Z})$ is a ring isomorphism which respects the quadratic forms q_X, $q_{X'}$ and the Hodge structures.*
- *The image $[\gamma]_*(\alpha)$ of α is a Kähler class on X'.*

It might be helpful to compare the situation with a well-known result for K3 surfaces. Let X be a K3 surface and let $\alpha \in \mathcal{C}_X$ such that $\alpha \cdot C \neq 0$ for any smooth rational curve $C \subset X$. Then there exists a finite number of smooth irreducible rational curves $C_i \subset X$ such that $s_{C_1}(\ldots(s_{C_n}(\alpha))\ldots)$ is a Kähler class on X. Here, s_C is the reflection in the hyperplane orthogonal to the (-2)-class $[C] \in H^2(X, \mathbb{Z})$, i.e. $s_C(\alpha) = \alpha + (C.\alpha)C$. In this case the birational correspondence Z is given by the diagonal $\Delta \subset X \times X$ and the two projections. Of course, in dimension two any birational map between K3 surfaces defines an isomorphism. The additional cycle Y is the product $C \times C$.

The second goal of this lecture is to prove a partial converse of Corollary 27.5. We shall see that two birational compact hyperkähler manifolds always define non-separated points in the moduli space of marked compact hyperkähler manifolds.

27.3 Deformation Equivalence of Birational Hyperkähler Manifolds

Let X and X' be projective varieties and let $f : X' \dashrightarrow X$ be a birational map. If f is an isomorphism outside of subvarieties of codimension at least two in X and X' then $\mathrm{Pic}(X) = \mathrm{Pic}(X')$. Thus, we can speak of the pull-back of a line bundle L on X to a line bundle $L' = f^*(L)$ on X'. Moreover, there is a natural isomorphism of graded rings $\bigoplus H^0(X, L^k) \cong \bigoplus H^0(X', L'^k)$.

Using this one easily sees that f can be extended to an isomorphism $X' \cong X$ if and only if there exists an ample line bundle L on X such that the pull-back $L' := f^*(L)$ is an ample line bundle on X'. The analogue of this result for Kähler manifolds was proved by Fujiki in [58]:

Proposition 27.6 *Let X and X' be compact Kähler manifolds and let f : $X' \dashrightarrow X$ be a birational map which is an isomorphism outside of subvarieties of codimension at least two. Then f extends to an isomorphism $X' \cong X$ if and only if there exists a Kähler class α on X such that $f^*(\alpha) \in H^2(X, \mathbb{R})$ is a Kähler class on X'.*

We will need the following slightly more precise version of this result:

Proposition 27.7 *Let $f : X' \dashrightarrow X$ be a birational map of irreducible holomorphic symplectic manifolds. If $\alpha \in H^2(X, \mathbb{R})$ is a class such that $\int_C \alpha > 0$ and $\int_{C'} f^*(\alpha) > 0$ for all rational curves $C \subset X$ and $C' \subset X'$, then f extends to an isomorphism $X \cong X'$.*

Proof. Let $\pi : Z \to X$ be a sequence of blow-ups resolving f, i.e. the birational map f is given by the correspondence $X \xleftarrow{\pi} Z \xrightarrow{\pi'} X'$. By Proposition 21.6 any exceptional divisor E_i of $\pi : Z \to X$ is also exceptional for $\pi' : Z \to X'$. Moreover, there exists a positive linear combination $\sum n_i E_i$, such that $-\sum n_i E_i$ is π-ample. As $H^2(Z, \mathbb{R})$ can be decomposed in two different ways $H^2(Z, \mathbb{R}) = \pi^* H^2(X, \mathbb{R}) \oplus \bigoplus \mathbb{R}[E_i]$ and $H^2(Z, \mathbb{R}) = \pi'^* H^2(X', \mathbb{R}) \oplus \bigoplus \mathbb{R}[E_i]$, any class $\beta \in H^2(X, \mathbb{R})$ can be written as $\pi^* \beta = \pi'^* \beta' + \sum a_i [E_i]$, where $\beta' = f^* \beta$ and $a_i \in \mathbb{R}$. Apply this to the class α as above. In a first step we show that all coefficients a_i are non-negative in this case. Interchanging the rôle of α and α' then yields $a_i = 0$ for all i, i.e. $\pi'^* \alpha' = \pi^* \alpha$.

Assume that this is not the case, i.e. $a_1, \ldots, a_k < 0$, $a_{k+1}, \ldots, a_\ell \geq 0$ for some $k \geq 1$. We may assume that $-(a_1/n_1) = \max_{i=1,\ldots k}\{-(a_i/n_i)\}$. Let $C' \subset E_1$ be a general rational curve contracted under $\pi : E_1 \to X$. As C' is general, one has $E_i \cdot C' \geq 0$ for all $i > 1$ and hence

$$-\sum_{i=1}^{\ell} a_i (E_i \cdot C') \leq -\sum_{i=1}^{k} a_i (E_i \cdot C')$$

$$\leq (-\frac{a_1}{n_1}) \sum_{i=1}^{k} n_i (E_i \cdot C') \leq (-\frac{a_1}{n_1}) \sum_{i=1}^{\ell} n_i (E_i \cdot C') < 0.$$

This yields a contradiction to

$$0 = \pi^* \alpha \cdot C' = (\pi'^* \alpha' + \sum_{i=1}^{\ell} a_i [E_i]) \cdot C' \geq \sum_{i=1}^{\ell} a_i (E_i \cdot C').$$

To conclude one shows that for all exceptional divisors E_i the two contractions $\pi : E \to X$ and $\pi' : E \to X'$ coincide. This immediately implies that the birational map $X' \dashrightarrow X$ extends to an isomorphism. Assume that there exists a rational curve $C' \subset E_i$ contracted by π, such that $\pi' : C' \to X'$ is finite. Since $\pi'^* \alpha' = \pi^* \alpha$, this yields the contradiction $0 < \pi'^* \alpha' \cdot C' = \pi^* \alpha \cdot C' = 0$. Analogously, one excludes the case that there exists a rational curve C contracted by π' with $\pi|_C$ finite. As the fibres of $\pi'|_{E_i}$ and $\pi|_{E_i}$ are covered by rational curves, this shows that the two projections do coincide. $\square$

Generalizing the result about deformation equivalence of two compact hyperkähler manifolds related by a Mukai flop (cf. Proposition 22.6) we can now prove the deformation equivalence of two arbitrary birational compact hyperkähler manifolds.

Proposition 27.8 *Let X and X' be compact hyperkähler manifolds and let*

$$f : X' \dashrightarrow X$$

be a birational map. Then there exist smooth proper families $\mathcal{X} \to S$ and $\mathcal{X}' \to S$ over a one-dimensional disk S with the following properties:

- *The special fibres are $\mathcal{X}_0 \cong X$ and $\mathcal{X}'_0 \cong X'$.*
- *There exists a birational map $\tilde{f} : \mathcal{X}' \dashrightarrow \mathcal{X}$ which is an isomorphism over $S \setminus \{0\}$, i.e. $\tilde{f} : \mathcal{X}'|_{S \setminus \{0\}} \cong \mathcal{X}|_{S \setminus \{0\}}$, and which coincides with f on the special fibre, i.e. $\tilde{f}_0 = f$.*

Proof. We will only sketch the main ideas of the proof and refer to [101] for details. First, choose a very general class $\alpha \in \mathcal{C}_X$ that corresponds to a very general class $\alpha' \in \mathcal{K}_{X'}$ under the isomorphism $H^{1,1}(X) \cong H^{1,1}(X')$ induced by the birational map. All we will need is $\int_{C'} \alpha' > 0$ for any (rational) curve $C' \subset X'$. Next apply Proposition 27.4 (and its proof) to X and α. Thus, it suffices to show that $\mathcal{X}'_0 \cong X'$ and that under this isomorphism $\tilde{f}_0$ and f coincide.

Let $\mathcal{Z} \subset \mathcal{X} \times_S \mathcal{X}'$ be the graph of the birational map, i.e. the closure of the graph of the isomorphism $\mathcal{X}|_{S \setminus 0} \cong \mathcal{X}'|_{S \setminus 0}$. The cycle $\gamma := \mathrm{im}(\mathcal{Z}_0 \to X \times \mathcal{X}'_0)$ decomposes into $\gamma = Z + \sum Y_i$, where the $Y_i \subset X \times \mathcal{X}'_0$ correspond to the exceptional divisors D_i of $\mathcal{X} \xleftarrow{\pi} \mathcal{Z} \xrightarrow{\pi'} \mathcal{X}'$ and $X \leftarrow Z \to \mathcal{X}'_0$ is a birational correspondence. If the codimension of the image of D_i under $\pi' : D_i \to \mathcal{X}'_0$ is at least two in $\mathcal{X}'_0$, then $[Y_i]_* : H^2(X) \to H^2(\mathcal{X}'_0)$ is trivial. If this is the case for all i, then $\beta := [\gamma]_*(\alpha) = [Z]_*(\alpha)$. Hence, under the birational correspondence $\mathcal{X}'_0 \leftarrow Z \to X \dashrightarrow X'$ the class α' on X' is mapped to the Kähler class β on $\mathcal{X}'_0$. Proposition 27.7 then shows that the birational map $\mathcal{X}'_0 \dashrightarrow X'$ can be extended to an isomorphism.

In order to show that the image of D_i in $\mathcal{X}'_0$ is of codimension at least two, one argues roughly as follows. If there exists a divisor D_i whose image in $\mathcal{X}'_0$ is a divisor, then also its image in X is a divisor. This is proven by using the holomorphic symplectic forms on $\mathcal{X}'_0$ and X which coincide on the complement of a codimension two subset. For codimension reasons, a general irreducible rational curve C in the fibre of $D_i \to \mathcal{X}_0'$ will not map entirely into the exceptional set of $f : X \dashrightarrow X'$. Thus, $\int_C \alpha > 0$. Eventually, a contradiction is derived by using arguments similar to those in the proof of Proposition 27.7 which yield $\int_C \alpha < 0$. $\square$

Corollary 27.9 *Let X and X' be birational compact hyperkähler manifolds. Then the Hodge structures of X and X' are isomorphic. In particular, X and X' have equal Hodge and Betti numbers. Moreover, $H^*(X, \mathbb{Z}) \cong H^*(X', \mathbb{Z})$ are isomorphic as graded rings.*

In fact, no invariant is known that could distinguish between non-isomorphic birational compact hyperkähler manifolds. Therefore, it usually is very difficult to actually decide whether two birational manifolds are really distinct (compare the examples of Debarre and Yoshioka in §21.3).

Corollary 27.9 and Proposition 27.8 have been successfully applied to concrete problems in [14] and [46]. Corollary 27.9 should be compared with results of Kontsevich, Batyrev, Wang, and Denef and Loeser:

Theorem 27.10 *Let X and X' be two birational projective manifolds with nef canonical bundle. Then X and X' have equal Hodge and Betti numbers.*

For Calabi-Yau manifolds the equality of the Betti numbers was first shown by Batyrev [7] by using p-adic integration. This was then generalized by Wang [199,200] to cases where the canonical bundles are nef on the exceptional sets proving also equality of the Hodge numbers. Later Kontsevich proposed the method of motivic integration. This was worked out by Batyrev [6] and independently and in more detail by Denef and Loeser [50]. Note however that there exist examples of birational projective Calabi–Yau threefolds with non-isomorphic (as abstract rings) cohomology rings [57]. This type of question can be lifted to the level of derived categories. It is still an open question whether the derived categories of two birational Calabi–Yau manifolds are equivalent. The only known results are for Calabi–Yau manifolds in dimension three ([28,25]) and for Mukai's elementary transformations in the case of hyperkähler manifolds [143,166].

Remark 27.11 Let us also note that Kaledin has developed techniques to show deformation equivalence in the local situation [125].

27.4   Exercises

27.1 Compare the K3 surface case with the general statement in Corollary 27.5.

27.2 Let $X \leftarrow Z \to X'$ be a Mukai flop as in Example 27.1. Show that the composition of $[Z]_* : H^*(X, \mathbb{Z}) \to H^*(X', \mathbb{Z})$ with $[Z]_* : H^*(X', \mathbb{Z}) \to H^*(X, \mathbb{Z})$ equals $\mathrm{id} + [P \times P']_*$.

28   The (Birational) Kähler Cone

The Kähler cone (or the ample cone) of a variety usually encodes crucial information about the geometry of the variety and can thus be a rather complicated object. For Calabi–Yau threefolds this was studied by Kawamata, Wilson and others ([129,205]). In the case of a compact hyperkähler manifold X we have seen that the Kähler cone is contained in the positive cone $\mathcal{C}_X \subset H^{1,1}(X, \mathbb{R})$, which is defined purely in terms of the Beauville–Bogomolov form and is, in particular, 'constant' in families. One would like to describe the Kähler cone $\mathcal{K}_X$ inside this positive cone. The Demailly–Paun theorem (cf. §26.2) says that the Kähler cone can be described as the set of classes that are positive on all subvarieties (or rather a connected component of this set). But in most cases it is very difficult to understand the set of all subvarieties of a given variety. Already for K3 surfaces, where only curves have to be tested, it turns out that testing very special ones suffices.

Proposition 28.1 *Let X be a K3 surface. Then a class α in the positive cone is Kähler if and only if α is positive on every smooth rational curve $C \subset X$.*

Note that an irreducible curve $C \subset X$ is smooth and rational if and only if C is a (-2)-curve, i.e. $C^2 = -2$. For a proof of this proposition see [16]. We will present an alternative proof by using the coarser description of the Kähler cone of an arbitrary compact hyperkähler manifold (cf. §28.1). A birational map between K3 surfaces can be extended to an isomorphism. In higher dimensions this is no longer the case and, therefore, also the birational Kähler cone has to be studied. The birational Kähler cone is by definition the set of all classes that are Kähler on some birational model. It will be studied in §28.3.

28.1   The Closure of the Kähler Cone

We will prove an anlogue of the above result for higher-dimensional hyperkähler manifolds. We will see that it suffices to test curves (and not subvarieties of arbitrary codimension) and that we can restrict ourselves again to rational curves. Unfortunately, we neither know whether smooth rational curves suffice nor can we bound the square of the curves that have to be taken into account. Note that the boundary of the Kähler cone is a complicated object. Questions whether it is locally finitely polyhedral etc. remain completely open. For the Hilbert scheme there is an interesting discussion of the ample cone in [95].

Proposition 28.2 *Let X be a compact hyperkähler manifold. The closure $\overline{\mathcal{K}}_X$ of the Kähler cone $\mathcal{K}_X \subset H^{1,1}(X, \mathbb{R})$ is the set of all classes $\alpha \in \overline{\mathcal{C}}_X$ with $\int_C \alpha \geq 0$ for all rational curves $C \subset X$.*

Proof. Let $\alpha \in \overline{\mathcal{K}}_X$. Then $\int_C \alpha \geq 0$ for any curve $C \subset X$. Since $\mathcal{K}_X \subset \mathcal{C}_X$, also $\alpha \in \overline{\mathcal{C}}_X$. This proves one direction.

Now, let $\alpha \in \overline{\mathcal{C}}_X$ and $\int_C \alpha \geq 0$ for all rationals curves $C \subset X$. Pick an arbitrary Kähler class ω. Then for $\varepsilon > 0$ one has $\alpha + \varepsilon\omega \in \mathcal{C}_X$ and $\int_C(\alpha + \varepsilon\omega) > 0$. For ω and ε general, the class $\alpha + \varepsilon\omega$ is general as well. In order to prove that $\alpha \in \overline{\mathcal{K}}_X$, it suffices to show that $\alpha + \varepsilon\omega \in \mathcal{K}_X$ for ω general and ε arbitrary small. Thus, it is enough to show that any general(!) class $\alpha \in \mathcal{C}_X$ with $\int_C \alpha > 0$ for all rational curves $C \subset X$ is contained in $\mathcal{K}_X$.

Applying Proposition 27.4 and Corollary 27.5 we see that there exists a cycle $\gamma = Z + \sum Y_i \subset X \times X'$ with $[\gamma]_*(\alpha) \in \mathcal{K}_{X'}$. In fact, revisiting the proof of Proposition 27.4 we find that $[\gamma]_*(\alpha)$ can be seen as the restriction to the special fibre over $0 \in S$ of the image of α under a birational map $\mathcal{X} \dashrightarrow \mathcal{X}'$. Then use Proposition 27.7 to conclude $\mathcal{X} \cong \mathcal{X}'$. Thus, $\alpha \in \mathcal{K}_X$.   □

Corollary 28.3 *Let X be a compact hyperkähler manifold not containing any rational curve. Then $\mathcal{C}_X = \mathcal{K}_X$.*

In dimension two this gives back the above theorem for K3 surfaces for which we can now give a proof. It is noteworthy that this result had originally been proved by using the Global Torelli theorem for K3 surfaces for which in higher dimensions there is not even a good conjecture at the moment.

Proof of Proposition 28.1.   Let C be an irreducible rational curve. Then $C^2 \geq 0$ or C is smooth and $C^2 = -2$. Let $\alpha \in \mathcal{C}_X$. If $C^2 \geq 0$, then $\alpha \cdot C \geq 0$. Hence, the proposition shows $\overline{\mathcal{K}}_X = \{\alpha \in \mathcal{C}_X | \alpha \cdot C \geq 0$ for every smooth rational curve$\}$. It suffices to show that for every class $\alpha \in \partial\overline{\mathcal{K}}_X \cap \mathcal{C}_X$ there exists a smooth rational curve C with $\alpha \cdot C = 0$. Let $\{C_i\}$ be a series of smooth irreducible rational curves, such that $\alpha \cdot C_i \to 0$. Let $\alpha_1 = \alpha, \alpha_2 \ldots, \alpha_{20}$ be an orthogonal base of $H^{1,1}(X, \mathbb{R})$. We can even require $\alpha_{i>1}^2 = -1$ and $\alpha_i.\alpha_{j\neq i} = 0$. For the coefficients of $[C_i] = \lambda_i\alpha + \sum \mu_{ij}\alpha_j$ one then concludes $\lambda_i \to 0$ and $\sum \mu_{ij}^2 \to 2$. Thus, the set of classes $[C_i]$ is contained in a compact ball and, hence, there is only a finite number of them. The assertion follows directly from this.   □

28.2   The Open Kähler Cone

In his article [26] Boucksom could prove that not only the closed, but also the (open) Kähler cone can be described in terms of rational curves as suggested by the above result. For his result one first has to prove the description of the closed Kähler cone in order to conclude Corollary 28.3. The latter will be needed in the proof of Lemma 28.4.

Recall that a class $\alpha \in \mathcal{C}_X$ is called general if α is not orthogonal to any integral class of type $(1,1)$. The last part of the proof of Proposition 27.4 shows that for a general class $\alpha \in \mathcal{C}_X$ the general fibre of $\mathcal{X}(\alpha) \to T(\alpha)$ satisfies $\mathcal{K}_{\mathcal{X}(\alpha)_t} = \mathcal{C}_{\mathcal{X}(\alpha)_t}$

Boucksom first observes that the same conclusion holds for any class $\alpha \in \mathcal{C}_X$ such that $\int_C \alpha \neq 0$ for any rational curve C.

Lemma 28.4 *If $\alpha \in \mathcal{C}_X$ is not orthogonal to any rational curve $C \subset X$ then the general fibre of the family $\mathcal{X}(\alpha) \to T(\alpha)$ over $T(\alpha) = \mathrm{Def}(X) \cap \mathbb{P}(\mathbb{C}\alpha \oplus \mathbb{C}\sigma \oplus \mathbb{C}\bar{\sigma})$ has maximal Kähler cone, i.e. $\mathcal{K}_{\mathcal{X}(\alpha)_t} = \mathcal{C}_{\mathcal{X}(\alpha)_t}$.*

Proof. Indeed, since $\int_C \alpha \neq 0$ for any rational curve, none of the rational curves $C \subset X$ will deform to the nearby fibres in $\mathcal{X}(\alpha) \to T(\alpha)$. Thus, the general fibre over $t \in T(\alpha)$ close to $0 \in T(\alpha)$ will not contain any rational curve. Thus, by Corollary 28.3 the general fibre will have maximal Kähler cone. □

Proposition 28.5 *Let X be a compact hyperkähler manifold. Then the Kähler cone $\mathcal{K}_X$ is the set of all classes $\alpha \in \mathcal{C}_X$ such that $\int_C \alpha > 0$ for all rational curves $C \subset X$.*

Proof. The proof of Proposition 28.2 applies literally. Under our assumption and by using Lemma 28.4 we can work directly with the class α without modifying it by $\varepsilon\omega$ as we did there. □

28.3 The Birational Kähler Cone

An analogous result can be proved for the birational Kähler cone. The birational Kähler cone was studied by Kawamata and Morrison in the case of Calabi–Yau threefolds ([129,160]). Let X be a compact hyperkähler manifold and let $f : X \dashrightarrow X'$ be a birational map to another compact hyperkähler manifold X'. Via the natural isomorphism $H^{1,1}(X', \mathbb{R}) \cong H^{1,1}(X, \mathbb{R})$ (cf. Proposition 25.14) the Kähler cone $\mathcal{K}_{X'}$ of X' can also be considered as an open subset of $H^{1,1}(X, \mathbb{R})$ and, as can easily be seen, as an open subset of $\mathcal{C}_X$. The union of all those open subsets is called the birational Kähler cone of X.

Definition 28.6 The *birational Kähler cone* is the union

$$\mathcal{BK}_X := \bigcup_{f : X \dashrightarrow X'} f^*(\mathcal{K}_{X'}),$$

where $f : X \dashrightarrow X'$ runs through all birational maps $X \dashrightarrow X'$ from X to another compact hyperkähler manifold X'.

Note that $\mathcal{BK}_X$ is in fact a disjoint union of the $f^*(\mathcal{K}_{X'})$, i.e. if $f_1^*(\mathcal{K}_{X_1})$ and $f_2^*(\mathcal{K}_{X_2})$ have a non-empty intersection then they are equal. This follows from Proposition 27.6.

In particular, the birational Kähler cone $\mathcal{BK}_X$ will rarely be a convex cone in $\mathcal{C}_X$. This can only happen if any birational map can be extended to an isomorphism. But it turns out, that its closure $\overline{BK}_X$ is a convex cone. Notice that this is far from being obvious. E.g. if X and X' are birational compact hyperkähler manifolds and α and α' are Kähler classes on X and X', respectively, there is a priori no reason why every generic class α'' contained in the segment joining α and α' should be a Kähler class on yet another birational compact hyperkähler manifold X''. In this point the theory resembles the theory of Calabi–Yau threefolds [129].

We will give a geometric description of the birational Kähler cone analogous to Proposition 28.2. The rational curves are replaced by uniruled divisors. There is also a 'form'-description which will not be treated here (cf. [101]).

Proposition 28.7 *Let X be a compact hyperkähler manifold. Then a class $\alpha \in H^{1,1}(X, \mathbb{R})$ is in the closure $\overline{BK}_X$ of the birational Kähler cone BK_X if and only if $\alpha \in \overline{C}_X$ and $q_X(\alpha, [D]) \geq 0$ for all uniruled divisors $D \subset X$.*

Proof. The idea of the proof is very simple and similar to the proof of Proposition 28.2. We will sketch the argument under a simplifying assumption and refer to [101] for the most general case.

Of course, if $\alpha \in \overline{BK}_X$, then $q(\alpha, [D]) \geq 0$ for any divisor. This is due to the observation that for a birational correspondence $Z \subset X \times X'$ the image $[Z]_*[D]$ is just the image D' of D under the rational map $X \dashrightarrow X'$. Hence, by Proposition 25.14 one has $q(\alpha, [D]) = q([Z]_*(\alpha), [Z]_*[D]) = q([Z]_*(\alpha), [D'])$ and the latter is positive if $[Z]_*(\alpha) \in \mathcal{K}_{X'}$.

For the other direction, we let $\alpha \in \mathcal{C}_X$ be a general class such that $q(\alpha, [D]) > 0$ for any uniruled divisor D. Applying Proposition 27.4 and Corollary 27.5 we obtain a cycle $\gamma = Z + \sum Y_i \subset X \times X'$ with $[\gamma]_*(\alpha) \in \mathcal{K}_{X'}$. As usual, $Z \subset X \times X'$ is a birational correspondence and the Y_i are irreducible exceptional, and thus uniruled, divisors of a birational map $\mathcal{X} \dashrightarrow \mathcal{X}'$. It suffices to show that $[Y_i]_*(\alpha) = 0$. Then $[Z]_*(\alpha) = [\gamma]_*(\alpha) \in \mathcal{K}_{X'}$ and, hence, $\alpha \in BK_X$. We prove this under the assumption that there is only one exceptional divisor, i.e. $\gamma = Z + Y$. If the image D' of $Y \to X'$ is of codimension at least two, then $[Y]_*(\alpha) = 0$. Let us assume that D' is a divisor. If the two projections $X \leftarrow Y \to X'$ were the same, the birational map $\mathcal{X} \dashrightarrow \mathcal{X}'$ could be extended over a larger open subset and the divisor Y could be avoided altogether. Hence, the projections are different and, therefore, D and D' are uniruled divisors. Also note that D' is the image of D under the birational correspondence Z. Moreover, $[Y]_*[D] = (Y \cdot C')[D']$, where C' is the general fibre of $Y \to D'$. Since Y is exceptional, $(Y \cdot C') < 0$. Similarly, $[Y]_*[D'] = (Y \cdot C)[D]$ with $(Y \cdot C) < 0$. A contradiction is obtained as follows: $q(\alpha, [D]) = q([\gamma]_*(\alpha), [\gamma]_*[D]) = q([\gamma]_*(\alpha), [Z]_*[D]) + (Y \cdot C')q([\gamma]_*(\alpha), [D']) < q([\gamma]_*(\alpha), [Z]_*[D])$, since $[\gamma]_*(\alpha)$ is Kähler. The last term equals $q([\gamma]_*[\gamma]_*(\alpha), [\gamma]_*[D']) = q(\alpha, [Z]_*[D']) + (Y \cdot C)q(\alpha, [D])$, which can be bounded from above by $q(\alpha, [D])$, since α is positive on uniruled divisors. $\square$

Corollary 28.8 *If X is a compact hyperkähler manifold not containing any uniruled divisor, then $\mathcal{BK}_X$ is dense in $\mathcal{C}_X$.*

The following result is due to Boucksom [27]. It predicts which divisors are uniruled.

Corollary 28.9 *Let D be an irreducible divisor in X with $q([D]) < 0$. Then D is uniruled.*

Proof. Pick a very general class $\alpha \in \mathcal{C}_X$ with $q(\alpha, [D]) < 0$. Then there exists by Proposition 27.4 and Corollary 27.5 a cycle $\gamma = Z + \sum Y_i \subset X \times X'$ with $\omega' := [\gamma]_*(\alpha) \in \mathcal{K}_{X'}$. As usual, $Z \subset X \times X'$ is a birational correspondence. Thus, $[Z]_*(\alpha) = \omega' - \sum [Y_i]_*(\alpha)$. Since $[Z]_*$ respects the Beauville–Bogomolov form, one has $0 > q(\alpha, [D]) = q([Z]_*(\alpha), [D']) = q(\omega', [D']) - \sum q([Y_i]_*(\alpha), [D']) > -\sum q([Y_i]_*(\alpha), [D'])$, where D' is the image of D under the birational correspondence Z. Thus, there exists at least one Y_j with $q([Y_j]_*(\alpha), [D']) > 0$. By construction, the Y_i are exceptional divisors of a birational correspondence $\mathcal{X} \dashrightarrow \mathcal{X}'$. Thus, either the image of $Y_i \to X'$ is of codimension at least two, and hence $[Y_i]_*(\alpha) = 0$, or it is a uniruled divisor $D_i \subset X'$. Moreover, in the latter situation $[Y_i]_*(\alpha)$ is a negative multiple of $[D_i]$. Indeed, this is a version of Proposition 27.7 applied to $\mathcal{X} \dashrightarrow \mathcal{X}'$. If α were positive on the general fibre of $Y_i \to X'$ with a codimension one image, then we would not have to blow-up $\mathcal{X}$ in this locus in order to make α positive. Hence, $q([D_j], [D']) < 0$, where D_j is as above. On the other hand, if E and E' are distinct irreducible divisors then $[E].[E']$ is given by the components of their codimension two intersection (with positive multiplicities). This shows, since $\sigma\bar\sigma$ is a positive form, that the integral $\int [E][E'](\sigma\bar\sigma)^{n-1}$ must be non-negative. Since $q([D_j], [D']) = \int [D_j][D'](\sigma\bar\sigma)^{n-1}$ up to a positive scalar, this shows that $D_j = D'$. Hence, D is uniruled. $\qquad\square$

28.4 Exercises

28.1 Show that $\mathcal{K}_X = \mathcal{C}_X$ if X is projective and the Picard number of X is one.

28.2 Could it be that for a uniruled irreducible divisor D one has $q([D]) > 0$?

References

1. F.J. Almgren, *Almgren's Big Regularity Paper*, World Scientific, River Edge, NJ, 2000, edited by V. Scheffer and J.E. Taylor.
2. J. Amoros, M. Burger, K. Corlette, D. Kotschick, and D. Toledo, *Fundamental groups of compact Kähler manifolds*, A.M.S. Mathematical Surveys and Monographs, vol. 44, A.M.S., Providence, RI, 1996.
3. S. Barannikov and M. Kontsevich, *Frobenius manifolds and formality of Lie algebras of polyvector fields*, International Mathematics Research Notices 4 (1998), 201–215.
4. W. Barth, C. Peters, and A. Van de Ven, *Compact Complex Surfaces*, Ergebnisse der Mathematik und ihrer Grenzgebiete, vol. 4, Springer–Verlag, Berlin, 1984.
5. V. Batyrev, *Dual polyhedra and mirror symmetry for Calabi–Yau hypersurfaces in toric varieties*, Journal of Algebraic Geometry 3 (1994), 493–535.
6. V. Batyrev, *Stringy Hodge numbers of varieties with Gorenstein canonical singularities*, Integrable systems and algebraic geometry (Kobe/Kyoto, 1997), World Scientific, River Edge, NJ, 1998, pp. 1–32.
7. V. Batyrev, *Birational Calabi–Yau n-folds have equal Betti numbers*, London Mathematical Society Lecture Notes, vol. 264, pp. 1–11, Cambridge University Press, Cambridge, 1999.
8. V. Batyrev and L. Borisov, *On Calabi–Yau complete intersections in toric varieties*, Higher-dimensional complex varieties (Trento, 1994), de Gruyter, Berlin, 1996, pp. 39–65.
9. V. Batyrev, I. Ciocan-Fontanine, B. Kim, and D. van Straten, *Conifold transitions and mirror symmetry for Calabi–Yau complete intersections in Grassmannians*, Nuclear Physics B514 (1998), 640–666.
10. V. Batyrev, I. Ciocan-Fontanine, B. Kim, and D. van Straten, *Mirror symmetry and toric degenerations of partial flag manifolds*, Acta mathematica 184 (2000), 1–39.
11. A. Beauville, *Some remarks on Kähler manifolds with $c_1 = 0$*, Classification of algebraic and analytic manifolds, Progress in Mathematics, vol. 39, Birkhäuser, Boston, MA, 1983, pp. 1–26.
12. A. Beauville, *Variétés Kähleriennes dont la première classe de Chern est nulle*, Journal of Differential Geometry 18 (1983), 755–782.
13. A. Beauville, *Systèmes hamiltoniens complètement intégrables associés aux surfaces K3*, Problems in the theory of surfaces and their classification (Cortona, 1988), Sympos. Math., vol. XXXII, Academic Press, London, 1991, pp. 25–31.
14. A. Beauville, *Counting rational curves on K3 surfaces*, Duke Mathematical Journal 97 (1999), 99–108.
15. A. Beauville, *Riemannian holonomy and algebraic geometry*, math.AG/9902110, 1999.
16. A. Beauville, J.-P. Bourguignon, and M. Demazure (eds.), *Géométrie des surfaces K3: modules et périodes*, Astérisque, vol. 126, 1985.
17. A. Beauville and R. Donagi, *La variété des droits d'une hypersurface cubique de dimension 4*, Comptes Rendus de l'Académie des Sciences. Série A, Sciences Mathematiques 301 (1985), 703–706.

18. M. Berger, *Sur les groupes d'holonomie homogène des variétés à connexion affine et des variétés riemanniennes*, Bull. Soc. Math. France **83** (1955), 225–238.

19. A. Bertram, *Another way to enumerate rational curves with torus actions*, Inventiones mathematicae **142** (2000), 487–512.

20. A. Besse, *Einstein Manifolds*, Springer, Berlin, 1987.

21. F. Bogomolov, *Hamiltonian Kählerian manifolds*, Dokl. Akad. Nauk SSSR **243** (1978), 1101–1104.

22. F. Bogomolov, *On Guan's example of simply connected non-Kähler compact complex manifolds*, American Journal of Mathematics **118** (1996), 1037–1046.

23. F. Bogomolov, *On the cohomology ring of a simple hyperkähler manifold (on the results of Verbitsky)*, Geometric and Functional Analysis **6** (1996), 612–618.

24. L. Bonavero, *Inégalités de Morse holomorphes singulières*, Comptes Rendus de l'Académie des Sciences. Série A, Sciences Mathematiques **317** (1993), 1163–1166.

25. A. Bondal and D. Orlov, *Semiorthogonal decomposition for algebraic varieties*, alg-geom/9506012, 1995.

26. S. Boucksom, *Le cône kählerien d'une variété hyperkählerienne*, Comptes Rendus de l'Académie des Sciences. Série A, Sciences Mathematiques **333** (2001), 935–938.

27. S. Boucksom, *An analytic approach to Zariski decomposition*, math.AG/0204336, 2002.

28. T. Bridgeland, *Flops and derived categories*, Inventiones mathematicae **147** (2002), 613–632.

29. M. Britze, *Generalized Kummer varieties*, Ph.D. thesis, University of Köln, 2002.

30. R.L. Bryant, *Metrics with exceptional holonomy*, Annals of Mathematics **126** (1987), 525–576.

31. R.L. Bryant, *Second order families of special Lagrangian 3-folds*, math.DG/0007128, 2000.

32. R.L. Bryant and P. Griffiths, *Some observations on the infinitesimal period relations for regular threefolds with trivial canonical bundle*, Arithmetic and geometry, Vol. II, Progress in Mathematics, vol. 36, Birkhäuser, Boston, MA, 1983, pp. 77–102.

33. R.L. Bryant and S.M. Salamon, *On the construction of some complete metrics with exceptional holonomy*, Duke Mathematical Journal **58** (1989), 829–850.

34. D. Burns, Y. Hu, and T. Luo, *Hyperkähler manifolds and birational transformation in dimension four*, math.AG/0004154, 2000.

35. F. Campana, *Densité des variétés Hamiltoniennes primitives projectives*, Comptes Rendus de l'Académie des Sciences. Série A, Sciences Mathematiques **297** (1983), 413–416.

36. P. Candelas, X. de la Ossa, P. Green, and L. Parkes, *A pair of Calabi–Yau manifolds as an exactly soluble superconformal theory*, Nuclear Physics **B359** (1991), 21–74.

37. P. Candelas, G. Horowitz, A. Strominger, and E. Witten, *Vacuum Configurations for Superstrings*, Nuclear Physics **B258** (1985), 46–74.

38. P. Candelas, M. Lynker, and R. Schimmrigk, *Calabi–Yau manifolds in weighted $\mathbb{P}_4$*, Nuclear Physics **B341** (1990), 383–402.

39. J. Carlson, S. Müller-Stach, and C. Peters, *Period maps and period domains*, Cambridge University Press, Cambridge, 2002.

40. I. Castro and F. Urbano, *New examples of minimal Lagrangian tori in the complex projective plane*, Manuscripta math. **85** (1994), 265–281.

41. S.-S. Chern, *Complex manifolds without potential theory*, second ed., Universitext, Springer–Verlag, New York, 1979.

42. K. Cho, Y. Miyaoka, and N. Shepherd-Barron, *Long extremal rays and symplectic resolutions*, Preprint.

43. E. Coddington and N. Levinson, *Theory of Ordinary Differential Equations*, McGraw-Hill, New York, 1955.

44. D. Cox and S. Katz, *Mirror Symmetry and Algebraic Geometry*, Mathematical Surveys and Monographs, vol. 68, A.M.S., Providence, RI, 1999.

45. O. Debarre, *Un contre-exemple au théorème de Torelli pour les variétés symplectiques irréductibles*, Comptes Rendus de l'Académie des Sciences. Série A, Sciences Mathematiques **299** (1984), 681–684.

46. O. Debarre, *On the Euler characteristic of generalized Kummer varieties*, American Journal of Mathematics **121** (1999), 577–586.

47. P. Deligne, *Théorème de Lefschetz et critères de dégénerescence de suites spectrales*, Inst. Hautes Études Sci. Publ. Math. **35** (1968), 259–278.

48. J.-P. Demailly, *Regularization of closed positive currents and intersection theory*, Journal of Algebraic Geometry **1** (1992), 361–409.

49. J.-P. Demailly and M. Paun, *Numerical characterization of the Kähler cone of a compact Kähler manifold*, math.AG/0105176, 2001.

50. J. Denef and F. Loeser, *Germs of arcs on singular algebraic varieties and motivic integration*, Inventiones mathematicae **135** (1999), 201–237.

51. I. Dolgachev, *Weighted projective varieties*, Group actions and vector fields (Vancouver, B. C., 1981), Springer Lecture Notes in Mathematics, vol. 956, Springer, Berlin, 1982, pp. 34–71.

52. G. Ellingsrud, L. Göttsche, and M. Lehn, *On the Cobordism Class of the Hilbert Scheme of a Surface*, Journal of Algebraic Geometry **10** (2001), 81–100.

53. G. Ellingsrud and S. Strømme, *The number of twisted cubic curves on the general quintic threefold*, Math. Scand. **76** (1995), 5–34.

54. I. Enoki, *Compact Ricci-Flat Kähler Manifolds*, Kähler metric and moduli spaces, Adv. Stud. Pure Math., vol. 18-II, Academic Press, Boston, 1990, pp. 229–256.

55. J.M. Figueroa-O'Farrill, C. Köhl, and B. Spence, *Supersymmetry and the cohomology of (hyper)kähler manifolds*, Nuclear Physics **B503** (1997), 614–626.

56. A.P. Fordy and J.C. Wood (eds.), *Harmonic Maps and Integrable Systems*, Aspects of Mathematics, vol. E23, Vieweg, Wiesbaden, 1994.

57. R. Friedman, *On threefolds with trivial canonical bundle*, Proc. Symp. Pure Math. **53** (1991), 103–134.

58. A. Fujiki, *A Theorem on Bimeromorphic Maps of Kähler Manifolds and Its Applications*, Publ. RIMS Kyoto Univ. **17** (1981), 735–754.

59. A. Fujiki, *On primitively symplectic compact Kähler V-manifolds of dimension four*, Classification of algebraic and analytic manifolds, Progress in Mathematics, vol. 39, Birkhäuser, Boston, MA, 1983, pp. 71–250.

60. A. Fujiki, *On the de Rham Cohomology Group of a Compact Kähler Symplectic Manifold*, Adv. Stud. Pure Math. **10** (1987), 105–165.

61. A. Fujiki and S. Nakano, *Supplement to 'On the inverse of monoidal transformation'*, Publ. RIMS Kyoto Univ. **7** (1971), 637–644.
62. K. Fukaya, *Multivalued Morse Theory, Asymptotic Analysis, and Mirror Symmetry*, Preprint, 2002.
63. W. Fulton, *Intersection theory*, second ed., Ergebnisse der Mathematik und ihrer Grenzgebiete, vol. 2, Springer–Verlag, Berlin, 1998.
64. A. Givental, *Equivariant Gromov-Witten invariants*, International Mathematics Research Notices **13** (1996), 613–663.
65. W. Goldman and M. Hirsch, *The radiance obstruction and parallel forms on affine manifolds*, Trans. Amer. Math. Soc. **286** (1984), 629–649.
66. W. Goldman and J. Millson, *The deformation theory of representations of fundamental groups of compact Kähler manifolds*, Inst. Hautes Études Sci. Publ. Math. **67** (1988), 43–96.
67. W. Goldman and J. Millson, *The homotopy invariance of the Kuranishi space*, Illinois Journal of Mathematics **34** (1990), 337–367.
68. E. Goldstein, *Minimal Lagrangian tori in Kähler-Einstein manifolds*, math.DG/0007135, 2000.
69. E. Goldstein, *Calibrated fibrations*, Communications in Analysis and Geometry **10** (2002), 127–150, math.DG/9911093.
70. L. Göttsche, *Hilbert Schemes of Zero-Dimensional Subschemes of Smooth Varieties*, Springer Lecture Notes in Mathematics, vol. 1572, Springer–Verlag, Berlin, 1994.
71. L. Göttsche and D. Huybrechts, *Hodge numbers of moduli spaces of stable bundles on K3 surfaces*, International Journal of Mathematics **7** (1996), 359–372.
72. D. Grayson and M. Stillman, *Macaulay 2: A computer program designed to support computations in algebraic geometry and computer algebra*, source and object code available from http://www.math.uiuc.edu/Macaulay2/.
73. B. Greene, *The Elegant Universe*, Norton, New York, 1998.
74. B. Greene and R. Plesser, *Duality in Calabi-Yau moduli space*, Nuclear Physics **B338** (1990), 15–37.
75. P. Griffiths, *On the periods of certain rational integrals, I and II*, Annals of Mathematics **90** (1969), 460–495 and 498–541.
76. P. Griffiths (ed.), *Topics in Transcendental Geometry*, Annals of Mathematics studies, vol. 106, Princeton University Press, Princeton, NJ, 1984.
77. P. Griffiths and J. Harris, *Principles of algebraic geometry*, Wiley, New York, 1978.
78. P. Griffiths and W. Schmid, *Recent developments in Hodge theory: a discussion of techniques and results*, Discrete Subgroups of Lie Groups and Applications to Moduli, Oxford University Press, Oxford, 1973.
79. V. Gritsenko, *Elliptic genus of Calabi-Yau manifolds and Jacobi and Siegel modular forms*, Algebra Anal. **11** (2000), 100–125, math.AG/9906190.
80. M. Gross, *A Finiteness Theorem for Elliptic Calabi-Yau Three-folds*, Duke Mathematical Journal **74** (1994), 271–299.
81. M. Gross, *Primitive Calabi-Yau Threefolds*, Journal of Differential Geometry **45** (1997), 288–318.
82. M. Gross, *Special Lagrangian Fibrations I: Topology*, Integrable Systems and Algebraic Geometry (M.-H. Saito, Y. Shimizu, and K. Ueno, eds.), World Scientific, Singapore, 1998, pp. 156–193, alg-geom/9710006.

83. M. Gross, *Connecting the Web: A Prognosis*, Mirror Symmetry III (D.H. Phong, L. Vinet, and S.-T. Yau, eds.), AMS/IP Studies in Advanced Mathematics, vol. 10, A.M.S., Providence, RI, 1999, pp. 157–169.

84. M. Gross, *Special Lagrangian Fibrations II: Geometry*, Differential Geometry inspired by String Theory, Surveys in Differential Geometry, vol. 5, International Press, Boston, MA, 1999, pp. 341–403, math.AG/9809072.

85. M. Gross, *Examples of Special Lagrangian Fibrations*, Symplectic Geometry and Mirror Symmetry, Proceedings of the 4th KIAS Annual International Conference (K. Fukaya, Y.-G. Oh, K. Ono, and G. Tian, eds.), World Scientific, Singapore, 2001, pp. 81–109, math.AG/0012002.

86. M. Gross, *Topological Mirror Symmetry*, Inventiones mathematicae **144** (2001), 75–137, math.AG/9909015.

87. M. Gross and B. Siebert, *Affine Structures and Mirror Symmetry*, in preparation.

88. D. Guan, *Examples of compact holomorphic symplectic manifolds which are not Kähler III*, International Journal of Mathematics **6** (1995), 709–718.

89. D. Guan, *On the Betti numbers of irreducible compact hyperkähler manifolds of complex dimension four*, Math. Res. Lett. **8** (2001), 663–669.

90. R. Hartshorne, *Ample Subvarieties of Algebraic Varieties*, Springer Lecture Notes in Mathematics, vol. 156, Springer, 1970.

91. R. Hartshorne, *Algebraic Geometry*, Springer-Verlag, 1977.

92. R. Harvey, *Spinors and calibrations*, Academic Press, San Diego, 1990.

93. R. Harvey and H.B. Lawson, *Calibrated geometries*, Acta Mathematica **148** (1982), 47–157.

94. M. Haskins, *Special Lagrangian Cones*, math.DG/0005164, 2000.

95. B. Hasset and Y. Tschinkel, *Rational curves on holomorphic symplectic fourfolds*, Geometric and Functional Analysis **11** (2001), 1201–1228.

96. N.J. Hitchin, *The moduli space of Special Lagrangian submanifolds*, Ann. Scuola Norm. Sup. Pisa Cl. Sci. **25** (1997), 503–515, dg-ga/9711002.

97. N.J. Hitchin, A. Karlhede, U. Lindström, and M. Rocek, *Hyperkähler metrics and supersymmetry*, Communications in Mathematical Physics **108** (1987), 535–589.

98. N.J. Hitchin and J. Sawon, *Curvature and characteristic numbers of hyperKähler manifolds*, Duke Mathematical Journal **106** (2001), 599–615.

99. T. Hübsch, *Calabi-Yau manifolds. A bestiary for physicists*, World Scientific, New Jersey, 1992.

100. D. Huybrechts, *Finiteness results for compact hyperkähler manifolds*, J. Reine Angew. Math. to appear.

101. D. Huybrechts, *The Kähler cone of a compact hyperkähler manifold*, Math. Ann. to appear.

102. D. Huybrechts, *Birational symplectic manifolds and their deformations*, Journal of Differential Geometry **45** (1997), 488–513.

103. D. Huybrechts, *Compact hyperkähler manifolds*, Habilitation, 1997, http://www.mi.uni-koeln.de/~huybrech/artikel.htm/HKhabmod.ps.

104. D. Huybrechts, *Compact Hyperkähler Manifolds: Basic Results*, Inventiones mathematicae **135** (1999), 63–113.

105. D. Huybrechts, *Erratum to the paper: 'Compact hyperkähler manifolds: basic results'*, math.AG/0106014, 2001.

106. D. Huybrechts, *Products of harmonic forms and rational curves*, Documenta math. **6** (2001), 227–239.

107. D. Huybrechts and M. Lehn, *The Geometry of Moduli Spaces of Sheaves*, Aspects of Mathematics, vol. E31, Vieweg, Braunschwieg, 1997.

108. S. Ji and B. Shiffman, *Properties of compact complex manifolds carrying closed positive currents*, J. Geom. Anal. **3** (1993), 37–61.

109. D.D. Joyce, *Compact Riemannian 7-manifolds with holonomy G_2. I*, Journal of Differential Geometry **43** (1996), 291–328.

110. D.D. Joyce, *Compact Riemannian 7-manifolds with holonomy G_2. II*, Journal of Differential Geometry **43** (1996), 329–375.

111. D.D. Joyce, *Compact Riemannian 8-manifolds with holonomy* Spin(7), Inventiones mathematicae **123** (1996), 507–552.

112. D.D. Joyce, *A new construction of compact 8-manifolds with holonomy* Spin(7), Journal of Differential Geometry **53** (1999), 89–130, math.DG/ 9910002.

113. D.D. Joyce, *Compact Manifolds with Special Holonomy*, Oxford University Press, Oxford, 2000.

114. D.D. Joyce, *Singularities of special Lagrangian fibrations and the SYZ Conjecture*, math.DG/0011179, 2000.

115. D.D. Joyce, *Constructing special Lagrangian m-folds in $\mathbb{C}^m$ by evolving quadrics*, Mathematische Annalen **320** (2001), 757–797, math.DG/0008155.

116. D.D. Joyce, *Evolution equations for special Lagrangian 3-folds in $\mathbb{C}^3$*, Annals of Global Analysis and Geometry **20** (2001), 345–403, math.DG/0010036.

117. D.D. Joyce, *U(1)-invariant special Lagrangian 3-folds. I. Nonsingular solutions*, math.DG/0111324, 2001.

118. D.D. Joyce, *U(1)-invariant special Lagrangian 3-folds. II. Existence of singular solutions*, math.DG/0111326, 2001.

119. D.D. Joyce, *U(1)-invariant special Lagrangian 3-folds. III. Properties of singular solutions*, math.DG/0204343, 2002.

120. D.D. Joyce, *U(1)-invariant special Lagrangian 3-folds in $\mathbb{C}^3$ and special Lagrangian fibrations*, math.DG/0206016, 2002.

121. D.D. Joyce, *On counting special Lagrangian homology 3-spheres*, hep-th/ 9907013, 1999. To appear in *Topology and Geometry: Commemorating SISTAG*, editors A.J. Berrick, M.C. Leung and X.W. Xu, A.M.S. Contemporary Mathematics series, 2002.

122. D.D. Joyce, *Ruled special Lagrangian 3-folds in $\mathbb{C}^3$*, Proceedings of the London Mathematical Society **85** (2002), 233–256, math.DG/0012060.

123. D.D. Joyce, *Special Lagrangian 3-folds and integrable systems*, math.DG/ 0101249, 2001. To appear in *Surveys on Geometry and Integrable Systems*, editors M. Guest, R. Miyaoka and Y. Ohnita, Advanced Studies in Pure Mathematics series, Mathematical Society of Japan, 2002.

124. D.D. Joyce, *Special Lagrangian m-folds in $\mathbb{C}^m$ with symmetries*, math.DG/ 0008021, 2000. To appear in the Duke Mathematical Journal, 2002.

125. D. Kaledin, *Symplectic resolutions: deformations and birational maps*, math.AG/0012008, 2000.

126. D. Kaledin and M. Verbitsky, *Partial resolutions of Hilbert type, Dynkin diagrams, and generalized Kummer varieties*, math.AG/9812078, 1998.

127. D. Kaledin and M. Verbitsky, *Trianalytic subvarieties of generalized Kummer varieties*, International Mathematics Research Notices **9** (1998), 439–461.

128. S. Katz, *On the finiteness of rational curves on quintic threefolds*, Compositio Math. **60** (1986), 151–162.

129. Y. Kawamata, *Crepant blowing-up of 3-dimensional canonical singularities and its application to degeneration of surfaces*, Annals of Mathematics **127** (1988), 93–163.

130. Y. Kawamata, *Unobstructed deformations – a remark on a paper of Z. Ran*, Journal of Algebraic Geometry **1** (1992), 183–190, Erratum J. Alg. Geom. **6** (1997), 803–804.

131. K. Kodaira, *Complex manifolds and deformation of complex structures*, Grundlehren der mathematischen Wissenschaften, vol. 283, Springer–Verlag, Berlin, 1986.

132. K. Kodaira and D. Spencer, *On deformations of complex analytic structures I, II*, Annals of Mathematics **67** (1958), 328–466.

133. K. Kodaira and D. Spencer, *On deformations of complex analytic structures III. Stability theorems for complex analytic structures*, Annals of Mathematics **71** (1960), 43–76.

134. J. Kollár and T. Matsusaka, *Riemann–Roch type inequalities*, American Journal of Mathematics **105** (1983), 229–252.

135. M. Kontsevich, *Homological Algebra of Mirror Symmetry*, Proceedings of the International Congress of Mathematicians (Zürich, 1994), Birkhäuser, Basel, 1994, pp. 120–139, alg-geom/9411018.

136. M. Kontsevich and Y. Soibelman, *Homological Mirror Symmetry and Torus Fibrations*, Symplectic Geometry and Mirror Symmetry, Proceedings of the 4th KIAS Annual International Conference (K. Fukaya, Y.-G. Oh, K. Ono, and G. Tian, eds.), World Scientific, Singapore, 2001, pp. 203–263, math.SG/0011041.

137. A.G. Kovalev, *Twisted connected sums and special Riemannian holonomy*, math.DG/0012189, 2000.

138. M. Kreuzer and H. Skarke, *Complete classification of reflexive polyhedra in four dimensions*, Advances in Theoretical and Mathematical Physics **4** (2000), 1209–1230, hep-th/0002240.

139. G. Lawlor, *The angle criterion*, Inventiones mathematicae **95** (1989), 437–446.

140. H.B. Lawson, *Lectures on Minimal Submanifolds*, vol. 1, Publish or Perish, Wilmington, Del., 1980.

141. H.B. Lawson and M.-L. Michelsohn, *Spin geometry*, Princeton Math. Series, vol. 38, Princeton University Press, 1989.

142. M. Lehn and C. Sorger, *The cup product of the Hilbert scheme for K3 surfaces*, math.AG/0012166, 2000.

143. N.C. Leung, *A general Plücker formula*, math.AG/0111179, 2001.

144. B. Lian, K. Liu, and S.-T. Yau, *Mirror principle. I*, Asian Journal of Mathematics **1** (1997), 729–763.

145. E. Looijenga and V. Lunts, *A Lie Algebra Attached to a Projective Variety*, Inventiones mathematicae **129** (1997), 361–412.

146. Hui Ma and Yujie Ma, *Totally real minimal tori in $\mathbb{CP}^2$*, math.DG/0106141, 2001.

147. E. Markman, *Brill–Noether duality for moduli of sheaves on a K3*, Journal of Algebraic Geometry **4** (2001), 623–694.

148. D. Markushevich, *Integrable symplectic structures on compact complex manifolds*, Math. USSR-Sb. **59** (1988), 459–469.

149. D. Markushevich, *Completely integrable projective symplectic 4-dimensional varieties*, Izv. Math. **59** (1995), 159–187.

150. D. Matsushita, *On fibre space structures of a projective irreducible symplectic manifold*, Topology **38** (1999), 79–83, Addendum in Topology 40 (2001), 431–432.

151. D. Matsushita, *Equidimensionality of Lagrangian fibrations on holomorphic symplectic manifolds*, Math. Res. Lett. **7** (2000), 389–391.

152. D. Matsushita, *Higher direct images of Lagrangian fibrations*, math.AG/0010283, 2000.

153. D. McDuff and D. Salamon, *J-holomorphic curves and quantum cohomology*, University Lecture Series, vol. 6, American Mathematical Society, Providence, RI, 1994.

154. I. McIntosh, *Special Lagrangian cones in $\mathbb{C}^3$ and primitive harmonic maps*, math.DG/0201157, 2002.

155. R.C. McLean, *Deformations of calibrated submanifolds*, Comm. Anal. Geom. **6** (1998), 705–747.

156. B.G. Moishezon, *A criterion for projectivity of complete algebraic abstract varieties*, Trans. Amer. Math. Soc. **63** (1967), 1–50.

157. F. Morgan, *Geometric Measure Theory, a Beginner's Guide*, Academic Press, San Diego, 1995.

158. D. Morrison, *Picard–Fuchs equations and mirror maps for hypersurfaces*, Essays on mirror manifolds (S.-T. Yau, ed.), International Press, Hong Kong, 1992, pp. 241–264.

159. D. Morrison, *Compactifications of moduli spaces inspired by mirror symmetry*, Astérisque **218** (1993), 243–271.

160. D. Morrison, *Beyond the Kähler cone*, Israel Math. Conf. Proc. **9** (1996), 361–376.

161. J. Morrow and K. Kodaira, *Complex manifolds*, Holt, Rinehart and Winston, New York, 1971.

162. S. Mukai, *Symplectic structure of the moduli space of sheaves on an abelian or K3 surface*, Inventiones mathematicae **77** (1984), 101–116.

163. S. Mukai, *Moduli of vector bundles on K3 surfaces and symplectic manifolds*, Sugaku Expositions **1(2)** (1988), 138–174.

164. Y. Namikawa, *Counter-example to global Torelli problem for irreducible symplectic manifolds*, math.AG/0110114, 2001.

165. Y. Namikawa, *Deformation theory of singular symplectic n-folds*, Mathematische Annalen **319** (2001), 597–623.

166. Y. Namikawa, *Mukai flops and derived categories*, math.AG/0203287, 2002.

167. M. Nieper-Wißkirchen, *Hirzebruch–Riemann–Roch formulae on irreducible symplectic Kähler manifolds*, J. Alg. Geom. to appear.

168. K. O'Grady, *The weight-two Hodge structure of moduli spaces of sheaves on a K3 surface*, Journal of Algebraic Geometry **6** (1997), 599–644.

169. K. O'Grady, *Desingularized moduli spaces of sheaves on a K3*, J. Reine Angew. Math. **512** (1999), 49–117.

170. K. O'Grady, *A new six-dimensional irreducible symplectic manifold*, math.AG/0010187, 2000.

171. T. Pacini, *Deformations of Asymptotically Conical Special Lagrangian Submanifolds*, math.DG/0207144, 2002.

172. Z. Ran, *Deformations of manifolds with torsion or negative canonical bundles*, Journal of Algebraic Geometry **1** (1992), 279–291.

173. M. Reid, *The moduli space of 3-folds with $K = 0$ may nevertheless be irreducible*, Mathematische Annalen **278** (1987), 329–334.

174. S.-S. Roan, *On Calabi-Yau orbifolds in weighted projective spaces*, International Journal of Mathematics **1** (1990), 211–232.
175. W.-D. Ruan, *Lagrangian tori fibration of toric Calabi-Yau manifold I*, math.DG/9904012, 1999.
176. W.-D. Ruan, *Lagrangian Tori Fibration of Toric Calabi-Yau Manifold III: Symplectic Topological SYZ Mirror Construction for General Quintics*, math.DG/9909126, 1999.
177. W.-D. Ruan, *Lagrangian torus fibration and mirror symmetry of Calabi-Yau hypersurface in toric variety*, math.DG/0007028, 2000.
178. M. Saito, *Decomposition theorem for proper Kähler morphisms*, Tohoku Math. J. **42** (1990), 127–148.
179. S. Salamon, *On the cohomology of Kähler and hyperkähler manifolds*, Topology **35** (1996), 137–155.
180. W. Schmid, *Variation of Hodge structure: the singularities of the period mapping*, Inventiones mathematicae **22** (1973), 211–319.
181. R.A. Sharipov, *Minimal tori in the five-dimensional sphere in* $\mathbb{C}^3$, Theor. Math. Phys. **87** (1991), 363–369, Revised version: math.DG/0204253.
182. T. Shioda, *The period map of abelian surfaces*, J. Fac. Sci. Univ. Tokyo **25** (1978), 47–59.
183. Y.-T. Siu, *Every K3 surface is Kähler*, Inventiones mathematicae **73** (1983), 139–150.
184. A. Strominger, S.-T. Yau, and E. Zaslow, *Mirror symmetry is T-duality*, Nuclear Physics **B479** (1996), 243–259, hep-th/9606040.
185. D. Sullivan, *Infinitesimal computations in topology*, Inst. Hautes Études Sci. Publ. Math. **47** (1977), 269–331.
186. B. Szendrői, *On an example of Aspinwall and Morrison*, math.AG/9911064. To appear in Proc. A.M.S., 1999.
187. B. Szendrői, *Calabi-Yau threefolds with a curve of singularities and counterexamples to the Torelli problem*, International Journal of Mathematics **11** (2000), 449–459.
188. B. Szendrői, *Calabi-Yau threefolds with a curve of singularities and counterexamples to the Torelli problem. II*, Math. Proc. Cambridge Philos. Soc. **129** (2000), 193–204.
189. G. Tian, *Smoothness of the universal deformation space of compact Calabi-Yau manifolds and its Peterson-Weil metric*, Mathematical Aspects of String Theory (S.-T. Yau, ed.), World Scientific, Singapore, 1987, pp. 629–646.
190. A.N. Todorov, *The Weil-Petersson geometry of the moduli space of SU($n \geq 3$) (Calabi-Yau) manifolds*, Communications in Mathematical Physics **126** (1989), 325–346.
191. J. Varouchas, *Kähler Spaces and Proper Open Morphisms*, Mathematische Annalen **283** (1989), 13–52.
192. M. Verbitsky, *On the action of a Lie algebra SO(5) on the cohomology of a hyperkähler manifold*, Funct. Anal. Appl. **24** (1990), 70–71.
193. M. Verbitsky, *Cohomology of compact hyperkähler manifolds*, alg-geom/9501001, 1995.
194. M. Verbitsky, *Algebraic structures on hyperkähler manifolds*, Math. Res. Lett. **3** (1996), 763–767.
195. M. Verbitsky, *Cohomology of compact hyperkähler manifolds and its applications*, Geometric and Functional Analysis **6** (1996), 602–611.

196. M. Verbitsky, *Trianalytic subvarieties of the Hilbert scheme of points on a K3 surface*, Geometric and Functional Analysis **8** (1998), 732–782.
197. C. Voisin, *Symétrie Miroir*, Panoramas et Synthèses, vol. 2, Soc. Math. France, Paris, 1996.
198. H. Wakakuwa, *On Riemannian manifolds with homogeneous holonomy group Sp(n)*, Tohoku Math. J. **10** (1958), 274–303.
199. C.-L. Wang, *On the topology of birational minimal models*, Journal of Differential Geometry **50** (1998), 129–146.
200. C.-L. Wang, *Cohomology theory in birational geometry*, preprint, 2000.
201. A. Weil, *Introduction à l'étude des variétés kählériennes*, Hermann, Paris, 1958.
202. J. Wierzba, *Contractions of Symplectic Varieties*, math.AG/9910130. J. Alg. Geom. to appear.
203. J. Wierzba and J. Wiesniewski, *Small Contractions of Symplectic 4-folds*, math.AG/0201028, 2002.
204. P.M.H. Wilson, *Calabi–Yau manifolds with large Picard number*, Inventiones mathematicae **98** (1989), 139–155.
205. P.M.H. Wilson, *The Kähler cone on Calabi–Yau threefolds*, Inventiones mathematicae **107** (1992), 561–583.
206. S.-T. Yau, *On the Ricci curvature of a compact Kähler manifold and the complex Monge–Ampère equations. I*, Communications on pure and applied mathematics **31** (1978), 339–411.
207. K. Yoshioka, *Irreducibility of moduli spaces of vector bundles on K3 surfaces*, math.AG/9907001, 1999.
208. K. Yoshioka, *Moduli spaces of stable sheaves on abelian surfaces*, Mathematische Annalen **321** (2001), 817–884.

Index

B-field, 94, 140, 143
J-holomorphic curve, 95, 99, 102
- regular, 97
- simple, 96
2875 lines, 95, 102, 139

$\hat{A}$-genus, 28
adjunction, 87, 89, 93
affine manifold with singularities, 147
affine structure, 141, 143–145, 147, 153,
    155, 156, 158
- integral, 141, 145, 146
almost Calabi–Yau m-folds, 53
almost complex structure, 95–98
- compatible, 97
- smooth homotopy of, 98
associative 3-folds, 35, 40, 55
- deformations, 40, 55
asymptotic cone, 43, 58
Atiyah–Singer Index Theorem, 55

Batyrev construction, 90
Beauville–Bogomolov form, 182
- of the generalized Kummer, 188
- of the Hilbert scheme, 187
- unnormalized, 212
Berger's theorem, 11–12
Bianchi identities, 8, 9
Bogomolov–Tian–Todorov theorem, 79
bubbling, 99

Calabi Conjecture, 20–22
Calabi–Yau m-folds, 13, 17–18, 71, 72
- almost, 53
calibrated geometry, 31–36
calibrated submanifolds, 33
- compact, 50
calibration, 17, 33
- maximal, 34
canonical bundle, 16, 72, 87
canonical coordinates, 119–123,
    134–137
Cauchy–Riemann equations, 96
Cayley 4-folds, 35, 41, 55

- deformations, 41, 55
Cheeger–Gromoll splitting Theorem, 22
Chern class, 17
coassociative 4-folds, 35, 40, 54
- deformations, 40, 55
complex manifold, 14
complex structure, 14
connection, 5
- Gauss–Manin, 105, 106
crepant resolution, 87–91, 93, 125, 157
crossed homomorphism, 142, 145
cubic intersection form, 85
- quantum deformation, 100
cubic surface, 102
- 27 lines on, 102
current, positive, 207
curvature, 6
- constant, 11

deformation, 73, 74, 81, 172
- universal, 73, 173
- unobstructed, 79, 174, 175, 203
Dehn twist, 106, 146
Dolbeault cohomology, 24

elliptic curve, 86, 105, 111, 113, 143,
    146
exceptional holonomy groups, 13, 26–31
- compact manifolds, 29–31

fibration, 171, 194
form, 5
- holomorphic volume, 17
- Kähler, 15
- Ricci, 16
formal semigroup ring, 100
framing, 94, 101, 121

G_2, 13, 27
G_2-manifold, 27
G_2-structure, 27
Geometric Measure Theory, 31, 32, 36,
    57
Gorenstein, 87

Griffiths transversality, 106, 108, 109, 137
Gromov compactness theorem, 99
Gromov–Witten invariants, 100, 122

Hilbert scheme, 167, 169
Hodge diamond, 88, 103, 115, 150
Hodge filtration, 103, 104, 106, 107, 109, 113
– limiting, 114, 115
Hodge number, 24, 88, 90, 91, 93
Hodge structures, 102–104, 115
– mixed, 102, 111, 113–115
Hodge theory
– annoying indices, 104
holomorphic symplectic manifold, 18, 85
– irreducible, 26
holomorphic symplectic structure, 18
holomorphic volume form, 17
holonomy group
– and constant tensors, 7
– and curvature, 7
– of a connection, 7
– of a metric, 9
– properties, 7
holonomy representation, 141, 142, 158, 159
HRR-formula, 187
hyperkähler geometry, 13, 18–19, 25–26

integral Mori semigroup, 100

Jacobi identity, 9
– graded, 76

K3 surface, 23, 165
– elliptic, 165, 171, 196
– Fermat quartic, 165
– Kummer surface, 165
Kähler class, 15, 93
Kähler cone, 84, 85, 143, 144, 179, 207, 221
– birational, 223
– of a very general, 201, 208
Kähler form, 15, 84, 99
Kähler geometry, 13–16
Kähler manifold, 72, 81, 83, 84, 103, 104, 111, 115

Kähler moduli space, 93, 108, 117, 122, 123, 140, 145
– complexified, 94
Kähler potential, 15
Kodaira–Spencer map, 82, 83, 93, 109
Kodaira–Spencer class, 173
– of the twistor space, 198
Kummer variety, generalized, 167, 188, 202, 209

Lagrangian submanifolds, 37
Legendrian index, 59
Levi-Civita connection, 8
Lie bracket, 5, 76
local system, 104–108, 111, 142, 148

manifold
– hyperkähler, 178
– irreducible holomorphic symplectic, 164
– – birational, 168, 170, 175, 202, 213, 217
marked manifold, 197
metric, 8
– Calabi–Yau, 13
– Hermitian, 15
– hyperkähler, 13, 18
– irreducible, 10
– Kähler, 13
– locally symmetric, 11
– nonsymmetric, 11
– product, 10
– quaternionic Kähler, 13
– reducible, 10
minimal submanifolds, 32–33
mirror map, 122
Mirror Symmetry, 63–64, 89, 91, 93, 101, 119, 121, 122, 140, 142, 147
moduli space, 197
moment map, 41, 44, 49
monodromy
– diffeomorphism, 105, 106, 146, 153
– maximally unipotent, 113
– quasi-unipotent, 111
– transformation, 105, 111, 116, 117, 126, 134, 135, 140, 146
– unipotent, 112, 115, 117, 126
monodromy weight filtration, 112–116, 153

Mukai flop, 168, 175, 214

Nijenhuis tensor, 14, 15
Nilpotent orbit theorem, 115–121

obstruction space, 82
octonions, 13

parallel transport, 7
period domain, 176, 197, 200
period map, 108–110, 113, 115, 116,
    118, 126, 177, 198
– differential of, 109
– surjectivity, 199
periods of quintic, 129
Picard–Fuchs equations, 129–134, 136,
    137

quaternions, 13, 18, 38
quintic 3-fold, 87–89, 91, 95, 96, 102,
    111, 121–123
– mirror, 90, 122, 125, 134

regular singular point, 133
Ricci curvature, 16
Ricci form, 16
Riemann curvature, 8
Riemannian metric, see metric

sheaf
– locally constant, 104
special Lagrangian m-folds, 35–40, 140
– affine structures on moduli space,
    52–53
– Asymptotically Conical, 58
– compact, 50–54
– constructions, 41–48
– deformations, 39–40, 50–52, 59–60
– examples, 46–48
– in almost Calabi–Yau m-folds, 53
– index of singularities, 61, 67
– obstructions, 50, 52

– ruled, 43
– singularities, 50, 57–62
– strongly AC, 58
– weakly AC, 58
– with phase $e^{i\theta}$, 37
special Lagrangian cones, 42, 57–60
– as an integrable system, 44
– examples, 46, 48
– rigid, 60
special Lagrangian fibration, 49
special Lagrangian fibrations, 64–68
– discriminant, 64, 66
– singularities, 65–68
Spin(7), 13, 28–29
Spin(7)-manifold, 28
Spin(7)-structure, 28
String Theory, 63
symmetric space, 10
SYZ Conjecture, 62–68, 140, 155

tangent cone, 36, 57
Teichmüller space, 73
tensors, 4
– constant, 7, 9
– Einstein summation convention, 5
– index notation, 5
Torelli theorem
– global, 201
– – counterexample, 202
– – for K3 surfaces, 201
– local, 109, 177
torsion, 6
twistor space, 196

vector bundles, 4

weighted projective space, 89

Yukawa coupling, 158
– $(1,1)$, 101, 117, 119, 122
– $(1,2)$, 103, 107, 108, 117, 119, 122,
    123, 134, 136–138

Universitext

Aksoy, A.; Khamsi, M. A.: Methods in Fixed Point Theory

Alevras, D.; Padberg M. W.: Linear Optimization and Extensions

Andersson, M.: Topics in Complex Analysis

Aoki, M.: State Space Modeling of Time Series

Audin, M.: Geometry

Aupetit, B.: A Primer on Spectral Theory

Bachem, A.; Kern, W.: Linear Programming Duality

Bachmann, G.; Narici, L.; Beckenstein, E.: Fourier and Wavelet Analysis

Badescu, L.: Algebraic Surfaces

Balakrishnan, R.; Ranganathan, K.: A Textbook of Graph Theory

Balser, W.: Formal Power Series and Linear Systems of Meromorphic Ordinary Differential Equations

Bapat, R.B.: Linear Algebra and Linear Models

Benedetti, R.; Petronio, C.: Lectures on Hyperbolic Geometry

Berberian, S. K.: Fundamentals of Real Analysis

Berger, M.: Geometry I, and II

Bliedtner, J.; Hansen, W.: Potential Theory

Blowey, J. F.; Coleman, J. P.; Craig, A. W. (Eds.): Theory and Numerics of Differential Equations

Börger, E.; Grädel, E.; Gurevich, Y.: The Classical Decision Problem

Böttcher, A; Silbermann, B.: Introduction to Large Truncated Toeplitz Matrices

Boltyanski, V.; Martini, H.; Soltan, P. S.: Excursions into Combinatorial Geometry

Boltyanskii, V. G.; Efremovich, V. A.: Intuitive Combinatorial Topology

Booss, B.; Bleecker, D. D.: Topology and Analysis

Borkar, V. S.: Probability Theory

Carleson, L.; Gamelin, T. W.: Complex Dynamics

Cecil, T. E.: Lie Sphere Geometry: With Applications of Submanifolds

Chae, S. B.: Lebesgue Integration

Chandrasekharan, K.: Classical Fourier Transform

Charlap, L. S.: Bieberbach Groups and Flat Manifolds

Chern, S.: Complex Manifolds without Potential Theory

Chorin, A. J.; Marsden, J. E.: Mathematical Introduction to Fluid Mechanics

Cohn, H.: A Classical Invitation to Algebraic Numbers and Class Fields

Curtis, M. L.: Abstract Linear Algebra

Curtis, M. L.: Matrix Groups

Cyganowski, S.; Kloeden, P.; Ombach, J.: From Elementary Probability to Stochastic Differential Equations with MAPLE

Dalen, D. van: Logic and Structure

Das, A.: The Special Theory of Relativity: A Mathematical Exposition

Debarre, O.: Higher-Dimensional Algebraic Geometry

Deitmar, A.: A First Course in Harmonic Analysis

Demazure, M.: Bifurcations and Catastrophes

Devlin, K. J.: Fundamentals of Contemporary Set Theory

DiBenedetto, E.: Degenerate Parabolic Equations

Diener, F.; Diener, M.(Eds.): Nonstandard Analysis in Practice

Dimca, A.: Singularities and Topology of Hypersurfaces

DoCarmo, M. P.: Differential Forms and Applications

Duistermaat, J. J.; Kolk, J. A. C.: Lie Groups

Edwards, R. E.: A Formal Background to Higher Mathematics Ia, and Ib

Edwards, R. E.: A Formal Background to Higher Mathematics IIa, and IIb

Emery, M.: Stochastic Calculus in Manifolds

Endler, O.: Valuation Theory

Erez, B.: Galois Modules in Arithmetic

Everest, G.; Ward, T.: Heights of Polynomials and Entropy in Algebraic Dynamics

Farenick, D. R.: Algebras of Linear Transformations

Foulds, L. R.: Graph Theory Applications

Frauenthal, J. C.: Mathematical Modeling in Epidemiology

Friedman, R.: Algebraic Surfaces and Holomorphic Vector Bundles

Fuks, D. B.; Rokhlin, V. A.: Beginner's Course in Topology

Fuhrmann, P. A.: A Polynomial Approach to Linear Algebra

Gallot, S.; Hulin, D.; Lafontaine, J.: Riemannian Geometry

Gardiner, C. F.: A First Course in Group Theory

Gårding, L.; Tambour, T.: Algebra for Computer Science

Godbillon, C.: Dynamical Systems on Surfaces

Goldblatt, R.: Orthogonality and Spacetime Geometry

Gouvêa, F. Q.: p-Adic Numbers

Gross, M.; Huybrechts, D.; Joyce, D.: Calabi-Yau Manifolds and Related Geometries

Gustafson, K. E.; Rao, D. K. M.: Numerical Range. The Field of Values of Linear Operators and Matrices

Hahn, A. J.: Quadratic Algebras, Clifford Algebras, and Arithmetic Witt Groups

Hájek, P.; Havránek, T.: Mechanizing Hypothesis Formation

Heinonen, J.: Lectures on Analysis on Metric Spaces

Hlawka, E.; Schoißengeier, J.; Taschner, R.: Geometric and Analytic Number Theory

Holmgren, R. A.: A First Course in Discrete Dynamical Systems

Howe, R., Tan, E. Ch.: Non-Abelian Harmonic Analysis

Howes, N. R.: Modern Analysis and Topology

Hsieh, P.-F.; Sibuya, Y. (Eds.): Basic Theory of Ordinary Differential Equations

Humi, M., Miller, W.: Second Course in Ordinary Differential Equations for Scientists and Engineers

Hurwitz, A.; Kritikos, N.: Lectures on Number Theory

Iversen, B.: Cohomology of Sheaves

Jacod, J.; Protter, P.: Probability Essentials

Jennings, G. A.: Modern Geometry with Applications

Jones, A.; Morris, S. A.; Pearson, K. R.: Abstract Algebra and Famous Inpossibilities

Jost, J.: Compact Riemann Surfaces

Jost, J.: Postmodern Analysis

Jost, J.: Riemannian Geometry and Geometric Analysis

Kac, V.; Cheung, P.: Quantum Calculus

Kannan, R.; Krueger, C. K.: Advanced Analysis on the Real Line

Kelly, P.; Matthews, G.: The Non-Euclidean Hyperbolic Plane

Kempf, G.: Complex Abelian Varieties and Theta Functions

Kitchens, B. P.: Symbolic Dynamics

Kloeden, P.; Ombach, J.; Cyganowski, S.: From Elementary Probability to Stochastic Differential Equations with MAPLE

Kloeden, P. E.; Platen; E.; Schurz, H.: Numerical Solution of SDE Through Computer Experiments

Kostrikin, A. I.: Introduction to Algebra

Krasnoselskii, M. A.; Pokrovskii, A. V.: Systems with Hysteresis

Luecking, D. H., Rubel, L. A.: Complex Analysis. A Functional Analysis Approach

Ma, Zhi-Ming; Roeckner, M.: Introduction to the Theory of (non-symmetric) Dirichlet Forms

Mac Lane, S.; Moerdijk, I.: Sheaves in Geometry and Logic

Marcus, D. A.: Number Fields

Martinez, A.: An Introduction to Semiclassical and Microlocal Analysis

Matsuki, K.: Introduction to the Mori Program

Mc Carthy, P. J.: Introduction to Arithmetical Functions

Meyer, R. M.: Essential Mathematics for Applied Field

Meyer-Nieberg, P.: Banach Lattices

Mines, R.; Richman, F.; Ruitenburg, W.: A Course in Constructive Algebra

Moise, E. E.: Introductory Problem Courses in Analysis and Topology

Montesinos-Amilibia, J. M.: Classical Tessellations and Three Manifolds

Morris, P.: Introduction to Game Theory

Nikulin, V. V.; Shafarevich, I. R.: Geometries and Groups

Oden, J. J.; Reddy, J. N.: Variational Methods in Theoretical Mechanics

Øksendal, B.: Stochastic Differential Equations

Poizat, B.: A Course in Model Theory

Polster, B.: A Geometrical Picture Book

Porter, J. R.; Woods, R. G.: Extensions and Absolutes of Hausdorff Spaces

Radjavi, H.; Rosenthal, P.: Simultaneous Triangularization

Ramsay, A.; Richtmeyer, R. D.: Introduction to Hyperbolic Geometry

Rees, E. G.: Notes on Geometry

Reisel, R. B.: Elementary Theory of Metric Spaces

Rey, W. J. J.: Introduction to Robust and Quasi-Robust Statistical Methods

Ribenboim, P.: Classical Theory of Algebraic Numbers

Rickart, C. E.: Natural Function Algebras

Rotman, J. J.: Galois Theory

Rubel, L. A.: Entire and Meromorphic Functions

Rybakowski, K. P.: The Homotopy Index and Partial Differential Equations

Sagan, H.: Space-Filling Curves

Samelson, H.: Notes on Lie Algebras

Schiff, J. L.: Normal Families

Sengupta, J. K.: Optimal Decisions under Uncertainty

Séroul, R.: Programming for Mathematicians

Seydel, R.: Tools for Computational Finance

Shafarevich, I. R.: Discourses on Algebra

Shapiro, J. H.: Composition Operators and Classical Function Theory

Simonnet, M.: Measures and Probabilities

Smith, K. E.; Kahanpää, L.; Kekäläinen, P.; Traves, W.: An Invitation to Algebraic Geometry

Smith, K. T.: Power Series from a Computational Point of View

Smoryński, C.: Logical Number Theory I. An Introduction

Stichtenoth, H.: Algebraic Function Fields and Codes

Stillwell, J.: Geometry of Surfaces

Stroock, D. W.: An Introduction to the Theory of Large Deviations

Sunder, V. S.: An Invitation to von Neumann Algebras

Tamme, G.: Introduction to Étale Cohomology

Tondeur, P.: Foliations on Riemannian Manifolds

Verhulst, F.: Nonlinear Differential Equations and Dynamical Systems

Wong, M. W.: Weyl Transforms

Zaanen, A.C.: Continuity, Integration and Fourier Theory

Zhang, F.: Matrix Theory

Zong, C.: Sphere Packings

Zong, C.: Strange Phenomena in Convex and Discrete Geometry

Printing: Mercedes-Druck, Berlin
Binding: Stein+Lehmann, Berlin